S0-FKN-756

BRYOPHYTE DEVELOPMENT: PHYSIOLOGY AND BIOCHEMISTRY

Editors

R. N. Chopra, Ph.D.
Professor
Department of Botany
University of Delhi
Delhi, India

Satish C. Bhatla, Ph.D.
Lecturer
Department of Botany
University of Delhi
Delhi, India

CRC Press
Boca Raton Ann Arbor Boston

Library of Congress Cataloging-in-Publication Data

Bryophyte development: physiology and biochemistry/editors, R.N.
Chopra, Satish C. Bhatla.
p. cm.
Includes bibliographical references.
ISBN 0-8493-5289-4
1. Bryophytes—Development. I. Chopra, R. N. II. Bhatla, Satish
C.
QK533.6.B77 1990
588′ .043—dc20 90-33329
CIP

This book represents information obtained from authentic and highly regarded sources. Reprinted material is quoted with permission, and sources are indicated. A wide variety of references are listed. Every reasonable effort has been made to give reliable data and information, but the author and the publisher cannot assume responsibility for the validity of all materials or for the consequences of their use.

Direct all inquiries to CRC Press, Inc., 2000 Corporate Blvd., N.W., Boca Raton, Florida 33431.

International Standard Book Number 0-8493-5289-4

Library of Congress Card Number 90-33329
Printed in the United States

PREFACE

Bryophytes are being used increasingly for the study of physiological and biochemical aspects of plant development. It is ease of handling, together with their simple organization and faster multiplication rate under controlled environmental conditions, that makes these plants excellent materials for investigations on developmental biology. An interesting aspect that makes bryophytes commercially important, as well, is the extraction of useful chemical constituents from their tissues.

With the accumulation of knowledge in the newer areas of investigations and the increasing degree of specialization, the most satisfactory way of bringing together authentic information in one volume is to invite the specialists in each area to contribute. The response to our invitation has been overwhelming, and we sincerely thank all the contributors for their cooperation. It has been a pleasure for us to prepare the manuscripts for press.

The topics presented in this volume are broadly of five categories: use of protoplasts, callus, and mutants for investigations on controlled growth and differentiation; physiological and biochemical aspects of moss protonema differentiation; physiology of reproduction; developmental physiology of liverworts; and production of physiologically active substances by bryophytes. In a work of this nature some amount of overlapping of information between different chapters is unavoidable.

It is hoped that this volume will be found useful by researchers in the field of developmental physiology and biochemistry of bryophytes.

R. N. Chopra
Satish C. Bhatla
Delhi
June 1988

THE EDITORS

Professor R. N. Chopra, Ph.D., F.B.S. (b. 1929), has been teaching and supervising research at the Department of Botany, University of Delhi, Delhi, India for over 33 years. He was the Chairman and also Director of the Center of Advanced Study in Botany for 3 years (1982 to 1985) in the same institution. He was Additional Secretary of the International Society of Plant Morphologists from 1975 to 1984. At present he is the President of the Indian Bryological Society and also a Councillor of the Indian Botanical Society.

Professor Chopra has traveled widely. He spent 1 year (1962 to 1963) at Duke University, Durham, North Carolina as a Postdoctoral Fellow and worked at Wye College, University of London as Senior UNESCO Fellow during 1972. He has participated in several symposia/conferences at the national and international levels and has also chaired sessions. He has over 150 research papers to his credit, as well as a book entitled "Biology of Bryophytes", recently published by Wiley Eastern Ltd., New Delhi. Professor Chopra's present interests include protonemal differentiation, bud induction, reproductive biology, apospory, apogamy, callus induction and its controlled differentiation, and effect of pollutants on bryophytes. Professor Chopra has built one of the finest schools of Bryology in India, the contributions of which are well cited in reviews and texts on the subject.

Satish C. Bhatla, Ph.D. (b. 1955) is on the teaching staff of the Department of Botany at the University of Delhi, Delhi, India. He has been a Fellow of the Alexander von Humboldt Foundation, West Germany, and received his Postdoctoral training at the University of Heidelberg, Heidelberg, West Germany. He has been participating in symposia/conferences and workshops at the national and international levels.

Dr. Bhatla has been engaged in research on the physiology and biochemistry of development in mosses since 1977. His present research interests include isolation and biochemical characterization of hormone-sensitive mutants and the mode of auxin and cytokinin action in cell differentation.

CONTRIBUTORS

Yoshinori Asakawa, Ph.D.
Dean and Professor
Faculty of Pharmaceutical Sciences
Tokushima Bunri University
Tokushima, Japan

Neil W. Ashton, Ph.D.
Associate Professor
Department of Biology
University of Regina
Regina, Saskatchewan
Canada

Dominick V. Basile, Ph.D.
Professor
Department of Biological Sciences
Lehman College, CUNY
Bronx, New York

S. C. Bhatla, Ph.D.
Lecturer
Department of Botany
University of Delhi
Delhi, India

Martin Bopp, Dr. rer. nat.
Professor and Director
Department of Botany
University of Heidelberg
Heidelberg, Federal Republic of Germany

Philip J. Boyd, Ph.D.
Department of Genetics
University of Leeds
Leeds, Yorkshire
England

R. N. Chopra, Ph.D.
Professor
Department of Botany
University of Delhi
Delhi, India

David J. Cove, Ph.D.
Professor
Department of Genetics
University of Leeds
Leeds, Yorkshire
England

Sadhana Dhingra-Babbar, Ph.D.
Department of Botany
University of Delhi
Delhi, India

Elmar Hartmann, Ph.D.
Professor
Institute for Plant Physiology, Cell Biology, and Microbiology
Free University Berlin
Berlin, Federal Republic of Germany

Kenji Kato, Ph.D.
Senior Research Associate
Suntory Institute for Bioorganic Research
Shimomoto, Osaka
Japan

Celia D. Knight, Ph.D.
Department of Genetics
University of Leeds
Leeds, Yorkshire
England

Manohar Lal, Ph.D.
Professor
Department of Botany
University of Delhi
Delhi, India

Barbara B. Lippincott, Ph.D.
Senior Research Associate
Department of Biochemistry, Molecular Biology, and Cell Biology
Northwestern University
Evanston, Illinois

James A. Lippincott, Ph.D.
Professor
Department of Biochemistry, Molecular Biology, and Cell Biology
Northwestern University
Evanston, Illinois

R. E. Longton, Ph.D.
Senior Lecturer
Department of Botany
University of Reading
Reading, England

M. K. C. Menon, Ph.D.
Department of Botany
Kirori Mal College
University of Delhi
Delhi, India

Yoshimoto Ohta, Ph.D.
Senior Research Associate
Suntory Institute for Bioorganic Research
Shimamoto, Osaka
Japan

Sarla, Ph.D.
Department of Botany
University of Delhi
Delhi, India

Mary Jane Saunders, Ph.D.
Associate Professor
Department of Biology
University of South Florida
Tampa, Florida

Walter W. Schwabe, Ph.D., D.Sc.
Professor
Department of Horticulture
Wye College
London University
Ashford, Kent
England

Luretta D. Spiess, Ph.D.
Senior Research Associate
Department of Biochemistry, Molecular Biology, and Cell Biology
Northwestern University
Evanston, Illinois

Reiji Takeda, Ph.D.
Senior Research Associate
Suntory Institute for Bioorganic Research
Shimamoto, Osaka
Japan

Martina Weber, Ph.D.
Institute for General Botany
University of Mainz
Mainz, Federal Republic of Germany

TABLE OF CONTENTS

Chapter 1

PROTOPLASTS AS TOOLS IN THE STUDY OF MOSS DEVELOPMENT

Neil W. Ashton, Philip J. Boyd, David J. Cove, and Celia D. Knight

TABLE OF CONTENTS

I. INTRODUCTION

In this review we shall, for the sake of brevity, adopt the now common usage of the word "protoplast" to refer to the isolated protoplast; we shall not confine our use of the term "protoplast" to isolated protoplasts which have been examined critically for the complete removal of cell wall material. We will therefore be using "protoplast" where more critically we should use the term "isolated spheroplast".

Although protoplasts have been isolated and regenerated from a number of different species of mosses,[1-6] they have only been used extensively for the study of development in *Physcomitrella patens,* and this review is therefore confined largely to this species. In *P. patens,* protoplasts regenerate directly and with high frequency to give rise to protonemal tissue.[7] This direct regeneration of protoplasts to form normal tissue is the key to the technical convenience of protoplasts in the study of moss development.

II. PROTOPLAST REGENERATION AND CELL POLARITY

Protoplasts of *P. patens,* isolated by the treatment of protonemal tissue with cellulolytic enzymes, regenerate rapidly with high frequency provided they are incubated in osmotically buffered medium and are supported within a matrix such as agar.[7] Regeneration frequencies of 50% are obtained routinely, and frequencies higher than 90% are common. Regeneration occurs in the normal culture medium except that, for maximum regeneration rates, a Ca^{2+} concentration of 10 m*M* is required, about 30 times more than is required for maximal protonemal growth.[8] Protoplast regeneration only occurs in light, although cell wall formation, as scored by calcofluor staining or the acquisition of osmotic resistance, occurs in darkness.[9] In a detailed study of the light requirements for protoplast regeneration in *P. patens*, Jenkins and Cove[9] showed that considerably higher intensities of light were required to induce the asymmetrical development of the regenerating protoplast than were needed for cell division or for the continued extension of the protonemal filament, once formed. Red and blue light were more or less equally effective in inducing asymmetry, but red light was much more effective than blue in allowing cell division and filament elongation. It therefore seems likely that there are at least two different processes of light detection involved in protoplast regeneration.

Protoplasts have also been of use in the study of the phototropic and polarotropic responses of protonemal cells. Spore germination in *P. patens* gives rise to filaments of primary chloronemal cells. The apical cells of these filaments show a complex pattern of phototropism and polarotropism.[10,11] Jenkins and Cove[10] showed that, in low intensities of red light, growth was toward an unpolarized light source, but at right angles to the direction of a source of polarized light and perpendicular to its E vector. At higher intensities of red light, growth is at right angles to the light source whether polarized or not; however, in polarized light the filaments grow oriented parallel to the E vector. Response to other wavelengths was also studied, and as a result it was suggested that the switch between the low and high light intensity responses was likely to be mediated by phytochrome. In addition, a blue light receptor may be involved in determining whether the high intensity response to unpolarized light is growth at right angles or away from the light source.[12] Secondary chloronemal filaments develop as side branches on caulonemal filaments and are morphologically similar to primary chloronemal filaments. However, they show a positive phototropic response even to high light intensities. Their polarotropic response is less easy to assess since the orientation of the caulonemal filament appears to influence their direction of growth, but secondary chloronemal filaments tend to grow perpendicular to the E vector in all light intensities.[11,13] Thus, the phototropic response of secondary chloronemal apical cells at all light intensities resembles that of the low light level response of primary chloronemal cells.

Protoplasts are usually isolated from young protonemal tissue which consists almost entirely of chloronemal cells. This tissue is grown from an inoculum of cell fragments which have been prepared by brief treatment of similar tissue in a blender. The polarotropic and phototropic responses of filaments developing from these regenerating protoplasts are similar to those of secondary chloronemata.[14] The chloronemal filaments developing from germinating spores therefore show a unique tropic response to high light intensities, and the possibility that this is caused by some substance which is present in the spore or only synthesized during spore germination has been investigated by isolating protoplasts from spores which have just germinated.[14] The frequency of regeneration from such protoplasts was found to be lower than from the usual protonemal tissue, so data are not extensive. Nevertheless, all observations from these protoplasts showed a pattern of phototropism and polarotropism similar to that of secondary chloronemata or of protoplasts isolated from protonemal tissue. While these studies need to be taken further, there appears to be no evidence from this work that the high-light tropic response of primary chloronemata involves spore- or germination-specific factors, although it is possible that the process of protoplasting and regeneration allows sufficient time for the decay or escape of such factors.

III. SOMATIC MUTAGENESIS

The production of mutant strains of an organism is particularly useful when studying development, as mutants that are blocked at a certain stage of development may provide data on the developmental process itself. There are many advantages in choosing to study development in haploid tissues, including simplicity of genetic analysis (but see Section IV below) and the ability to directly observe mutant phenotypes caused by the presence of mutant alleles recessive to the wild-type. Mutagenic treatments are best carried out on uniform unicellular material. The usual technique for mutagenesis in *P. patens* has involved treating spores with mutagen and then selecting survivors for the mutant phenotype of choice.[15,16] However, when developmentally abnormal strains are produced, many are sexually sterile since development has been arrested before the production of gametes. Thus, further mutagenic treatment of these strains utilizing spores (e.g., for reversion studies) has not been possible. A technique of somatic mutagenesis involving protoplast isolation has been developed for *P. patens*[17] which has overcome this problem.

Young protonemal tissue was first treated with mutagen; then protoplasts were isolated and regenerated so that the regenerated mutant strain originated from a single mutagenized cell. Experiments involving the regeneration of mechanically prepared tissue fragments after mutagenesis gave poor regeneration rates (as few as 2% of fragments containing one or more whole cells regenerated). The most consistent results have been obtained using *N*-methyl-*N'*-nitro-*N*-nitrosoguanidine (MNNG) as the mutagen, but other chemicals such as ethyl methanesulfonate (EMS), ethylnitrosourea (ENUA), and methylnitrosourea (MNUA) have been tried, as well as ultraviolet (UV) irradiation.[18] EMS was found to be lethal to protoplasts at the concentration used for spore mutagenesis (5% v/v) and even at lower concentrations (0.2 to 1% v/v) was toxic, so this mutagen was not used further. MNNG was preferred to other alkylating agents (ENUA and MNUA), as in a series of experiments MNNG produced more morphologically abnormal strains for a given amount of mutagen over the range of concentrations tested.[18]

Reliable kill curves have been difficult to obtain for the somatic mutagenesis technique, as it was difficult to standardize the amount of tissue used for protoplast isolation; therefore, the most reliable indication of successful mutation is the percentage of morphologically abnormal clones produced. Recently, the mutation rates of resistance to the purine analogue 8-azaguanine (8-AG) and to the amino acid analogue *p*-fluorophenylalanine (PFPA) have been used as an indication of the reliability of the technique, and rates comparable to those obtained with spore mutagenesis have been obtained.[17] The technique has been used to isolate a mutant which overproduces gametophores,[19] and revertants of this class of mutants (originally produced by mutagenic treatment of spores) have also been isolated.[18] Somatic mutagenesis is now the preferred method of mutant isolation in *P. patens*, as it requires neither the production of sporophytes, a process which takes 3 to 4 months under laboratory conditions, nor the harvesting of spores, a laborious task if large numbers of spores are required.

IV. SOMATIC HYBRIDIZATION

The advantages of studying development in haploid tissues were discussed in Section III. The ease with which mutants blocked at various stages of development can be isolated has allowed considerable progress to be made in the analysis of the signals and processes involved.[19] Although haploidy makes the isolation of developmentally abnormal mutants straightforward, their conventional genetic analysis is problematic if the abnormality leads to infertility. Genetic analysis of mutants is desirable for two reasons. First, the complexity of a process can be assessed by determining the number of genes which can mutate to affect the process. Second, the construction of double mutant strains is important to establish the epistatic relationship of the mutations concerned and, hence, to determine the order in which the gene products act.

To determine the number of genes whose products are involved in a process it is necessary to carry out complementation analysis, a procedure which normally requires diploid or heterokaryotic tissue. Although mosses produce diploid sporophytic tissue as a natural part of their life cycle, this cannot be obtained from mutants which are blocked early in their development. To circumvent this problem a method for the production of somatic hybrids has been devised which involves the fusion of protoplasts.[7,20] In order to carry out a complementation test between two developmentally abnormal mutants it is usually necessary that the mutants have been isolated in different and complementary auxotrophic strains, since the selection of the somatic hybrid involves plating a mixture of protoplasts isolated from the two strains onto a selective medium on which neither strain can grow. Details of the methods used are given by Ashton et al.[21] Protoplasts are induced to fuse either by treatment with polyethylene glycol and Ca^{2+} or by electrofusion. Once selected, the hybrids can then be examined for their developmental phenotypes. It is also possible to establish the dominance status of a developmental mutant by producing a hybrid between it and a developmentally normal strain using the same procedure.

Complementation analysis using somatic hybrids has now been carried out on a number of different classes of developmentally abnormal mutants of *P. patens*, and the results are summarized in Table 1. Because in these analyses somatic hybrids have been selected by using strains with complementing auxotrophic requirements, it has not been possible to select all possible hybrids. In addition, it has sometimes proved difficult to isolate hybrids between certain strains, even though they have suitable genetic backgrounds. Complementation analyses are therefore seldom complete.

The construction of double mutants between different developmental mutants has not been achieved so far in *P. patens* except by the induction of a second mutation in an already-mutant strain (see Section III). Complementing somatic hybrids are usually sexually fertile, although they take much longer to complete their life cycle. Only a limited amount of genetic analysis has been carried out on such hybrids which give rise to diploid progeny;[22] however, they should provide a method of constructing diploid strains which are homozygous for two different mutant alleles.

TABLE 1
Complementation Analyses of Developmentally Abnormal Mutants of *Physcomitrella patens*

Mutant genes[a]	Capacity for sexual reproduction[b]	Recombination with[c]	Dominance status with respect to wild-type alleles[d]	Complementation demonstrated with[e]	Noncomplementation demonstrated with[e]
BAR 1	S		R	*NAR* 113	
BAR 2	S		R		
BAR 297	S		R		
BAR 130	NT		R		
BAR 61	NT		R		
BAR 330	NT		I		
GAD 48	NT		D		
*gad*A33	S		R	*GAD*B284	*GAD*A66, *GAD*A186
*GAD*A66	S		R	*GAD*B74	*gad*A33, *GAD*A91
*GAD*A91	S		R	*GAD*B284	*GAD*A66, *GAD*A186
*GAD*A186	S		R	*GAD*B74	*gad*A33, *GAD*A91
*GAD*B74	S		I	*GAD*A66, *GAD*A186	*GAD*B284
*GAD*B284	S		R	*gad*A33, *GAD*A91	*GAD*B74
gam 1	F		R		
*GAM*A7/13	S		R	*GAM* 121,196,224,280	*GAM*A173, *GAM*A295
*GAM*A173	S		R	*GAM* 26,7/10,363	*GAM*A7/13
*GAM*A295	S		R	*GAM* 26,7/10,363	*GAM*A7/13
GAM 26	S		R	*GAM* 121,139,A173,224,280,A295	
GAM 7/10	S		R	*GAM* 121,139,A173,196,224,280,A295	
GAM 355	S		I	*GAM* 139,196	
GAM 363	S		R	*GAM* 121,139,A173,196,224,280,A295	
GAM 121	S		R	*GAM* 26,7/10,A7/13,363	
GAM 139	S		R	*GAM* 26,7/10,355,363	
GAM 196	S		R	*GAM* 7/10,A7/13,355,363	
GAM 224	S		R	*GAM* 26,7/10,A7/13,363	
GAM 280	S		R	*GAM* 26,7/10,A7/13,363	

TABLE 1 (continued)
Complementation Analyses of Developmentally Abnormal Mutants of *Physcomitrella patens*

Mutant genes[a]	Capacity for sexual reproduction[b]	Recombination with[c]	Dominance status with respect to wild-type alleles[d]	Complementation demonstrated with[e]	Noncomplementation demonstrated with[e]
*GTR*A2	S		R	*GTR*B1, *GTR*C5	*gtr*A106
*GTR*A3	S		R	*GTR*B1	*gtr*A106
*gtr*A106	S		R	*GTR*C5	*GTR*A2, *GTR*A3
*GTR*B1	S		R	*GTR*A2, *GTR*A3, *GTR*C5	
*GTR*C5	NT		R	*GTR*B1, *gtr*A106	
NAR 91	S		D		
NAR 112	S		D		
NAR 113	S		I	*BAR* 1	
NAR 87	NT		I		
NAR 429	NT		I		
NAR 171	NT		R		
NAR 180	NT		R		
*OVE*A78	S		R	*ove*C200	*ove*A201, *OVE*A409
OVE 79	S		R	*ove*A201	
*ove*B100	S	*ove*A201 (NL)	R	*OVE* 130, *ove*C200, *ove*A201, *OVE*A409	*OVE*B300, *OVE*B302
*OVE*A102	S		R		*ove*A201
OVE 130	S		R	*ove*B100	
OVE 131	S		R		
OVE 133	S		R		
OVE 134	S		R	*ove*A201	
*ove*C200	S		R	*OVE*A78, *ove*B100	
*ove*A201	S	*ove*B100 (NL)	R	*OVE* 79, *ove*B100, *OVE* 134, *OVE* 202	*OVE*A78, *OVE*A102
OVE 202	S		R	*ove*A201	
*OVE*B300	S		R		*ove*B100
OVE 301	S		R		
*OVE*B302	S		R		*ove*B100

*OVE*A409	S		R	*ove*B100	*OVE*A78
OVE 701	S		D		
*OVE*K57	S		R	*OVE*J90, *OVE*J97, *ove*J99	
*OVE*J90	S		R	*OVE*K57	*OVE*J107
*OVE*J97	S		R	*OVE*K57	*OVE*J107
*ove*J99	S		R	*OVE*K57	*OVE*J107
*OVE*J107	S		R		*OVE*J90, *OVE*J97, *ove*J99
*ptr*A1	SS	*ptr*B2 (56%), *ptr*B3 (68%)	R	*ptr*B2, *ptr*B3, *PTR*C4, *PTR*C5	
*ptr*B2	F	*ptr*A1 (56%), *ptr*B3 (0%)	R	*ptr*A1, *PTR*C4, *PTR*C5	*ptr*B3
*ptr*B3	F	*ptr*A1 (68%), *ptr*B2 (0%)	R	*ptr*A1, *PTR*C5	*ptr*B2
*PTR*C4	F		R	*ptr*A1, *ptr*B2	*PTR*C5
*PTR*C5	S		R	*ptr*A1, *ptr*B2, *ptr*B3	*PTR*C4

[a] Mutant alleles, which have been established to segregate in crosses in a Mendelian manner, are represented by italicized lower-case three-letter symbols, followed by a single upper-case letter indicating the complementation group (if known), and then the allele number (e.g., *ptr*A1). If italicized upper-case symbols are used instead (e.g., *PTR*C4) these represent putative alleles which result in a stable mutant phenotype. Here it is simplest to assume that the abnormal phenotype is caused by the possession of a single mutant allele, but this condition has not yet been established critically. In most cases, the Mendelian behavior of genes has been demonstrated by examining the pattern of segregation of wild-type and mutant phenotypes among either haploid progeny grown from spores produced by meioses within a crossed diploid sporophyte[15,64,66] or the diploid progeny produced by meioses within a tetraploid sporophyte generated by self-fertilization of a diploid gametophyte which was itself obtained either by somatic hybridization[7,65,66] or by aposporous generation of diploid tissue from an immature crossed sporophyte.[15] Occasionally the segregation pattern among plants grown from spores from a sporophyte which appears to have arisen apogamously or parthenogenetically on a diploid somatic hybrid has been used to show that certain genes are behaving in a Mendelian fashion.[22,68] Examples of each of the segregation patterns observed in *P. patens* among progeny developing from spores of different origins can be found in Reference 22. **Abbreviations:** *BAR*, 6-benzylaminopurine resistant; *BAR* 1, 2, and 297 belong to category 4, *BAR* 130 belongs to category 5, and *BAR* 61 and 330 belong to category 6.[67] *GAD*, abnormal gametophore development; all *GAD* strains listed are ''narrow-leaved'' mutants.[68] *GAM*, exhibiting an absence of or reduced and/or delayed production of gametophores. *GTR*, gravitropically abnormal. *NAR*, 1-naphthaleneacetic acid resistant; *NAR* 91, 112, and 113 belong to category 1, *NAR* 87 and 429 belong to category 2, and *NAR* 171 and 180 belong to category 3.[67] *OVE*, gametophore overproducing; *OVE*J90, J97, J99, J107, and K57 belong to category 10, and all other *OVE* strains listed belong to category 9.[70] *PTR*, phototropically abnormal.

TABLE 1 (continued)
Complementation Analyses of Developmentally Abnormal Mutants of *Physcomitrella patens*

[b] **Abbreviations:** F, fertile, i.e., capable of self- and cross-fertilization; S, sterile, i.e., incapable of self- or cross-fertilization; SS, self-sterile, but capable of cross-fertilization; NT, not tested. Most mutants which appear to be completely sterile are so because they are blocked or grossly abnormal at early stages of gametophyte development.

[c] Only recombination frequencies between genes which can mutate to have a similar phenotypic effect are given.[15,66,69] When only one recombinant genotype can be recognized phenotypically the frequency of this recombinant class has been doubled to obtain the recombination frequency. Recombination frequencies between genes having dissimilar effects on the phenotype are available elsewhere.[15,69] The recombination frequencies quoted in this table were obtained by performing conventional sexual crosses. By examining the phenotypes of progeny plants obtained by self-fertilization of an *ove*A201/*ove*B100 somatic hybrid it has been demonstrated that there is no linkage (NL) between the *ove*A and *ove*B genes.[65]

[d] **Abbreviations:** D, dominant; R, recessive; I, incompletely dominant; NT, not tested. The dominance status of the mutant alleles with respect to their wild-type alleles has been ascertained from the phenotypes of heterozygous somatic hybrids[7,20,65-67,71-73] and/or aposporous diploids.[15,64]

[e] The ability or inability of pairs of independently isolated mutants with identical or similar phenotypes to complement one another has been established in all cases from the phenotypes of somatic hybrids made between the strains concerned by protoplast fusion.[20,65-67,71-73]

TABLE 2
Frequencies of Recovery of Kanamycin-Resistant Regenerants from *P. patens* Protoplasts Treated With Plasmid DNA

		Frequency of kanamycin-resistant regenerants after 13—19 days	
Experiment[a]	Plasmid[b]	Per recoverable protoplast	Per μg DNA
A	pSS1[62]	5.6×10^{-4}	6
B	pSS1[62]	7.9×10^{-4}	5
C	pSS1[62]	2.3×10^{-4}	2
	pBin19[36]	1.6×10^{-3}	11
	pKC7[35]	1.5×10^{-4}	1

[a] A, B, and C are independent experiments. Selection was applied at day 0 in all experiments, but in experiments A and B at 12.5 μg/ml and in experiment C at 25 μg/ml kanamycin sulfate (Sigma). Protoplasts were regenerated for 7 days on osmotically buffered medium (Section II) containing kanamycin and then transferred to standard medium containing fresh kanamycin at weekly intervals.

[b] The vector-only control in each experiment was plasmid pBR322.[63]

V. GENETIC TRANSFORMATION

The ability to transform prokaryotes has made a significant contribution to our understanding of gene expression and regulation. More recently, transformation has been reported in a number of plant[23-25] and animal species, and similar advances in the study of eukaryotes are already apparent.[26,27] This technique is likely to be of key importance in determining the genes responsible for development.

Ideally, plant cell transformation should involve the uptake of DNA into a single cell from which a new plant develops. Protoplasts are therefore an ideal vehicle, preferable in this respect to the use of whole plants or tissue explants, but their use is dependent upon their efficient isolation and regeneration. In *P. patens*, protoplasts are derived from chloronemal cells and are capable of regenerating directly into a new plant (see Section II). In higher plant species, protoplast manipulation is more complex and time-consuming. Protoplasts must first regenerate into undifferentiated callus tissue, which must then be induced to differentiate. Techniques to achieve this are not yet available for many plant species.

Because protoplast isolation and regeneration is straightforward in *P. patens* we have used protoplasts to transform cells to kanamycin resistance by the direct gene transfer method.[28] The protocol used involves incubating plasmid DNA with protoplasts in the presence of polyethylene glycol and exposing them to a mild heat shock.[21,29-31] Experiments have been performed using a number of different plasmids, each containing the aminoglycoside phosphotransferase gene which is responsible for inactivating the antibiotic kanamycin. Data are presented in Table 2 which show that kanamycin-resistant regenerants, not present in controls, can be recovered reproducibly after treatment with plasmid DNA.[32] Each experiment involved two controls, one in which the protoplasts were not exposed to DNA and one in which DNA from a plasmid containing no gene for kanamycin resistance was used. On no occasion have any kanamycin-resistant regenerants occurred in either of these control treatments. Strains of *P. patens* which carry a mutation resulting in auxotrophy were chosen for transformation so that the fate of the introduced gene could be followed after sexual crossing. Our preliminary data show that reproducible frequencies of regenerants are obtained which are comparable to the transformation frequencies obtained in other plant species.[33] Higher rates have now been obtained in other plants by the use of electroporation and other modifications,[34] and we are in the process of attempting some of these. These data also indicate that kanamycin-resistant regenerants may arise even if the transforming plasmid does not carry a eukaryotic promoter (Table 2).[30] The kanamycin resistance gene in plasmid pKC7 is derived from the transposon Tn5 and is preceded by a prokaryotic promoter.[35] The same kanamycin resistance gene in plasmids pBin19 and pSS1 has the promoter from the *nos* gene of the Ti plasmid of *Agrobacterium tumefaciens,* which functions in eukaryotic cells.[36] Although there has recently been a report of fungal cell transformation by a gene having a bacterial promoter,[25] the plasmid in this case also carries an insert of fungal DNA.

Since our intention is to use the transformation technique to study development in mosses, we are anxious to transform *P. patens* using heterologous cloned genes. One of the plasmids (pSS1, see Table 2) used to transform *P. patens* to kanamycin resistance also carries the cytokinin biosynthetic gene from *A. tumefaciens.* In an experiment involving the transformation of a morphologically wild-type strain, one kanamycin-resistant regenerant was obtained which also showed a phenotype similar to mutants known to overproduce cytokinin.[29] We are currently using the same plasmid to transform a developmentally mutant strain which may be defective in cytokinin biosynthesis in an attempt to restore the wild-type phenotype.

This technique has further potential in the mutagenesis of *P. patens*, either by the process of "transposon tagging"[37,38] or by simple insertional inactivation by the transforming plasmid DNA. The advantage of this form of mutagenesis is that the inactivated gene could be recovered for sequence analysis.

We are currently engaged in confirming *P. patens* transformation by the Southern hybridization technique.[34,39] There is one peculiarity of the selection of kanamycin-resistant transformants in *P. patens* which has so far prevented us from obtaining sufficient tissue from a number of regenerants for DNA isolation. After primary selection on kanamycin-containing medium, strongly resistant colonies are unable to regenerate normally after subculture. However, transfer of undisturbed regenerants to fresh kanamycin does not inhibit growth.[32] Because of this and because many inocula are also unable to survive transfer to minimal medium it is unlikely that the transformants are genetically unstable. Instead it seems that kanamycin or the presence of a kanamycin resistance gene may interfere with regeneration of chloronemal cells. Although this may indicate that selection for kanamycin resistance is unsuitable for *P. patens*, there may be ways of overcoming this problem by allowing caulonemal cells to develop before application of the antibiotic. Alternative selectable marker genes are also under consideration,[40] as are alternative transformation procedures.

VI. AUXIN ACCUMULATION AND TRANSPORT

Indole-3-acetic acid (IAA) has been detected in *P. patens*[41] and *Funaria hygrometrica*,[42] and auxins have been demonstrated to have significant roles in the growth and development of moss gametophytes.[19] Consequently, there is considerable interest in the nature of auxin accumulation and transport mechanisms in mosses.

In higher plants, auxin accumulation by cells and auxin transport between cells have been investigated most frequently using either tissue segments,[43] suspension-culture cells,[44] or membrane vesicles obtained from homogenized cells and containing a substantial population of outside-out sealed plasma membrane (PM) vesicles.[45-48] In mosses the same phenomena have been studied in whole moss gametophytes,[49] excised rhizoids,[50] protonemal fragments,[51,52] and, more recently, protoplasts.[51-53] Protoplasts offer several advantages over both whole-cell systems and membrane vesicles. In the whole-cell systems, accumulation within the cytoplasm has to be detected against a background of binding to the cell wall and sequestration within the cell wall solution as well as the solution in the free space between cells. Of these problems, which are either nonexistent or minimized when protoplasts are used, binding to the cell wall is probably the most problematic since it has been demonstrated that nitrocellulose, a nitrated derivative of cellulose, binds IAA, 1-naphthaleneacetic acid (NAA), and 2,4-dichlorophenoxyacetic acid (2,4-D) in a pH-dependent fashion which mimics the pH-dependent uptake of auxin by plant cells.[54] Protoplasts also offer an advantage over membrane vesicles. There appears to be a time limit on the ability of the vesicles to maintain the transmembrane pH difference (ΔpH) and/or inside-negative transmembrane electrical potential difference ($\Delta\Psi$) which are the driving forces responsible for auxin accumulation.[46,55] No such limitation seems to exist in the case of protoplasts, at least during the time frame of the experiments performed so far. Another advantage of protoplasts of *P. patens* for this kind of study is that almost all of the protoplasts isolated with the commercial enzyme mixture Driselase are derived from chloronemal cells, since the cell walls of caulonemata and gametophores are refractory to this enzyme solution. In many other systems a mixture of cell types which may differ with respect to their auxin accumulation and transport characteristics is present. In *P. patens*, chloronemal protoplasts may be compared with protonemal fragments consisting of both chloronemal and caulonemal cells, presenting the possibility of discovering whether these two cell types have identical or dissimilar auxin accumulation and transport properties. One disadvantage of protoplasts is that they cannot be used to investigate the polar nature of auxin transport.

The methods for investigating auxin accumulation and transport by moss protoplasts are basically of two kinds. In type A experiments, protoplasts are incubated for specified times in osmotically and pH-buffered solutions containing a low concentration of radioactively labeled auxin, typically 10 n*M* ^{3}H-IAA or 1 μ*M* ^{14}C-IAA, and other appropriate additions, incubation being terminated by the separation of protoplasts and medium by centrifugation. Auxin accumulation is then assayed by transferring the protoplast pellet to scintillation fluid followed by liquid scintillation counting. If C_i/C_o (where C_i refers to the concentration of auxin inside the protoplasts and C_o is the concentration in the incubation medium) is to be computed, the protoplast pellet volume must be measured and a measured volume of incubation medium must be assayed to estimate the concentration of auxin remaining in it at the end of the incubation period. In type B experiments, net auxin efflux is examined by preloading protoplasts with labeled auxin by incubating them as described for type A experiments, followed by removal of the incubation medium and its replacement with "efflux medium" containing no radioactive auxin. The radioactivity remaining in the protoplasts as well as that appearing in the efflux medium is then monitored periodically. In both approaches, if the experiments are performed within closed thermodynamic systems, such as microcentrifuge tubes, the same steady-state C_i/C_o will be attained eventually since, in both cases, at any given time C_i/C_o is determined by the relative rates of influx and efflux, which will be equal when the steady-state position is reached.

The following unpublished results, some of which have already been presented at conferences,[51,52] have been obtained using *P. patens* protoplasts.

1. **Method A:** the extent of accumulation of IAA and 2,4-D is time dependent (Table 3), with net influx continuing to increase for at least 70 min, although the rate of accumulation has decreased greatly by this time. **Method B:** retention of IAA by preloaded protoplasts decreases with time. The greatest loss, more than 80% of the IAA originally present, occurs within the first 15 min, but net efflux continues for at least 65 min (Table 4).

TABLE 3
Accumulation of IAA and 2,4-D by Protoplasts of Wild-Type *P. patens*

pH of incubation medium	Incubation time					
	7 min		35 min		65 min	
	IAA	2,4-D	IAA	2,4-D	IAA	2,4-D
4	5.3	8.8	10	13	—	—
5	2.5	2.8	—	—	5.9	6.5
6	0.80	1.1	—	—	1.6	1.0
7	0.92	0.97	—	—	0.87	0.88

Note: The data, given to two significant figures, represent net influx of IAA or 2,4-D, expressed in pmol per μl of packed protoplasts, at the times indicated when protoplasts are incubated in media initially containing 1 μ*M* ^{14}C-IAA or 1 μ*M* ^{14}C-2,4-D.[58]

TABLE 4
The Effect of 2,3,5-TIBA on IAA Retention by *P. patens* Protoplasts

Additions	Time in efflux medium		
	15 min	40 min	65 min
None	15	8.8	7.8
10 μ*M* TIBA	17	9.6	8.2
100 μ*M* TIBA	20	10	7.8

Note: *P. patens* protoplasts were preloaded with IAA by incubating them in 1 μ*M* ^{14}C-IAA; then the protoplasts were pelleted by gentle centrifugation and the medium replaced by pH 5 efflux medium containing the additions indicated. The data, given to two significant figures, are the percentages of labeled IAA remaining within the protoplasts.

2. **Method A:** accumulation of IAA and 2,4-D is dependent upon the pH of the external medium. The extent of net influx can be ordered as follows, where numbers refer to medium pH: 4 > 5 > 6 > or ≃ 7. The magnitude of 2,4-D net influx is greater than that of IAA at pH 4 and 5, but approximately the same at pH 6 and 7 (Table 3). **Method B:** net efflux of IAA is pH dependent also, being greater in media of higher pHs.
3. **Method A:** at medium pHs of 4, 5, and 5.5, accumulation of IAA is decreased by osmotic shock, freeze-thawing, and 1 m*M* 2,4-dinitrophenol (DNP), the size of the effect increasing with decreasing pH of the medium. At pH 6 and 7, net auxin influx is unaffected or slightly increased by these treatments. **Method B:** retention of IAA is decreased by 1 m*M* DNP at pH 4, 5, and 6, with the effect being greater at lower pHs. At pH 7, DNP increases retention slightly. Retention is decreased by freeze-thawing at pH 4, 5, 6, and 7.
4. **Method A:** the net influx of radioactively labeled IAA from medium initially containing either 1 μ*M* ^{14}C-IAA or 10 n*M* ^{3}H-IAA is affected little or not at all by the presence of an excess of unlabeled IAA. A wide range of incubation times (5 to 73 min), medium pHs (4 to 7), and IAA concentrations (2 μ*M* to 1 m*M*) have been tested.

5. **Method A:** net influx of IAA is unaffected or increased only very slightly by *N*-1-naphthylphthalamic acid (NPA) tested in the concentration range 500 n*M* to 10 μ*M*. It is unaffected by 2,3,5-triiodobenzoic acid (TIBA) between 500 n*M* and 5 μ*M*.
6. **Method B:** in most experiments, net efflux of IAA into medium of pH 5 was reduced by 1 m*M* unlabeled IAA, 10 and 50 μ*M* NPA, 1, 10, and 100 μ*M* 2,3,5-TIBA, 10 and 100 μ*M* 3,4,5-TIBA, and 10 and 100 μ*M* *p*-chlorophenoxyisobutyric acid (PCIB). In no experiment did any of these additions stimulate net efflux. The effect could be detected only within the first 15 min following transfer of preloaded protoplasts to efflux medium. The data in Table 4, while showing specifically the effect of 2,3,5-TIBA on IAA retention, are fairly representative of the effect of any of the substances mentioned above.

In higher plant systems the ability of cells to accumulate and transport auxin can be explained satisfactorily by the chemiosmotic model of auxin transport postulated by Rubery and Sheldrake,[56] supplemented by including a specific influx carrier. The most recent data suggest that three components are involved in auxin transport:

1. Non-carrier-mediated, simple diffusion of undissociated IAA (IAAH) through the PM followed by its ionization and trapping of the IAA anions (IAA^-) produced within the more alkaline cytosol, the extent of accumulation depending upon ΔpH
2. Carrier-mediated transport by an $IAA^-/2H^+$ symport, influx being driven by both ΔpH and ΔΨ[47,48]
3. Facilitated IAA^- efflux by an IAA^- uniport, efflux being driven by the ΔΨ and $\Delta[IAA^-]$ across the PM.[48] It is generally accepted that polar auxin transport inhibitors increase net auxin influx by inhibiting this IAA^- efflux carrier.[57]

The results reported here using *P. patens* protoplasts are consistent with most findings obtained from higher plants and with the model outlined above. However, it is necessary to account for the failure to demonstrate a saturable component to auxin accumulation. There are a number of possible explanations, including:

1. The influx carrier may be inactivated or not synthesized under the culture conditions employed. Rose et al.[49] have shown that moss grown in high light intensity (2.0 W m^{-2}), HLM, does not exhibit saturable auxin uptake, whereas moss grown in low light intensity (0.50 to 0.56 W m^{-2}) does. The *P. patens* tissue used for isolating protoplasts was grown at light intensities considerably higher than those employed by Rose et al.[49] for HLM. However, it was grown in the presence of ammonium as an additional nitrogen source which preferentially favors the production of chloronemata; the low light conditions of Rose et al.[49] favored the production of the same tissue type. This explanation of the absence of saturable auxin accumulation also seems unlikely because protonemal fragments made from tissue grown under the same high light intensity conditions do exhibit saturable uptake.
2. Driselase is a crude preparation of mixed enzymes, including cellulase and pectinase, and may contain proteases. These might degrade any portion of the influx carrier protein(s) which protrude on the outer surface of the PM. The action of such proteases upon IAA^- efflux carriers could also account for the relatively weak effect of polar auxin transport inhibitors on auxin accumulation and retention by *P. patens* protoplasts.
3. The influx carrier may be absent from chloronemal cells but present in other cell types, e.g., caulonemal cells and/or rhizoids. This would enable some cell types to accumulate IAA preferentially at the expense of others in which simple IAAH diffusion is the only or at least major uptake mechanism, as has been suggested in the case of higher plants.[57]

The observed, albeit small, reduction in net auxin efflux in the presence of the polar auxin transport inhibitors NPA and 2,3,5- and 3,4,5-TIBA, as well as PCIB and 1 m*M* IAA, is consistent with the presence of an IAA^- efflux carrier. However, the stimulatory effect of NPA upon auxin accumulation is noticeably more pronounced when *P. patens* protonemal fragments, rather than protoplasts, are utilized as the experimental system.[52,58] Consequently, comments similar to those made earlier with reference to the influx carrier may be equally valid for the efflux carrier. The observed effect of PCIB upon net IAA efflux is interesting, since it has been demonstrated that PCIB strongly inhibits polar auxin transport in *Funaria hygrometrica* rhizoids[50] and that it can prevent the differentiation of chloronema into caulonema[59] and induce the dedifferentiation of caulonema into chloronema.[60]

The finding that chloronemal protoplasts can accumulate 2,4-D to a greater extent than IAA from medium with pH 4 or 5 and to a similar extent when the medium pH is 6 or 7 (Table 2) is significant, since 2,4-D is much less effective than IAA in inducing the differentiation of chloronemal into caulonemal cells in *P. patens*.[61] It is now impossible or, at least, very difficult to account for the observed differential sensitivity to these two substances in terms of differential ability of moss cells to accumulate them, and another explanation must be sought.

REFERENCES

1. **Batra, A. and Abel, W. O.,** Development of moss plants from isolated and regenerated protoplasts, *Plant Sci. Lett.*, 20, 183, 1981.
2. **Binding, H.,** Regeneration und Verschmelzung nackter Laubmoos Protoplasten, *Z. Pflanzenphysiol.*, 55, 305, 1966.
3. **Gwozdz, E. A. and Waliszewska, B.,** Regeneration of enzymatically isolated protoplasts of the moss *Funaria hygrometrica*, Sibth, *Plant Sci. Lett.*, 15, 41, 1979.
4. **McClelland, D. J.,** Genetical Studies of Gametophyte Development in the Moss *Physcomitrella patens,* Ph.D. thesis, University of Leeds, Leeds, England, 1988.
5. **Stumm, I., Meyer, Y., and Abel, W. O.,** Regeneration of the moss *Physcomitrella patens* (Hedw.) from isolated protoplasts, *Plant Sci. Lett.*, 5, 113, 1975.
6. **Saxena, P. K. and Rashid, A.,** Development of gametophytes from isolated protoplasts of the moss *Anoectangium thomsonii* Mitt., *Protoplasma,* 103, 401, 1980.
7. **Grimsley, N. H., Ashton, N. W., and Cove, D. J.,** The production of somatic hybrids by protoplast fusion in the moss, *Physcomitrella patens, Mol. Gen. Genet.*, 154, 97, 1977.
8. **Boyd, P. J. and Knight, C. D.,** unpublished data.
9. **Jenkins, G. I. and Cove, D. J.,** Light requirements for the regeneration of protoplasts of the moss, *Physcomitrella patens, Planta,* 157, 39, 1983.
10. **Jenkins, G. I. and Cove, D. J.,** Phototropism and polarotropism of primary chloronemata of the moss, *Physcomitrella patens:* responses of the wild-type, *Planta,* 158, 357, 1983.
11. **Cove, D. J. and Knight, C. D.,** Gravitropism and phototropism in the moss, *Physcomitrella patens*, in *Developmental Mutants in Higher Plants,* Thomas, H. and Grierson, D., Eds., Cambridge University Press, London, 1987, 181.
12. **Jenkins, G. I. and Cove, D. J.,** unpublished data.
13. **Knight, C. D. and Cove, D. J.,** The genetic analysis of tropic responses, *Environ. Exp. Bot.*, 29, 57, 1989.
14. **Henderson, D. J., Jenkins, G. I., and Cove, D. J.,** unpublished data.
15. **Ashton, N. W. and Cove, D. J.,** The isolation and preliminary characterisation of auxotrophic and analogue resistant mutants of the moss, *Physcomitrella patens, Mol. Gen. Genet.*, 154, 87, 1977.
16. **Knight, C. D., Cove, D. J., Boyd, P. J., and Ashton, N. W.,** The isolation of biochemical and developmental mutants in *Physcomitrella patens*, in *Methods in Bryology,* Glime, J. M., Ed., Hattori Botanical Laboratory, Nichinan, Miyazaki, Japan, 1988, 47.
17. **Boyd, P. J., Grimsley, N. H., and Cove, D. J.,** Somatic mutagenesis of the moss, *Physcomitrella patens, Mol. Gen. Genet.*, 211, 545, 1988.
18. **Grimsley, N. H. and Grimsley, J. M.,** unpublished data.

19. **Cove, D. J. and Ashton, N. W.,** The hormonal regulation of gametophytic development in bryophytes, in *The Experimental Biology of Bryophytes,* Dyer, A. F. and Duckett, J. G., Eds., Academic Press, London, 1984, 177.
20. **Grimsley, N. H., Ashton, N. W., and Cove, D. J.,** Complementation analysis of auxotrophic mutants of the moss, *Physcomitrella patens*, using protoplast fusion, *Mol. Gen. Genet.,* 155, 103, 1977.
21. **Ashton, N. W., Boyd, P. J., Cove, D. J., and Knight, C. D.,** Genetic analysis in *Physcomitrella patens*, in *Methods in Bryology,* Glime, J. M., Ed., Hattori Botanical Laboratory, Nichinan, Miyazaki, Japan, 1988, 59.
22. **Cove, D. J.,** Genetics of Bryophyta, in *New Manual of Bryology,* Schuster, R. M., Ed., Hattori Botanical Laboratory, Nichinan, Miyazaki, Japan, 1983, 222.
23. **An, G., Watson, B. D., and Chiang, C. C.,** Transformation of tobacco, tomato, potato and *Arabidopsis thaliana* using a binary Ti vector system, *Plant Physiol.,* 81, 301, 1986.
24. **Hasnain, S. E., Manavathu, E. K., and Leung, W. C.,** DNA-mediated transformation of *Chlamydomonas rheinhardii* cells: use of aminoglycoside 3′-phosphotransferase as a selectable marker, *Mol. Cell. Biol.,* 5, 3647, 1985.
25. **Suarez, T. and Eslava, A. P.,** Transformation of *Phycomyces* with a bacterial gene for kanamycin resistance, *Mol. Gen. Genet.,* 212, 120, 1988.
26. **Moses, P. B. and Chua, N.-H.,** Light switches for plant genes, *Sci. Am.,* 258, 64, 1988.
27. **Schell, J. S.,** Transgenic plants as tools to study the molecular organisation of plant genes, *Science,* 237, 1176, 1987.
28. **Paszkowski, J., Shillito, R. D., Saul, M., Mandak, V., Hohn, T., Hohn, B., and Potrykus, I.,** Direct gene transfer to plants, *EMBO J.,* 3, 2717, 1984.
29. **Knight, C. D.,** Gravitropism in the Moss *Physcomitrella patens,* Ph.D. thesis, University of Leeds, Leeds, England, 1987.
30. **Long, Z.,** Genetical and Biochemical Studies of Photosynthetic Mutants in the Moss *Physcomitrella patens,* Ph.D. thesis, University of Leeds, Leeds, England, 1986.
31. **Mohammed, T.,** Studies on Cloning Genes from *Neurospora crassa* in Heterologous Hosts, Ph.D. thesis, University of Leeds, Leeds, England, 1987.
32. **Knight, C. D.,** unpublished data.
33. **Cocking, E. C. and Davey, M. R.,** Gene transfer in cereals, *Science,* 236, 1259, 1987.
34. **Shillito, R. D. and Potrykus, I.,** Direct gene transfer to protoplasts of dicotyledonous and monocotyledonous plants by a number of methods including electroporation, *Methods Enzymol.,* 153, 313, 1987.
35. **Rao, R. N. and Rogers, S. G.,** Plasmid pKC7: a vector containing ten restriction endonuclease sites suitable for cloning DNA segments, *Gene,* 7, 79, 1979.
36. **Bevan, M.,** Binary *Agrobacterium* vectors for plant transformation, *Nucleic Acids Res.,* 12, 8711, 1984.
37. **Martin, C., Carpenter, R., Sommer, H., Saedler, H., and Coen, E. S.,** Molecular analysis of instability in flower pigmentation of *Antirrhinum majus* following isolation of the *palida* locus by transposon tagging, *EMBO J.,* 4, 1625, 1985.
38. **Peterson, P. A.,** Mobile elements in plants, *Crit. Rev. Plant Sci.,* 6(2), 105, 1987.
39. **Southern, E. M.,** Detection of specific sequences among DNA fragments separated by gel electrophoresis, *J. Mol. Biol.,* 98, 503, 1975.
40. **Rogers, S. G., Klee, H. J., Horsch, R. B., and Fraley, R. T.,** Improved vectors for plant transformation. Expression cassette vectors and new selectable markers, *Methods Enzymol.,* 153, 253, 1987.
41. **Ashton, N. W., Schulze, A., Hall, P., and Bandurski, R. S.,** Estimation of indole-3-acetic acid in gametophytes of the moss, *Physcomitrella patens, Planta,* 164, 142, 1985.
42. **Jayaswal, R. K. and Johri, M. M.,** Occurrence of biosynthesis of auxin in protonema of the moss *Funaria hygrometrica, Phytochemistry,* 24, 1211, 1985.
43. **Sussman, M. R. and Goldsmith, M. H. M.,** Auxin uptake and action of *N*-1-naphthylphthalamic acid in corn coleoptiles, *Planta,* 150, 15, 1981.
44. **Rubery, P. H.,** The specificity of carrier-mediated auxin transport by suspension-cultured crown gall cells, *Planta,* 135, 275, 1977.
45. **Jacobs, M. and Hertel, R.,** Auxin binding to subcellular fractions from *Cucurbita* hypocotyls: in vitro evidence for an auxin transport carrier, *Planta,* 142, 1, 1978.
46. **Hertel, R., Lomax, T. L., and Briggs, W. R.,** Auxin transport in membrane vesicles from *Cucurbita pepo* L., *Planta,* 157, 193, 1983.
47. **Benning, C.,** Evidence supporting a model of voltage-dependent uptake of auxin into *Cucurbita* vesicles, *Planta,* 169, 228, 1986.
48. **Sabater, M. and Rubery, P. H.,** Auxin carriers in *Cucurbita* vesicles. II. Evidence that carrier-mediated routes of both indole-3-acetic acid influx and efflux are electroimpelled, *Planta,* 171, 507, 1987.
49. **Rose, S., Rubery, P. H., and Bopp, M.,** The mechanism of auxin uptake and accumulation in moss protonemata, *Physiol. Plant.,* 58, 52, 1983.

50. **Rose, S. and Bopp, M.,** Uptake and polar transport of indoleacetic acid in moss rhizoids, *Physiol. Plant.*, 58, 57, 1983.
51. **Ashton, N. W.,** The role of medium acidification and auxin transport in the morphogenesis of *Physcomitrella patens, Br. Bryol. Soc. Bull.*, 49, 16, 1987.
52. **Ashton, N. W. and Bridge, P.,** Accumulation of radioactively labelled indole-3-acetic acid by wild type and auxin insensitive mutants of the moss, *Physcomitrella patens,* Abstract 2-10-5, presented at XIV Int. Botanical Congr., Berlin, July 24 to August 1, 1987.
53. **Bopp, M., Geier, V., and Kessler, V.,** Moss-protonema: protoplasts and transport of auxin, Abstracts 28 and 29, presented at Bryological Methods Workshop, Satellite Conf. XIV Int. Botanical Congr., Mainz, West Germany, July 17 to 23, 1987.
54. **Ashton, N. W. and Bridge, P.,** unpublished data.
55. **Sabater, M. and Sabater, F.,** Auxin carriers in membranes of lupin hypocotyls, *Planta,* 167, 76, 1986.
56. **Rubery, P. H. and Sheldrake, A. R.,** Carrier-mediated auxin transport, *Planta,* 118, 507, 1974.
57. **Goldsmith, M. H. M.,** A saturable site responsible for polar transport of indole-3-acetic acid in sections of maize coleoptiles, *Planta,* 55, 68, 1982.
58. **Ashton, N. W.,** unpublished data.
59. **Sood, S. and Hackenberg, D.,** Interaction of auxin, antiauxin and cytokinin in relation to the formation of buds in moss protonema, *Z. Pflanzenphysiol.*, 91, 385, 1979.
60. **Bopp, M.,** The hormonal regulation of morphogenesis in mosses, in *Proceedings in Life Sciences. Plant Growth Substances 1979,* Skoog, F., Ed., Springer-Verlag, Berlin, 1980, 351.
61. **Ashton, N. W., Lomax-Reichert, T., and Ray, P. M.,** Studies on the mode of action of auxin in *Physcomitrella patens,* Abstract p. 14, presented at 206th Conf. Society for Experimental Biology, Trinity College, Dublin, Ireland, 1982.
62. **Schofield, S.,** unpublished data.
63. **Maniatis, T., Fritsch, E. F., and Sambrook, J.,** *Molecular Cloning, a Laboratory Manual,* Cold Spring Harbor Laboratory, Cold Spring Harbor, NY, 1982, 479.
64. **Engel, P. P.,** The induction of biochemical and morphological mutants in the moss *Physcomitrella patens, Am. J. Bot.*, 55, 438, 1968.
65. **Featherstone, D. R.,** Studies of Mutants of the Moss, *Physcomitrella patens,* which Over-Produce Gametophores, Ph.D. thesis, University of Cambridge, Cambridge, England, 1980.
66. **Courtice, G. R. M.,** Developmental Genetic Studies of *Physcomitrella patens,* Ph.D. thesis, University of Cambridge, Cambridge, England, 1979.
67. **Ashton, N. W., Grimsley, N. H., and Cove, D. J.,** Analysis of gametophytic development in the moss, *Physcomitrella patens,* using auxin and cytokinin resistant mutants, *Planta,* 144, 427, 1979.
68. **Courtice, G. R. M. and Cove, D. J.,** Mutants of the moss, *Physcomitrella patens,* which produce leaves of altered morphology, *J. Bryol.*, 12, 595, 1983.
69. **Courtice, G. R. M., Ashton, N. W., and Cove, D. J.,** Evidence for the restricted passage of metabolites into the sporophyte of the moss *Physcomitrella patens* (Hedw.) Br. Eur., *J. Bryol.*, 10, 191, 1978.
70. **Ashton, N. W., Cove, D. J., and Featherstone, D. R.,** The isolation and physiological analysis of mutants of the moss, *Physcomitrella patens,* which over-produce gametophores, *Planta,* 144, 437, 1979.
71. **Grimsley, N. H., Featherstone, D. R., Courtice, G. R. M., Ashton, N. W., and Cove, D. J.,** Somatic hybridization following protoplast fusion as a tool for the genetic analysis of development in the moss *Physcomitrella patens,* in *Advances in Protoplast Research — Proc. 5th Int. Protoplast Symp. 1979,* Akadémiai Kiadó, Budapest, Hungary, 1980, 363.
72. **Grimsley, N. H.,** Genetic Analysis of the Moss *Physcomitrella patens* Using Protoplast Fusion, Ph.D. thesis, University of Cambridge, Cambridge, England, 1978.
73. **Futers, T. S., Wang, T. L., and Cove, D. J.,** Characterisation of a temperature-sensitive gametophore over-producing mutant of the moss, *Physcomitrella patens, Mol. Gen. Genet.*, 203, 529, 1988.

Chapter 2

MUTANTS AS TOOLS FOR THE ANALYTICAL DISSECTION OF CELL DIFFERENTIATION IN *PHYSCOMITRELLA PATENS* GAMETOPHYTES

Neil W. Ashton and David J. Cove

TABLE OF CONTENTS

I. INTRODUCTION

The amenability of *Physcomitrella patens* to genetic, biochemical, and physiological investigations of its development has been well documented,[1-26] and the most significant traits of this plant contributing to its suitability for such studies have recently been summarized.[27]

Our studies of *P. patens* development, and those of our co-workers, have been concerned primarily with two aspects of gametophytic morphogenesis:

1. The roles and interaction of light and plant hormones in regulating cell differentiation[1,3,4,7,9,10,12,13,15,17,24-28]
2. The phototropic and gravitropic responses of protonemal filaments and gametophores[6,11,12,18-21,29]

In both cases, developmentally abnormal mutants have proven to be immensely valuable analytical tools. In most instances the mutants have been induced by treating haploid spores with a chemical mutagen, usually *N*-methyl-*N'*-nitro-*N*-nitrosoguanidine (NTG) or ethyl methanesulfonate (EMS), although occasionally other mutagens, including ultraviolet (UV) light, have been utilized.[3,4,12,14,22,30] In the future it may prove more convenient to induce developmental mutations by somatic mutagenesis, a technique recently perfected in *P. patens*.[7,22,31]

The majority of tropically abnormal *P. patens* strains[11,12,19-21] have been obtained, following mutagenesis, by means of nonselective (i.e., total) isolation procedures.[22,30] Mutants affected in cell differentiation have been isolated both nonselectively[3,4,22] and selectively.[4,22] In the nonselective isolation procedure mutagenized spores have been allowed to germinate and grow into gametophytes, which have been examined visually to detect mutants with abnormal morphologies. Many morphologically abnormal strains obtained in this way have been shown subsequently to have altered sensitivities to exogenous auxin and/or cytokinin.[4,22] Many mutants isolated selectively by their resistance to concentrations of the synthetic auxin 1-naphthaleneacetic acid (NAA) or the synthetic cytokinin 6-benzylaminopurine (BAP), which cause profound changes in the growth and development of the wild-type, have been found to be altered developmentally even when grown in the absence of exogenous hormones.[4,22]

Philosophical considerations underlying the isolation of abnormal strains, possessing genes which have been mutated by conventional means[30] or by insertional inactivation with foreign DNA introduced by a genetic transformation procedure,[6,27,32] and their utilization for studying developmental processes are discussed in the first chapter of this volume and elsewhere.[6,7,10,21]

Tropisms, including the use of mutants for their study, in *P. patens* have been reviewed very recently and, therefore, will not be considered further in this chapter.[11,12] Instead, we shall examine what has been discovered so far about the hormonal and photoregulation of cell differentiation in this moss. Emphasis will be placed upon what has been ascertained through the cytological and physiological analysis of auxin and/or cytokinin sensitivitity mutants. We wish at this point to stress that, while it is simplest to assume that the developmentally abnormal phenotype of each of these mutants is caused by the possession of a single mutant allele, in most cases we have not yet demonstrated this critically. Failure of the genetic analysis of the most interesting mutants affected in cell differentiation to keep pace with studies of their cytology, physiology, and biochemistry has resulted mainly from two of their characteristics:

1. Many of the mutants are arrested at an early stage of gametophytic development and

they do not make gametangia. Genetic analysis by sexual crossing[6,30,33] is precluded in such strains.

2. Genetic analysis by somatic hybridization following protoplast fusion[6,16,34-36] has been hampered by the discovery that strains belonging to some categories of auxin and/or cytokinin sensitivity mutants possess mutant alleles which are dominant or incompletely dominant to their respective wild-type alleles.[4]

In this chapter, wherever they augment information obtained with the aid of mutants, data acquired without the use of mutants will be referred to also.

II. CELL DIFFERENTIATION IN WILD-TYPE *PHYSCOMITRELLA PATENS*

The following description refers to the gametophytic development of wild-type *P. patens* when cultured on a completely defined medium[22,30] containing nitrate as the sole nitrogen source and solidified with agar, at 25°C, under continuous standard (>60 μmol m^{-2} s^{-1}) white light (WL) or high levels (>16 μmol m^{-2} s^{-1}) of monochromatic red (665 nm) light (MRL).

Under these conditions spores germinate and protoplasts, in the presence of a suitable osmotic buffer, as well as isolated protonemal fragments regenerate to produce filaments of chloronemal cells, primary chloronemata, which grow by apical cell division. Subapical cells divide once or twice to produce further apical cells from which side branches, also composed of branching chloronemata, are derived. The cross walls formed by the division of primary chloronemal apical cells, about every 20 h,[37] are perpendicular to the long axis of the chloronema, the cells of which contain numerous chloroplasts.

After 5 or 6 days some chloronemal apical cells divide to produce cells of a second type, caulonemal apical cells, which divide more frequently than chloronemal apical cells, about every 5.5 to 6 h.[23] The occurrence of this transition probably requires several cell divisions for its completion, as has been reported in *Funaria hygrometrica*.[38] The caulonemal filaments, generated by the extension and division of caulonemal apical cells, comprise large cells which contain fewer chloroplasts and have cross walls at an oblique angle to the long axis of the filaments. Older caulonemal cells have walls impregnated with a red pigment. Most subapical cells of caulonemata divide once to form side branches which, at their inception, consist of single-cell side branch initials (SBIs). This process does not occur until the subapical cell becomes, as a result of the continued division of the apical cell, the third or fourth cell in the caulonema.[23] A small proportion (about 1%) of subapical caulonemal cells never form SBIs.[23] Between 85 and 90% of SBIs give rise to compound filaments of secondary chloronemal cells, which are morphologically similar to primary chloronemata.[23] However, these two kinds of chloronemata cannot be physiologically identical since they exhibit different phototropic characteristics.[39] Between 5 and 6% of SBIs grow into further caulonemal filaments.[23] The remaining SBIs either show no further development or they develop into gametophore buds which subsequently become the leafy shoots bearing the gametangia. In the wild-type grown on medium lacking exogenous hormones no more than one gametophore arises on each caulonema.

The probability of commitment of SBIs to these various developmental fates is affected by many factors, including nutritional status, plant hormones, and the quality and quantity of light. Nutritional status, which has been dealt with at least briefly elsewhere, will not be considered in this chapter.[23]

With respect to the influences of exogenous plant hormones, the qualitative results obtained by Ashton et al.[4] and the elegant quantitative data of McClelland[23] have revealed that:

1. NAA increases the proportion of SBIs developing into secondary caulonemal filaments while reducing the proportion of secondary chloronemata by a similar amount as well as inhibiting the extent of branching of remaining chloronemata, but has no effect upon the number of gametophores which arise from SBIs.
2. BAP increases the proportion of SBIs giving rise to buds, produces a corresponding decrease in the proportion of secondary chloronemata, and has no effect on the number of SBIs developing into secondary caulonemata. Like NAA, BAP also inhibits the branching of any secondary chloronemata which are formed.

Most of our studies on the regulation of SBI fate by light have been of two types:

1. Cultures have been grown from the outset in a variety of light regimes including, for example, continuous WL,[23] continuous polychromatic red light (PRL) (Table 1), and various levels of MRL (Table 2).
2. Dark-grown caulonemal filaments which do not have SBIs arising from their subapical cells have been exposed to MRL. Very low doses of light, e.g., 20 s of MRL at 25 $\mu mol\ m^{-2}\ s^{-1}$, induce the majority of subapical cells in such filaments to form a side branch.[39] The developmental fates of side branches arising from dark-grown caulonemata upon transfer to continuous illumination with various levels of MRL have been examined in detail,[39] and a preliminary account of the data obtained has been presented elsewhere.[27]

In this chapter, reference will be made to the results of these investigations on the photodetermined fates of SBIs wherever it is appropriate to do so.

III. MUTANTS AS TOOLS FOR STUDYING THE ROLES OF HORMONES, AND THEIR INTERACTION WITH LIGHT, IN THE REGULATION OF CELL DIFFERENTIATION IN *P. PATENS*

Cytological and physiological analysis of mutants, recognized originally by their resistance to high concentrations of either NAA (5 to 25 μM) or BAP (50 to 500 μM) in the culture medium when grown in standard WL, has enabled us to discern several phenotypic categories. We have also been able to build a model of the roles of auxin and cytokinin in regulating cell differentiation during *P. patens* gametophytic development.[4,10] Briefly, the morphology of BAP-resistant (BAR) category 4 mutants, which on medium lacking exogenous hormones consists of primary chloronemal filaments, is repaired to near normality by low levels (e.g., 500 n*M*) of NAA, suggesting that the differentiation of primary chloronemal apical cells into caulonemal apical cells requires auxin. Category 5 BAR strains, which we believe to be a leaky version of category 4 mutants, produce a normal or nearly normal number of caulonemata, overproduce secondary chloronemal filaments, but form no or few gametophores. Category 5 mutants are slightly less resistant to exogenous cytokinin than category 4 strains, strengthening our view that they represent less extreme versions of the latter mutant type. Like category 4 mutants, category 5 strains are repaired by low levels of NAA, indicating that the formation of gametophores and the suppression of the overproliferation of secondary chloronemata also require auxin, probably at a somewhat higher level than is needed for the differentiation of caulonemal apical cells. The morphology of category 2, NAA-resistant (NAR) mutants resembles that of category 5 BAR strains. However, category 2 mutants are resistant to exogenous NAA and are repaired by low levels (5 to 50 n*M*) of exogenous BAP, suggesting that, in addition to auxin, cytokinin is required for bud formation and to inhibit secondary chloronemal growth. It can also be deduced from our

TABLE 1
The Phenotypes of Wild-Type and a Selection of Hormone Sensitivity Mutants Grown for 4 Weeks in Continuous Polychromatic Red Light

Category[a]	Representative strain	Morphologies of gametophytes grown under PRL[b]	
		Control medium[c]	Medium containing 100 nM BAP
	WT	*Upper surface of colony:* caulonemata give rise to long, unbranched, or poorly branching secondary chloronemata. *Lower surface of colony:* caulonemata give rise to many secondary caulonemata. Each gametophyte also produces 100 to 200 etiolated leafy shoots with many basal rhizoids as well as rhizoids emanating from all regions of the stems.	*Sensitive:* only a small amount of protonemal tissue, comprised mainly of caulonemata or caulonema-like filaments. Thousands of small leafy shoots are present with many on each protonemal filament.
1 (1A)	NAR 91	Primary chloronemata only.	*Insensitive*
1 (1B)	NAR 185	Primary chloronemata with only a few or no caulonemata. No gametophores.	*Strongly resistant:* growth and development almost identical to that on medium containing no BAP.
1 (2A)	NAR 113	Primary chloronemata only.	*Sensitive and repaired:* production of chloronemata inhibited; formation of caulonemata and thousands of small, leafy shoots induced.

Note: All the mutants described in this table, as well as others belonging to these categories, have also been cultured under PRL in the presence of exogenous auxins: all of them are insensitive to or very strongly resistant to both NAA and IAA at concentrations ranging from 200 nM to 5 μM. Under PRL the wild-type exhibits morphological changes in the presence of IAA or NAA which are very similar to those induced by NAA in WL-grown cultures.[4] However, in PRL the sensitivity of wild-type tissues to NAA is greater than in WL. Morphologically abnormal category 1 mutants, which are resistant to both auxin and cytokinin, have been subdivided into two groups, 1A and 1B, on the basis of their differing phototropic responses in unilateral PRL and other distinguishing traits, further discussion of which will not be attempted here.

[a] The original categories[4] to which these mutants were assigned are given together with (in brackets) new category designations which seem more appropriate in the light of more recent data, some of which we have reported in this chapter and elsewhere.[27]

[b] Polychromatic red light (PRL) was obtained using the same light sources as for standard while light (WL), details of which have been published already,[30,34] and covering the petri plates with one layer of red (Roscolux No. 27) filter. Photon flux of PRL was approximately 6 μmol m^{-2} s^{-1} at the surface of the culture medium.

[c] Control medium was ABC medium supplemented with thiamine HCl, *p*-aminobenzoic acid, nicotinic acid,[22] and 0.5% sucrose.

findings that sensitivity of *P. patens* protonemal cells to each type of hormone requires the presence of the other type. This has been demonstrated directly in the case of cytokinin sensitivity using a category 4 mutant, BAR 1.[4] This EMS-induced morphologically abnormal strain, which was obtained through a total isolation procedure, is resistant to both the synthetic cytokinin BAP and to exogenous N^6-(Δ^2-isopentenyl) adenine (i^6Ade), which also appears to be the major endogenous cytokinin of *P. patens*.[24-26] However, in the presence of approximately 500 nM NAA, BAR 1 is about as sensitive as wild-type to exogenously supplied i^6Ade.[4]

TABLE 2

The Phenotypes of Wild-Type and a Selection of Auxin-Insensitive, Cytokinin-Repairable Morphological Mutants Cultured for 5 Weeks, in the Absence and Presence of Exogenous Cytokinin, in Various Levels of Continuous Monochromatic Red Light

Strain	Category[a]	100 n*M* BAP in culture medium	Photon flux of continuous monochromatic red light[b]			
		No	34—78 nmol m^{-2} s^{-1}	302—526 nmol m^{-2} s^{-1}	974 nmol—1.65 μmol m^{-2} s^{-1}	3.71—8.21 μmol m^{-2} s^{-1}
		Yes	17—101 nmol m^{-2} s^{-1}	297—532 nmol m^{-2} s^{-1}	946 nmol—1.47 μmol m^{-2} s^{-1}	2.67—8.76 μmol m^{-2} s^{-1}
WT		No	Primary caulonemata without secondary filaments. Most subapical caulonemal cells have side-branch initials. Ca. 10 leafy shoots per gametophyte at the highest photon flux in this range; <5 at the lowest photon flux.		*Upper surface:* Primary caulonemata with secondary chloronemata. *Lower surface:* primary caulonemata with secondary caulonemata and/or secondary unbranched filaments "intermediate" in appearance between caulonemata and chloronemata. Ca. 10—20 leafy shoots per gametophyte.	Primary caulonemata with secondary branched chloronemata. Ca. 50 leafy shoots per gametophyte.
		Yes	Caulonemata and/or rhizoids with no or few side-branches. Ca. 1000 small leafy shoots and buds; several gametophores per protonemal filament.		Ca. 1000 small leafy shoots emanating from a small amount of protonemal tissue.	Primary caulonemata with secondary caulonemata and secondary unbranched chloronemata together with several thousands of small leafy shoots.

NAR 112	1(2A)	No	Primary chloronemata only and less than at higher intensities.	Primary chloronemata only.		
		Yes	Caulonemata and rhizoids and some chloronemata. 500—1000 small shoots.	Central primary chloronemata, many primary and secondary caulonemata and/or rhizoids and several thousands of small leafy shoots and callus buds per gametophyte.		
NAR 113	1(2A)	No		Primary chloronemata only.		
		Yes	Primary chloronemata only.	Central primary chloronemata with some peripheral caulonemata. Thousands of normal leafy shoots present only at the periphery. Some buds observed on the central chloronemal tissue.		
NAR 87	2(2B)	No	Central primary chloronemata and peripheral abnormal caulonema-like filaments without side branches. No gametophores.	Central primary chloronemata and primary caulonemata with secondary chloronemata. No gametophores.	Central primary chloronemata. *Upper surface:* caulonemata with secondary chloronemata. *Lower surface:* primary caulonemata with secondary caulonemata and "intermediate" filaments. No gametophores.	Primary chloronemata only.
		Yes	100—200 small shoots. Some chloronemal tissue.	200—1000 small shoots, sometimes in clusters, emanating from caulonemata and/or "intermediate" filaments on which side-branch initials, but no or few filaments, have also developed.		No chloronemata apparent. Many caulonemata, on each of which there are many side-branch initials and several gametophores. Several thousands of small leafy shoots per gametophyte.

TABLE 2 (continued)
The Phenotypes of Wild-Type and a Selection of Auxin-Insensitive, Cytokinin-Repairable Morphological Mutants Cultured for 5 Weeks, in the Absence and Presence of Exogenous Cytokinin, in Various Levels of Continuous Monochromatic Red Light

[a] The original categories[4] to which these mutants were assigned are given together with (in brackets) new category designations which seem more appropriate in the light of more recent data, some of which we have reported in this chapter and elsewhere.[27]

[b] Monochromatic red (657—669 nm) light was obtained by passing light, provided by Aldis® Tutor 2 projectors (Rank Audio Visual Ltd., Brentford, England) with thyristor dimmer controls and 250 W tungsten-halogen bulbs (type M36; Thorn Lighting, Birmingham, England), through interference filters (DAL type; Schott, Mainz, F.R.G.) into otherwise light-proof culture chambers.

Recently it has been reported by Bhatla and Bopp[40] that a mutant of *Funaria hygrometrica*, which apparently has an almost identical phenotype to that of BAR 1, is deficient in endogenous IAA, containing less than one third of the amount present in wild-type *F. hygrometrica*. IAA oxidase activity in this mutant is nearly three times that in the wild-type. When grown under standard WL culture conditions for 3 weeks, category 1 mutants consisting only of primary chloronemal filaments are resistant to both NAA and BAP. Since we had no evidence that cytokinin is needed for the differentiation of caulonemal apical cells, but had shown that auxin is involved in this developmental transition, we concluded that category 1 strains might be affected either in auxin reception or in the process of signal transduction following hormone reception and leading to the hormone-induced response.[4-10]

Recently we presented data obtained by further analysis of category 1 and 2 strains.[27] Here we will discuss these data at greater length and also report additional results from our continued studies on other categories of hormone sensitivity mutants. Using this and other additional information we have constructed a more elaborate model of the control of cell differentiation in *P. patens* (Figure 1).

In order to discover whether NAR strains are insensitive to high levels of exogenous indole-3-acetic acid (IAA), which is known to be an endogenous auxin in *P. patens*,[5] as well as to exogenous NAA, and whether BAR strains may be repaired with low levels of IAA as well as with NAA, representative mutants have been grown under continuous PRL on medium containing IAA.[27] Under PRL, IAA in the growth medium is not photodegraded, whereas in standard WL it is degraded rapidly.[41] Sensitivity to exogenous NAA and to BAP has also been examined in PRL.[27] All category 1 and 2 strains, when cultured in PRL, are insensitive or strongly resistant to both IAA and NAA. A surprising result has been that some category 1 strains which we had previously demonstrated to be BAP resistant in WL[4] are, in PRL, as sensitive to BAP as the wild-type, and their abnormal phenotype is essentially repaired by low levels of the cytokinin[27] (Table 1). We now suggest, therefore, that cytokinin-repairable category 1 (redesignated category 2A) strains are relatively nonleaky versions of category 2 (redesignated 2B) mutants. Other category 1 strains are insensitive to BAP in WL and in PRL[27] (Table 1). Recently the sensitivity of category 2A and 2B mutants to exogenous cytokinin has also been demonstrated under continuous MRL[27] (Table 2). Also, these same category 2A strains are now known to exhibit some sensitivity to cytokinin, although less than wild-type, when cultured for an extended period of time in WL.[41] Since all category 2A mutants appear to consist exclusively or almost exclusively of primary chloronemal filaments and, indeed, form many more chloronemal cells in both PRL and WL than wild-type, and since their development is repaired at least in PRL (Table 1) and MRL (Table 2) by a low level of BAP, we have concluded that, in addition to auxin, cytokinin is needed for the differentiation of primary chloronemal apical cells into caulonemal apical cells as well as for inhibition of the overproliferation of primary chloronemal cells.

The existence and phenotype of category 2B mutants indicate that the formation of primary caulonemata probably requires less cytokinin than the production of gametophores and suppression of secondary chloronemal growth. Cytokinin-repairable category 2A and 2B strains may, therefore, be deficient in endogenous cytokinin to differing extents as a consequence of either defective synthesis or accelerated degradation of cytokinin. An alternative explanation is that they overproduce, to differing extents, a cytokinin antagonist. In view of these findings we are now able to suggest that category 1 mutants which are insensitive to high levels of both hormone types in PRL as well as WL may be defective in either auxin or cytokinin reception or signal transduction processes.

Another surprising discovery has been that, in continuous PRL, the production of gametophores by the majority of category 4 (redesignated 4A) and category 5 (redesignated 4B) BAR mutants is not increased significantly by the presence of either NAA or IAA in the

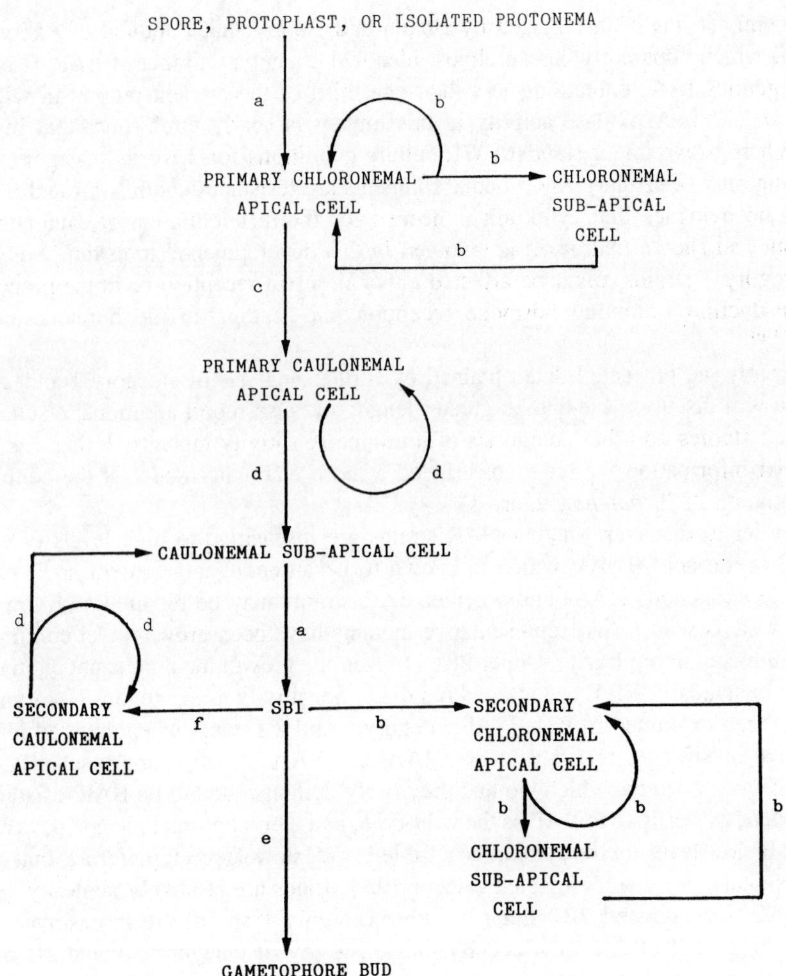

FIGURE 1. The roles of auxin and cytokinin, and their interaction with light, in regulating protonemal cell differentiation in *P. patens*. a: light (+); b: light (+), auxin (−), cytokinin (−); c: light (−), auxin (+), cytokinin (+); d: light (o), auxin (+), cytokinin (+); e: light (+), auxin (+), cytokinin (+); f: light (−), auxin (+). + indicates that the relevant factor is required for and promotes the transition indicated; − indicates that the factor inhibits the transition; o indicates that the factor has no apparent effect on the process indicated.

medium, although protonemal growth and differentiation of these strains are as sensitive in PRL to exogenous auxin as in wild-type.[41] Also, under PRL, the same category 4A and 4B mutants remain insensitive or strongly resistant to 100 n*M* BAP even in the presence of 200 n*M* to 1 μ*M* IAA.[41] These results are quite different from those obtained in WL and described earlier. At this time we are unable to suggest a satisfactory explanation of these findings.

In wild-type *P. patens* fewer chloronemal (especially secondary) apical and subapical cells are produced in PRL (Table 1) than in WL.[4] Thus, in PRL fewer SBIs differentiate into secondary chloronemal apical cells, and the secondary chloronemata (which arise by the growth and division of these cells) themselves form no or few side branches. The number of SBIs which differentiate into secondary caulonemal apical cells, however, increases in PRL (Table 1). Studies with MRL (Table 2) indicate that a photon flux in excess of approximately 2 μmol m^{-2} s^{-1} is required for the maximal production of compound secondary chloronemata which resemble those seen in cultures grown in standard WL and that somewhat lower levels of MRL favor the formation of secondary caulonemata.

Auxin-sensitive category 4A and 4B BAR mutants make many more chloronemal cells when grown in a fairly low level (about 6 μmol m^{-2} s^{-1}) of PRL on medium containing no exogenous hormones than does the wild-type.[41] We have already described similar observations made on cytokinin-repairable category 2A and 2B NAR strains cultured in continuous PRL (Table 1) or MRL (Table 2). The data obtained with MRL show clearly that category 2A and 2B mutants, in the absence of exogenous cytokinin, require lower light levels than wild-type for the production of chloronemal cells. These findings, when considered together, indicate that auxin and cytokinin may be required to antagonize a light-promoted proliferation of chloronemal cells.

IV. ADDITIONAL OBSERVATIONS WHICH SUPPORT OR AUGMENT OUR MODEL OF THE REGULATION OF CELL DIFFERENTIATION IN *P. PATENS*

A. EVIDENCE THAT LIGHT IS REQUIRED FOR THE FORMATION AND DIVISION OF PRIMARY CHLORONEMAL APICAL CELLS

1. Light is needed for spore germination[12] and for protoplast regeneration;[17] both processes lead to the production of a primary chloronema. In the case of protoplast regeneration, the first stage in this process, cell wall synthesis, occurs independently of light. However, the next stage, the production of an asymmetrical cell, requires a high level of light (i.e., approximately 7 μmol m^{-2} s^{-1} of MRL or about 3 μmol m^{-2} s^{-1} of monochromatic blue [442 nm] light), while the subsequent formation of a perpendicular cross wall giving rise to the first primary chloronemal apical cell, as well as the extension and division of this cell, can occur in much lower levels of light (e.g., 200 nmol m^{-2} s^{-1} of MRL).[17]
2. Further confirmation of a light requirement for the elongation and/or division of primary chloronemal apical cells has been obtained by transferring very young, WL-grown primary chloronemal cultures of wild-type *P. patens* into darkness. Such cultures exhibit little or no further growth even if incubated for several weeks and provided with an elaborated carbon source, usually 0.5% sucrose, in the medium.[42]

B. FURTHER EVIDENCE CONCERNING LIGHT AND HORMONAL REQUIREMENTS FOR THE DIFFERENTIATION OF CAULONEMAL APICAL CELLS AND FOR THEIR CONTINUED GROWTH AND DIVISION

1. If older WL-grown wild-type cultures, in which caulonemal apical cells have already differentiated, are transferred into darkness, the caulonemal apical cells continue to elongate and divide at about the same rate as in the light; the best estimate of the rate of division of caulonemal apical cells in the dark, 5.7 h, has been obtained using time-lapse video microscopy.[21] Most or all of the subapical cells formed by division of caulonemal apical cells in the dark do not divide; thus, dark-grown protonemal filaments, which exhibit negative gravitropism[3,12,20] and are morphologically similar (although not identical) to light-grown caulonemata, have no side branches or initials.[3,12,20] At this time we do not know with certainty whether or not some chloronemal apical cells in older cultures differentiate, upon transfer to darkness, into caulonemal apical cells which by their subsequent growth contribute to the population of dark-grown filaments.
2. Preliminary data, obtained by transferring WL-grown category 1, 2A, 2B, 4A, and 4B mutants into darkness, reveal that many of these strains, including some which

appear to consist exclusively of primary chloronemal filaments and certainly contain no normal caulonemata at the time of transfer, produce at least some caulonemal or caulonemal-like apical cells in darkness.[27,43] In some mutant strains, these apical cells divide in the dark and give rise to filaments which in most cases grow more slowly and are not morphologically identical to the dark-grown protonemal filaments of the wild-type.[41,42] In general, category 2A mutants grow better in darkness than category 1, 4A, or 4B strains.[41] However, in only one case (i.e., the category 2B mutant NAR 87, which makes a normal number of caulonemata in WL) have we observed the production, growth, and morphology of protonemal filaments generated in the dark which closely resemble those of dark-grown caulonemal-like filaments of the wild-type.[41,43] These findings enable us to make the following tentative suggestions:

a. The differentiation of primary chloronemal apical cells into caulonemal apical cells may be light independent, at least under some circumstances, and in darkness this transition may require no (or only very low levels of) auxin and cytokinin. These conclusions are not inconsistent with our earlier observation that a significant role of these two hormone types may be to antagonize a light-promoted maintenance and proliferation of chloronemal apical cells.
b. The extension and division of caulonemal apical cells can occur in the absence of light, but one or both of these processes need auxin and probably a low level of cytokinin.

C. FURTHER OBSERVATIONS CONCERNING LIGHT AND HORMONAL REQUIREMENTS FOR THE PRODUCTION OF SBIs AND THEIR SUBSEQUENT DIFFERENTIATION INTO SIDE BRANCHES OF VARIOUS TYPES

1. As indicated earlier, dark-grown caulonemata do not have SBIs, but very low levels of light (e.g., MRL at 3 nmol m^{-2} s^{-1}) induce the majority of subapical caulonemal cells to divide, giving rise to side branches.[27,39]
2. The differentiation of SBIs into secondary caulonemal apical cells requires a low level of light (e.g., in MRL at 3 nmol m^{-2} s^{-1}, about 50% of SBIs differentiate into caulonemal cells). Higher light levels inhibit this transition. Thus, in MRL at 200 nmol m^{-2} s^{-1}, only about 3% of SBIs have this fate.[27,39]
3. The differentiation of SBIs into secondary chloronemal apical cells requires higher light levels than are needed for the differentiation of caulonemal apical cells; e.g., in MRL some secondary chloronemata are formed at about 200 nmol m^{-2} s^{-1}, but maximal production (i.e., from 95% of SBIs) occurs in MRL only at 550 nmol m^{-2} s^{-1} or higher photon fluence rates.[27,39]
4. The development of some SBIs into gametophore buds occurs in MRL at about 500 nmol m^{-2} s^{-1} or possibly even at somewhat lower photon fluence rates (Table 2). However, maximal gametophore production, corresponding to approximately 1 to 2% of SBIs, occurs only at much higher photon fluence rates, e.g., >16 μmol m^{-2} s^{-1} of MRL.[27,39]

V. THE LIGHT REQUIREMENT IN *P. PATENS* FOR CYTOKININ-INDUCED GAMETOPHORE BUD FORMATION

Exogenous cytokinin alone[3] or in combination with exogenous auxin[44] cannot replace the light requirement of *P. patens* for gametophore production. Preliminary data indicate that the light requirement for this developmental transition is phytochrome mediated,[45] as appears to be the case in most other light-dependent aspects of *P. patens* morphogenesis[12,17,18,21] studied so far, with the exception of light-induced asymmetry in regenerating protoplasts.[17]

P. patens can be cultured in a regime of continuous liquid medium replacement which makes the wild-type moss dependent on an external supply of hormones for the formation of caulonemata and gametophores.[9,10,13] Experiments of this kind, in which the moss has been illuminated with standard WL, have provided data which confirm conclusions derived from the analysis of mutants.[10] Recently it has been demonstrated that the continuous feeding of liquid medium containing both auxin and cytokinin to dark-grown cultures does not initiate bud formation.[37,44] However, when medium containing both hormones is fed to a WL-grown culture, collected, and then fed to a dark-grown culture, bud formation is induced on the latter.[37,44] When auxin-containing medium is fed to a WL-grown culture, collected, mixed with cytokinin, and then fed to a dark-grown culture, no buds are formed.[44] A plausible interpretation of these results is that *P. patens* requires light and cytokinin simultaneously to produce another leachable substance, designated Factor L, which together with auxin and perhaps also cytokinin is needed (at least in the dark) to bring about bud formation.

VI. AUXIN ACCUMULATION BY WILD-TYPE AND AUXIN SENSITIVITY MUTANTS

In the absence of exogenous hormones, category 3 NAR mutants are morphologically identical (or at least similar) to the wild-type.[4,10] In WL, all of them are about as sensitive to exogenous cytokinin as the wild-type, but they are strongly resistant to high levels of exogenous NAA.[4,10] Since we have shown that endogenous auxin is needed for the differentiation of primary chloronemal apical cells into caulonemal apical cells and also for the formation of buds from SBIs, and since category 3 strains produce approximately normal numbers of caulonemata and gametophores, it seems reasonable to conclude that these mutants are neither deficient in endogenous auxin nor defective in sensitivity to it. A plausible explanation of their abnormal phenotype, therefore, appeared to be that they are altered in their ability to accumulate exogenous auxin.[4,10] Recently, however, direct investigation of auxin accumulation by wild-type and by a selection of category 3 NAR mutants using radioactively labeled IAA has revealed that all of the strains tested are similar to wild-type with respect to both total auxin uptake and the saturable, carrier-mediated component of auxin uptake.[28,46] All of a selection of category 1, 2A, and 2B auxin-insensitive mutants included in this study have also been found to possess a normal IAA uptake capability.[28,46]

An analysis of category 3 mutants cultured in PRL suggests that this is a relatively heterogeneous group. Also, most of these strains, all of which are strongly resistant to NAA in WL, exhibit a less pronounced resistance to NAA in PRL and a somewhat weaker resistance or, in some cases, no detectable resistance to IAA.[41] Although we cannot yet fully account for these results, a possible explanation for the phenotype of at least some category 3 mutants is that a part of their auxin uptake mechanism is altered so that it cannot interact normally with NAA, but still functions fairly well with IAA.

REFERENCES

1. **Ashton, N. W.,** The role of medium acidification and auxin transport in the morphogenesis of *Physcomitrella patens, Br. Bryol. Soc. Bull.,* 49, 16, 1987.
2. **Ashton, N. W.,** Medium acidification by *Physcomitrella patens*, in *Methods in Bryology,* Glime, J. M., Ed., Hattori Botanical Laboratory, Nichinan, Miyazaki, Japan, 1988, 119.

3. **Ashton, N. W., Cove, D. J., and Featherstone, D. R.,** The isolation and physiological analysis of mutants of the moss, *Physcomitrella patens*, which over-produce gametophores, *Planta*, 144, 437, 1979.
4. **Ashton, N. W., Grimsley, N. H., and Cove, D. J.,** Analysis of gametophytic development in the moss, *Physcomitrella patens*, using auxin and cytokinin resistant mutants, *Planta*, 144, 427, 1979.
5. **Ashton, N. W., Schulze, A., Hall, P., and Bandurski, R. S.,** Estimation of indole-3-acetic acid in gametophytes of the moss, *Physcomitrella patens, Planta*, 164, 142, 1985.
6. **Ashton, N. W., Boyd, P. J., Cove, D. J., and Knight, C. D.,** Genetic analysis in *Physcomitrella patens*, in *Methods in Bryology*, Glime, J. M., Ed., Hattori Botanical Laboratory, Nichinan, Miyazaki, Japan, 1988, 59.
7. **Ashton, N. W., Boyd, P. J., Cove, D. J., and Knight, C. D.,** Protoplasts as tools in the study of moss development, in *Bryophyte Development: Physiology and Biochemistry*, Chopra, R. N. and Bhatla, S. C., Eds., CRC Press, Boca Raton, FL, 1990, chap. 1.
8. **Courtice, G. R. M. and Cove, D. J.,** Mutants of the moss, *Physcomitrella patens*, which produce leaves of altered morphology, *J. Bryol.*, 12, 595, 1983.
9. **Cove, D. J.,** The role of cytokinin and auxin in protonemal development in *Physcomitrella patens* and *Physcomitrium sphaericum, J. Hattori Bot. Lab.*, 55, 79, 1984.
10. **Cove, D. J. and Ashton, N. W.,** The hormonal regulation of gametophytic development in bryophytes, in *The Experimental Biology of Bryophytes*, Dyer, A. F. and Duckett, J. G., Eds., Academic Press, London, 1984, 177.
11. **Cove, D. J. and Knight, C. D.,** Gravitropism and phototropism in the moss, *Physcomitrella patens*, in *Developmental Mutants in Higher Plants*, Thomas, H. and Grierson, D., Eds., Cambridge University Press, London, 1987, 181.
12. **Cove, D. J., Schild, A., Ashton, N. W., and Hartmann, E.,** Genetic and physiological studies of the effect of light on the development of the moss, *Physcomitrella patens, Photochem. Photobiol.*, 27, 249, 1978.
13. **Cove, D. J., Ashton, N. W., Featherstone, D. R., and Wang, T. L.,** The use of mutant strains in the study of hormone action and metabolism in the moss, *Physcomitrella patens*, in *Proc. 4th John Innes Symp.*, Davies, D. R. and Hopwood, D. A., Eds., The John Innes Charity, Norwich, England, 1980, 231.
14. **Engel, P. P.,** The induction of biochemical and morphological mutants in the moss *Physcomitrella patens, Am. J. Bot.*, 55, 438, 1968.
15. **Futers, T. S., Wang, T. L., and Cove, D. J.,** Characterisation of a temperature-sensitive gametophore over-producing mutant of the moss, *Physcomitrella patens, Mol. Gen. Genet.*, 203, 529, 1988.
16. **Grimsley, N. H., Featherstone, D. R., Courtice, G. R. M., Ashton, N. W., and Cove, D. J.,** Somatic hybridization following protoplast fusion as a tool for the genetic analysis of development in the moss *Physcomitrella patens*, in *Advances in Protoplast Research — Proc. 5th Int. Protoplast Symp.*, Akadémiai Kiadó, Budapest, 1980, 363.
17. **Jenkins, G. I. and Cove, D. J.,** Light requirements for the regeneration of protoplasts of the moss *Physcomitrella patens, Planta*, 157, 39, 1983.
18. **Jenkins, G. I. and Cove, D. J.,** Phototropism and polarotropism of primary chloronemata of the moss, *Physcomitrella patens*: responses of the wild type, *Planta*, 158, 357, 1983.
19. **Jenkins, G. I. and Cove, D. J.,** Phototropism and polarotropism of primary chloronemata of the moss, *Physcomitrella patens*: responses of mutant strains, *Planta*, 159, 432, 1983.
20. **Jenkins, G. I., Courtice, G. R. M., and Cove, D. J.,** Gravitropic responses of wild type and mutant strains of the moss, *Physcomitrella patens, Plant Cell Environ.*, 9, 637, 1986.
21. **Knight, C. D. and Cove, D. J.,** The genetic analysis of tropic responses, *Environ. Exp. Bot.*, 29, 57, 1989.
22. **Knight, C. D., Cove, D. J., Boyd, P. J., and Ashton, N. W.,** The isolation of biochemical and developmental mutants in *Physcomitrella patens*, in *Methods in Bryology*, Glime, J. M., Ed., Hattori Botanical Laboratory, Nichinan, Miyazaki, Japan, 1988, 47.
23. **McClelland, D. J.,** Protonemal branching patterns in *Physcomitrella patens, Br. Bryol. Soc. Bull.*, 49, 20, 1987.
24. **Wang, T. L., Cove, D. J., Beutelmann, P., and Hartmann, E.,** Isopentenyladenine from mutants of the moss, *Physcomitrella patens, Phytochemistry*, 19, 1103, 1980.
25. **Wang, T. L., Beutelmann, P., and Cove, D. J.,** Cytokinin biosynthesis in mutants of the moss *Physcomitrella patens, Plant Physiol.*, 68, 739, 1981.
26. **Wang, T. L., Horgan, R., and Cove, D. J.,** Cytokinins from the moss *Physcomitrella patens, Plant Physiol.*, 68, 735, 1981.
27. **Ashton, N. W., Cove, D. J., Wang, T. L., and Saunders, M. J.,** Developmental studies of *Physcomitrella patens* using auxin and cytokinin sensitivity mutants, in *Plant Growth Substances 1988*, Pharis, P. and Rood, S. B., Eds., Springer-Verlag, Berlin, in press.

28. **Ashton, N. W. and Bridge, P.,** Accumulation of radioactively labelled indole-3-acetic acid by wild type and auxin insensitive mutants of the moss, *Physcomitrella patens,* Abstract 2-10-5, presented at XIV Int. Botanical Congr., Berlin, July 24 to August 1, 1987.
29. **Knight, C. D.,** Gravitropism in the Moss, *Physcomitrella patens,* Ph.D. thesis, University of Leeds, Leeds, England, 1987.
30. **Ashton, N. W. and Cove, D. J.,** The isolation and preliminary characterisation of auxotrophic and analogue resistant mutants of the moss, *Physcomitrella patens, Mol. Gen. Genet.,* 154, 84, 1977.
31. **Boyd, P. J., Grimsley, N. H., and Cove, D. J.,** Somatic mutagenesis of the moss, *Physcomitrella patens, Mol. Gen. Genet.,* 211, 545, 1988.
32. **Long, Z.,** Genetical and Biochemical Studies of Photosynthetic Mutants in the Moss, *Physcomitrella patens,* Ph.D. thesis, University of Leeds, Leeds, England, 1986.
33. **Courtice, G. R. M., Ashton, N. W., and Cove, D. J.,** Evidence for the restricted passage of metabolites into the sporophyte of the moss, *Physcomitrella patens* (Hedw.) Br. Eur., *J. Bryol.,* 10, 191, 1978.
34. **Grimsley, N. H., Ashton, N. W., and Cove, D. J.,** The production of somatic hybrids by protoplast fusion in the moss, *Physcomitrella patens, Mol. Gen. Genet.,* 154, 97, 1977.
35. **Grimsley, N. H., Ashton, N. W., and Cove, D. J.,** Complementation analysis of auxotrophic mutants of the moss, *Physcomitrella patens,* using protoplast fusion, *Mol. Gen. Genet.,* 155, 103, 1977.
36. **Watts, J. W., Doonan, J. H., Cove, D. J., and King, J. M.,** Production of somatic hybrids of moss by electrofusion, *Mol. Gen. Genet.,* 199, 349, 1985.
37. **McClelland, D. J. and Cove, D. J.,** unpublished data.
38. **Knoop, B.,** Development in bryophytes, in *The Experimental Biology of Bryophytes,* Dyer, A. F. and Duckett, J. G., Eds., Academic Press, London, 1984, 143.
39. **Jenkins, G. I. and Cove, D. J.,** unpublished data.
40. **Bhatla, S. C. and Bopp, M.,** Use of naphthaleneacetic acid-resistant mutants of *Funaria hygrometrica* for investigating the regulatory sites in the auxin-induced cell differentiation, Abstract 48, presented at 13th Int. Conf. Plant Growth Substances, Calgary, Alberta, Canada, July 17 to 22, 1988.
41. **Ashton, N. W.,** unpublished data.
42. **Cove, D. J.,** unpublished data.
43. **Courtice, G. R. M.,** Developmental Genetic Studies of *Physcomitrella patens,* Ph.D. thesis, University of Cambridge, Cambridge, England, 1979.
44. **Cove, D. J., Ashton, N. W., McClelland, D. J., and Wang, T. L.,** Bud formation in *Physcomitrella patens,* Abstract 287, poster presentation at 13th Int. Conf. Plant Growth Substances, Calgary, Alberta, Canada, July 17 to 22, 1988.
45. **Futers, T. S.,** The Mode of Action of Cytokinin in the Moss, *Physcomitrella patens,* Ph.D. thesis, University of Leeds, Leeds, England, 1984.
46. **Ashton, N. W. and Bridge, P.,** unpublished data.

Chapter 3

PHOTOMODULATION OF PROTONEMA DEVELOPMENT

Elmar Hartmann and Martina Weber

TABLE OF CONTENTS

I. INTRODUCTION

The specific developmental patterns throughout the ontogeny of plants depend on regulation by environmental factors. The most important external factor is light. Two fundamentally different aspects of light action on green plants have to be distinguished. On the one hand, light supplies energy necessary for photosynthesis; on the other hand, it acts as an external signal in the regulation of plant development. The specific influence of photosynthesis on morphogenetic processes is not yet quite clear. Of course, photosynthetic activity is essential to maintain growth in green plants and is therefore a prerequisite of development. Its involvement in photomorphogenesis could be excluded in several cases; nevertheless, there are certain experimental conditions wherein it seems impossible to separate trophic and regulatory effects of light exactly.

The processing of light signals in plants starts with the perception of the stimulus by the photoreceptor, resulting in initiation of primary biochemical interactions at the cellular level. The cascade of the proceeding physiological reactions finally results in the constitution of the morphogenetic response.[1] At present, much is known about the initial step of the reaction sequence, i.e., the perception of the light stimulus by a photoreceptor. However, there are no concrete data on how signals from photomorphogenetically active light qualities are transduced in plants. The fundamental step of coupling the photochemical reaction to a primary biochemical process is still obscure.

With respect to the different light qualities of the natural spectrum, red and blue light are most effective in the control of plant development. Thus, the discussion about signal perception is focused on red light and blue light photoreceptors.

The red light receptor, responsible for a variety of photomorphogenetic responses in lower and higher plants, is phytochrome. The chromoprotein exists in two convertible forms with different absorption maxima in the red region of the visible light spectrum (P_r, 660 nm and P_{fr}, 730 nm).[2] As the absorption spectra overlap in the red spectral range, the two pigment forms always coexist in a wavelength-dependent photoequilibrium.

Two types of phytochrome-mediated reactions can be distinguished under experimental and natural conditions: low irradiance responses (LIR) and high irradiance responses (HIR). Several photomorphogenetic responses are induced by short-term irradiations with low intensities of red light, sufficient to establish the effective amount of the physiologically active form, P_{fr}. Illumination with far-red light after the inductive red irradiation converts P_{fr} back to P_r and suppresses the response. Far-red reversibility is the criterion to regard phytochrome as a morphogenetic factor. Photomorphogenetic responses such as germination, growth regulation, formation of flowers, or synthesis of anthocyanins are defined as differentiations. They are characterized by the irreversible induction of a developmental process which continues under noninductive conditions. The response is obviously related to differential gene expression. The second category of light effects — photomodulations — are rapidly reversible and are only processed during the presence of the inductive light factor.

The HIR response is not yet completely understood. HIR responses need continuous irradiation to reach a significant level and show action maxima in the red, far-red, blue, and ultraviolet (UV) spectral range.[3] The responsibility of phytochrome for HIRs was demonstrated by Hartmann[4] and Schäfer.[5] Besides phytochrome, a specific blue light photoreceptor system probably participates in the generation of HIRs.[6,7] As P_r and P_{fr} absorb in the blue and UV spectral range, too,[8] the short wavelength region also might contribute to phytochrome-mediated photomorphogenesis.[9] Nevertheless, the involvement of phytochrome can be excluded in those responses, which are strictly blue light dependent. Flavins and/or carotenoids are discussed as photoreceptors (cryptochrome).

In bryophytes the influence of both blue- and red-light-mediated photomorphogenesis is evident, especially in the early stages of the ontogeny. The life cycle of bryophytes is completed in distinct developmental steps, and the transition from one stage to the other depends on the light environment. As an experimental advantage, the protonemal stage of individual strains can be preserved by vegetative culture under defined conditions and, thus, supplies a genetically homogeneous material for developmental studies. Therefore, the filamentous moss protomena became a standard object for photomorphogenetic research in lower plants. This chapter deals with the influence of light on all aspects of protonemal growth and development in mosses and liverworts, including light effects on spore germination, protoplast differentiation, and the phototropic responses.

II. LIGHT-MODULATED SPORE GERMINATION

The first and probably the most fundamental step in protonemal development is the constitution of polarity in the course of spore germination. The formation of a main axis becomes obvious when a filamentous germ tube protrudes from the spore. Spore germination up to this stage has been described to be affected by light.[10]

As germination is a continuous process, the problem arises to define its limits. Based on morphological changes during spore germination in mosses and liverworts, three main steps can be distinguished: (1) absorption of water and swelling of the spore; (2) greening of the spore content and bursting of the exosporium; and (3) protrusion of the germ tube, indicating the beginning of protonemal growth. In studies on the light dependence of spore germination either the second or the third stage of this series served as germination criterion. As a common result of all investigations the importance of the initial swelling phase has to be emphasized. The absorption of water seems to be based on physical processes and does not depend on light.[12] Water uptake by the dry spores leads to the rehydration of cell components and the gradual increase in germinability.[13,14]

The further progress of spore germination is controlled by light. Greening of the spore content and bursting of the exosporium could be induced in *Funaria hygrometrica* by 3 h of irradiation with white light of low intensity.[12] Bauer and Mohr[15] demonstrated the involvement of phytochrome in this process. A red-light-induced stimulation of germination could be reversed by a subsequently applied far-red pulse of a few minutes. Apart from the red light stimulation, a lower but significant amount of germination was reported in the far-red control sample. Sophisticated studies by Schild[16] and Cove et al.[17] demonstrated a complicated involvement of phytochrome in moss spore germination, which is different from light induction in seeds. Krupa[18] compared the influence of blue (425 nm) and red (654 nm) light on *Funaria* spore germination and found that the first light-dependent stage was induced by both spectral ranges. His study, however, revealed a significantly higher effectiveness of red light. Blue light, given after an inductive red irradiation, even caused a decrease in the germination ratio. All spectral regions affected early germination processes in the moss species *Ceratodon purpureus, Dicranum scoparium,* and *F. hygrometrica,* showing a lower significance of blue and far-red light for *Funaria.*[19]

The available data provide evidence that the first light-dependent germination step in mosses is controlled by phytochrome. Spore germination up to the moment when the spore coat ruptures is a low-energy induction phenomenon with red/far-red reversibility. The inductive activity of blue light so far may be attributed to absorption by phytochrome and generation of a sufficient amount of P_{fr}. The inhibitory effect of blue light on a red-light-induced germination stimulus is a well-known phenomenon in fern spores.[20-22] In *Lygodium japonicum* a close relationship between the P_{fr} content and the percentage of germination has been demonstrated.[23] As P_{fr} is converted to P_r, to a certain extent, by the absorption of blue light, a decrease in the P_{fr} content might be an explanation for the observed inhibition effect in moss spores. In fern spores, however, phytochrome-dependent germination is completely inhibited by a brief exposure to blue light. For *Lygodium* this result was explained by the suppression of a specific physiological step in the germination process, performed by a photoreactive blue-absorbing pigment system.[23] Thus, besides the direct influence of the short wavelength region on phytochrome in moss spores, the possible interaction of a specific blue light photoreceptor system might be involved.

A remarkable fact shown by bryophytes as well as ferns is that a prolonged preswelling of the spores in darkness increases their sensitivity to light.[18,20,23-25] Thus, much shorter irradiation times are required to give a certain response if spores are pretreated in this way. The initial swelling phase, therefore, has been related to some activation processes which control the sensitivity for light. Cove et al.[17] reported strong evidence for a specific factor which is only operational with P_{fr}. In fern spores Tomizawa et al.[26] measured an increase in the phytochrome content during the imbibition period, parallel to an increase in sensitivity to light. This effect has been related to a mobilization of stored phytochrome[27] as well as to *de novo* synthesis.[28] These findings underline some of the physiological conditions that up to now have been covered by terms like ''maturity'' or ''ripeness''. In any case, the responsiveness of spores to light depends strictly on their pretreatment in darkness,[24,25] and much attention should be paid to this fact in quantitative studies on light action.

Most investigators consider a spore as germinated when a part of the filamentous protonema or rhizoid is visible. According to this criterion, germination in liverworts has been proved to be obligatory light dependent. In Marchantiales the influence of different light qualities on the germination ratio in long-term experiments revealed the following scheme of decreasing effectiveness: red, far-red, yellow-green, blue.[13,29] In *Sphaerocarpus,* too, all light qualities contribute to spore germination, although different kinetics were observed in the red and the blue spectral range.[24]

In mosses, in contrast to liverworts, a significant degree of germination in darkness has been observed in some species. In particular, a stimulation of germination in darkness, even to the germ tube stage, has been obtained by the addition of exogenous carbon sources such as sucrose to the nutrient medium.[11,30,31] However, these results could not be confirmed by other investigators.[17,32] Considering the inductive efficiency of low-energy, short-term irradiations, the ''contamination'' with inadvertent or nonspecific light exposures in the course of such experiments consequently must be ruled out.

In mosses, as in liverworts, light of all wavelengths was shown to be effective in promoting germination to its final stage.[19] In Valanne's study, far-red reversibility of germination in *Ceratodon* and *Dicranum* was described, which could not be confirmed for *Funaria*. Action spectra of germination in *Funaria*[32] and *Physcomitrella*[17] reflect the high effectiveness of the red spectral range, but a reversion effect of far-red light was not observed. Krupa,[18] in his comparative study on the effects of blue and red light in *Funaria* spore germination, strictly discriminated the two light-dependent germination stages as criteria I and II. He described an analogous effect of blue and red light in that the percentage of germination in general increased with the exposure time. In red light the dependence on the irradiation time was striking with regard to criterion II, whereas the criterion I stage was induced to a high percentage by only brief light exposures. In blue light, the germination ratio gradually increased and spore development was promoted continuously from the criterion I to the criterion II phase without any obvious differences in the mode of light action on the two stages.

The view on the interaction of different light qualities in later germination stages reveals a rather complex situation. The red light dependence of spore development now shows some features of an HIR, as the germination ratio clearly depends on the duration of red irradiation and shorter exposure times with higher intensities are less effective.[18] In *Physcomitrella* and *Funaria*, the high efficiency of the red spectral range with a lack of simple reversibility by far-red light points to the same direction. In contrast to some classical HIR phenomena, usually studied with etiolated or green higher plant seedlings, blue and far-red light exhibit lower effectiveness in moss spore germination.

The influence of blue and far-red light in the later germination stage is mainly discussed with regard to either HIR or photosynthesis.[18,19,24,32] In fact, time and irradiance dependence of the responses provide a strong argument for HIR.

Additionally, the stimulating effect of the photosynthetically relevant wavelength regions might be due to the energy requirement of growth processes in the spore distension phase. Spore germination in early stages might proceed, supported by the reserve substances of the spore.[33] In later stages, however, photosynthesis as well as exogenously supplied carbon sources provide the conditions for further growth and development. The energy supply during elongation of the germ tube may be regarded as the main reason for the influence of photosynthesis on spore germination. This idea is supported by the result that 3-(3,4-dichlorophenyl)-1,1-dimethylurea (DCMU), an inhibitor of photosynthetic electron transport, prevents germ tube growth in *Physcomitrella patens,* whereas it has no effect on earlier germination stages.[17]

In conclusion, phytochrome seems to play the major part in the photocontrol of spore germination, acting via different regulatory mechanisms either subsequently or simultaneously. The problem of complexity is, in part, conditioned by the choice of definitions. The so-called germination criteria are stages in a continuous developmental process, used for practical reasons. A variety of physiological processes, such as polarization, chloroplast development, switching from lipid to carbohydrate metabolism, or nucleic acid and protein synthesis, participate in the regulation of distinct developmental stages. Thus, our criteria are not differentiated enough to deal with all aspects of light action on germination.

III. LIGHT-MODULATED PROTONEMA DEVELOPMENT

A. PROTONEMA GROWTH AND DEVELOPMENT

The filamentous protonema originates from the elongating germ tube providing the structural pattern for the subsequent formation of gametophores.

In liverworts the filamentous protonema is reduced to a few cells and is soon succeeded by the leafy or thallose gametophore. Lesser importance of the protonemal stage in liverworts and, additionally, some disagreement about definitions of germination resulted in poor interest of researchers in this field. Inoue[13] presented a detailed study on morphogenesis in Marchantiales, including the role of light in different developmental stages.

The first, unequal cell division, usually occurring within the spore coat, delimits the initials for rhizoid and germ tube cells. The pattern of rhizoid formation seems to be characteristic for a species or genus, but whether it is expressed or not depends on light. In some species rhizoid formation was suppressed in low light intensities, whereas other species showed much higher sensitivity, leading to a high percentage of rhizoid formation under low light intensity conditions. The effect of different light qualities was investigated with *Reboulia hemisphaerica.* Blue and red light were most effective in promoting rhizoid development; far-red and green light showed less activity. In contrast, in *Marchantia polymorpha*, rhizoid formation was completely suppressed in blue light.[13]

Germ tube elongation in white light generally increased with decreasing light intensity. In *Reboulia,* germ tube elongation revealed the opposite tendency of rhizoid development. In the blue and the red spectral range, germ tubes were relatively short, whereas they elongated remarkably in far-red and green light. In *Marchantia,* the germ tube elongation was maximal in blue light and no rhizoid growth was observed. In *Plagiochasma intermedium,* germ tubes elongated greatly in blue and far-red light, but failed to develop in green light.

Thus, while rhizoid development (if it takes place) follows a stable and species-typical pattern, protonema development in liverworts is markedly affected by light. Considering the inconsistency of different species in the reaction to light, a general conception for photoregulation of protonema development in liverworts can hardly be worked out.

In mosses, in contrast to liverworts, a large and extensively branched protonema develops which consists of two main cell types, chloronema and caulonema.

The first protonemal type of tissue derived from the developing germ tube is the primary chloronema. The filaments are separated by vertical cross walls and contain numerous round chloroplasts. In some species a germ rhizoid emerges that, however, does not develop further.

After a certain culture period the chloronema stage is followed by the formation of caulonema cells. The caulonemal type of filament consists of longer cells with a few spindle-shaped chloroplasts and oblique cross walls. With proceeding filament growth, caulonema cells at a certain distance from the tip begin to form lateral ramifications. Depending on external and internal conditions, these initials develop into either side branches or gametophytic buds. The side branches may be of the caulonema or chloronema type, called secondary chloronema in the latter case. Subsequently, the moss caulonema develops into a richly ramified filament system with side branches of second order or higher.

The above-described normal pattern of protonema development is expressed best in white light. Dark-grown protonema cultures exhibit typical features of etiolation. Mitra et al.[34] reported that dark-grown cultures of *Pohlia nutans* form slender, unbranched filaments with elongated cells and colorless or pale chloroplasts. The growth of this species in the dark depended strictly on the external supply of sucrose in the nutrient medium. These results were confirmed in principle for *C. purpureus,*[35] *F. hygrometrica,*[36] and *Physcomitrella patens.*[17] A comparative investigation of the influence of different light qualities on protonema development revealed the effectiveness of photosynthetic active wavelength in stimulating growth processes. Mitra et al.[34] observed remarkable cell elongation in the blue and the red wavelength range, but only poor growth in green light. A further stimulation was obtained by the addition of sugar. The elongation of protonemal filaments requires carbohydrate sources, which may be provided by either photosynthetic activity or by the addition of sugars. Thus, protonema development generally depends on light, by which photosynthesis supplies the metabolic basis for growth processes.

The formation of a specific developmental pattern requires the regulative involvement of specific light qualities. The red spectral range is most effective in bringing about the chloronemal cell type. Jahn[37] reported for *F. hygrometrica* a nearly normal protonemal development in red light, while in the blue spectral range the chloronemal stage was reduced and much-elongated, caulonema-like cells were produced. In *Physcomitrella patens* normal development occurred in red light, whereas the blue and far-red region had effects similar to darkness.[17] The formative influence of white light on the chloronemal stage seems to be due to the activity of the red spectral range and probably can be attributed to phytochrome.

Chloronema filaments exhibit irregular branching that also seems to depend on phytochrome. Ramification in *F. hygrometrica* and *Physcomitrella patens* was stimulated by red irradiation, but only few branches were formed in far-red light.[17,38] Larpent and Jaques[36] even demonstrated far-red reversibility for the induction of lateral branches in chloronema filaments. Unfortunately, not much work has been done to elaborate the effects of phytochrome on branching of moss protonema.

Besides the determinative influence on the constitution of the chloronemal filament type, red light also has an effect on chloronema growth activity, independent of photosynthesis. The action spectrum of the growth of secondary chloronemata in submerged cultures of *Physcomitrium turbinatum* shows a distinct peak in the red wavelength region, but no comparable effectiveness of blue light.[39] Evidence for a photosynthesis-independent growth regulation was also found in chloronemata of *F. hygrometrica*[19] and *Physcomitrella patens.*[40] Finally, Bittisnich and Williamson[41] demonstrated a far-red reversible phytochrome control of extension growth in chloronemata of *F. hygrometrica.*

The initiation of the development of caulonema cells depends on hormonal regulation. An involvement of a photomorphogenetic pigment system has not yet been demonstrated.

The achievement of a "critical size" of the chloronema has been considered a prerequisite for the conversion from the chloronema to the caulonema stage.[42] This phenomenon was attributed to a minimum assimilatory capacity of the chloronemal system, necessary to provide enough energy and precursors for hormone synthesis. Bopp[42] pointed out that the caulonemal character of moss protonemata is not stable and must be maintained by favorable culture conditions (e.g., high light intensities). In low light intensities no caulonema develops and the protonema is preserved in the chloronema stage over a long period. The modulation effect of light intensities on the developmental condition of protonemata is further emphasized by switching from high to low irradiance. This change induces caulonema cells to form chloronema. Knoop[43] related this effect to a "critical growth velocity" of the tip cells, obtained under favorable conditions, that seems to be connected with the generation of an intracellular hormone signal. Besides the low light intensity, all influences that retard growth velocity in the tip cells initiate chloronema regeneration in *F. hygrometrica*. Thus, there seems to exist some sort of feedback mechanism between growth processes at the tip and the hormonal regulation modus. Nevertheless, there are remarkable species-specific differences. The moss *C. purpureus* preferentially forms caulonemal filaments in darkness on a medium supplemented with a sugar source (Figure 1).[44]

Branching in the caulonema occurs regularly and seems to depend on the cell cycle. In *F. hygrometrica* new side branches are initiated in the third cell of each filament simultaneously with cell division at the tip. While the apical cells divide every 6 to 7 h, the subapical cells divide in a rhythm of 12 to 14 h.[45] Formation of the side branches always occurs at the "right time", even if tip growth is retarded by inhibitors.[46] Branching can be induced by cytokinins,[47] indicating that there may be some regulation by internal hormone levels.

The inductive effect of light on the formation of caulonemal branch initials has not been proven, although some authors discussed the promotive influence of light on branching.[17,48] We found in *C. purpureus* a strong influence of blue light in increasing branching of the caulonema compared with red light of equal quantum flux densities, as shown in Figure 2C. In all experiments with high intensities of blue light, which induces a negative phototropic response (see Section IV.C), a new side branch was initiated just below the bending bow (Figure 1B). Branching of caulonema filaments is certainly also controlled by an endogenous rhythm and perhaps by phytochrome or cryptochrome. However, the mode of light action remains unclear.

B. BUD INDUCTION

The first sign of gametophore development in mosses is the induction of buds and the formation of a three-cutting-faced tip cell. By this fundamental step occurs the transition from one-dimensional filament extension to three-dimensional gametophore growth.

In most moss species, buds are derived exclusively from lateral branch initials of the caulonema. The unequal cell division of the initial cell is followed by the outgrowth of a lateral cell. The first observable events in the bud induction process are a swelling of this newly derived initial cell and retardation of its growth. Bud induction is strictly light dependent, and the light requirement cannot be replaced by the addition of carbon sources in darkness.[34,49-51] Red light has been proven to be the effective light quality in the induction process.[34,37,49,50] Simon and Naef[51] demonstrated far-red reversibility of bud induction, providing a convincing argument for phytochrome involvement. Nevertheless, as already mentioned for phytochrome-mediated branching, its effect on bud formation also has to be studied in more detail.

Bud formation was greatly suppressed in blue light, but the block could be removed by a change of the substrate.[37] After a certain culture time, however, inhibition was observed again. These results indicated the production and release of an inhibitory substance in blue light which accumulated in the nutrient medium and suppressed bud induction. This might

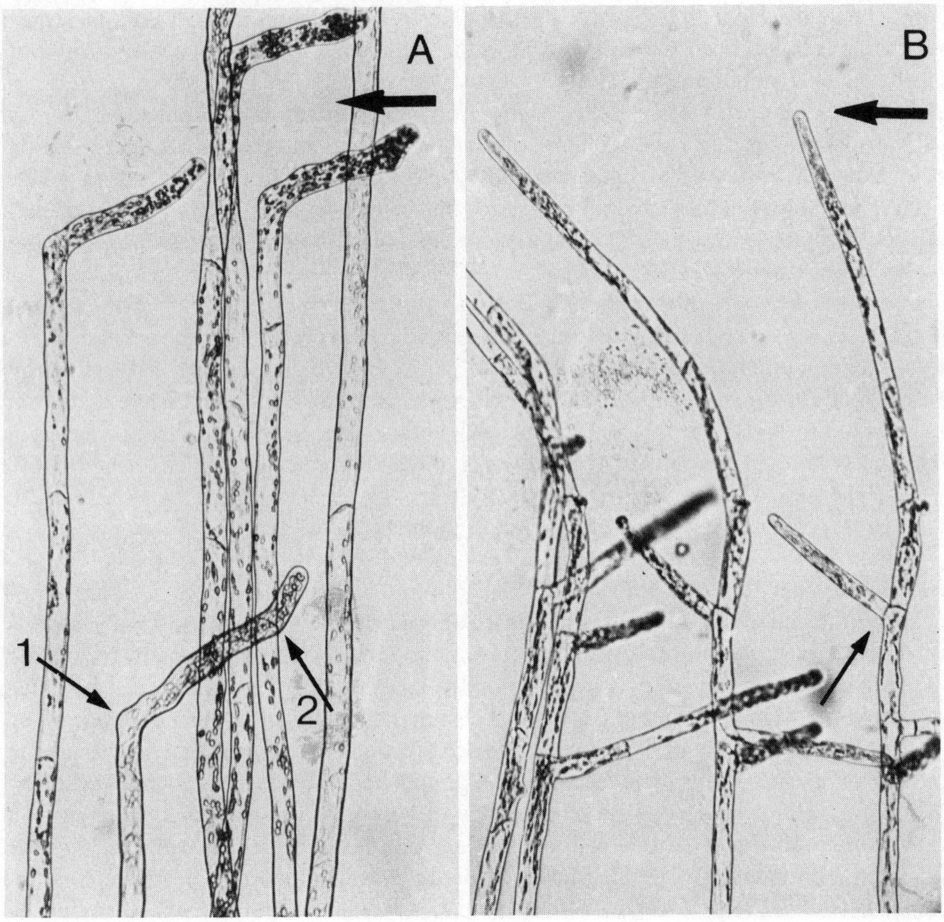

FIGURE 1. *Ceradon purpureus.* (A) Positive phototropic response of 4-day dark-adapted protonemal tissue. Continuous unilateral red light (664 nm, 15 W m^{-2}) was applied for 5 h to vertically positioned test samples.[44] The light direction is marked with a thick arrow. The small arrow (1) shows the beginning of phototropic growth with a sharp bending. The second arrow (2) marks the end of irradiation and the beginning of the negative gravitropic ''back-bending'' response. (B) Negative phototropic response after 22 h irradiation with blue light (420 nm, 8 W m^{-2}). The growth response is a typical ''bowing''. The beginning of the bow is characterized by a side branch (arrow).

be an explanation for the lack of gametophore development in blue light, described for *Pohlia nutans*[34] and *Amblystegium riparium.*[49]

It is well known that the application of cytokinins accelerates and enhances bud formation in the presence of light, and cytokinins could partly overcome the blue-light-derived suppression of gametophore development.[37,49,52] Nevertheless, cytokinins cannot substitute for the light requirement of bud production on caulonema filaments, although induction in darkness has been described for gametophores.[30,31] Considering this fact and the synergistic effect of light and cytokinins, a phytochrome-dependent hormone action on bud induction might be discussed as a possible step in the transduction process.

IV. PHOTO- AND POLAROTROPISM

A. GENERAL ASPECTS

Unilateral irradiation of germinating spores, protoplasts, and elongating protonema fil-

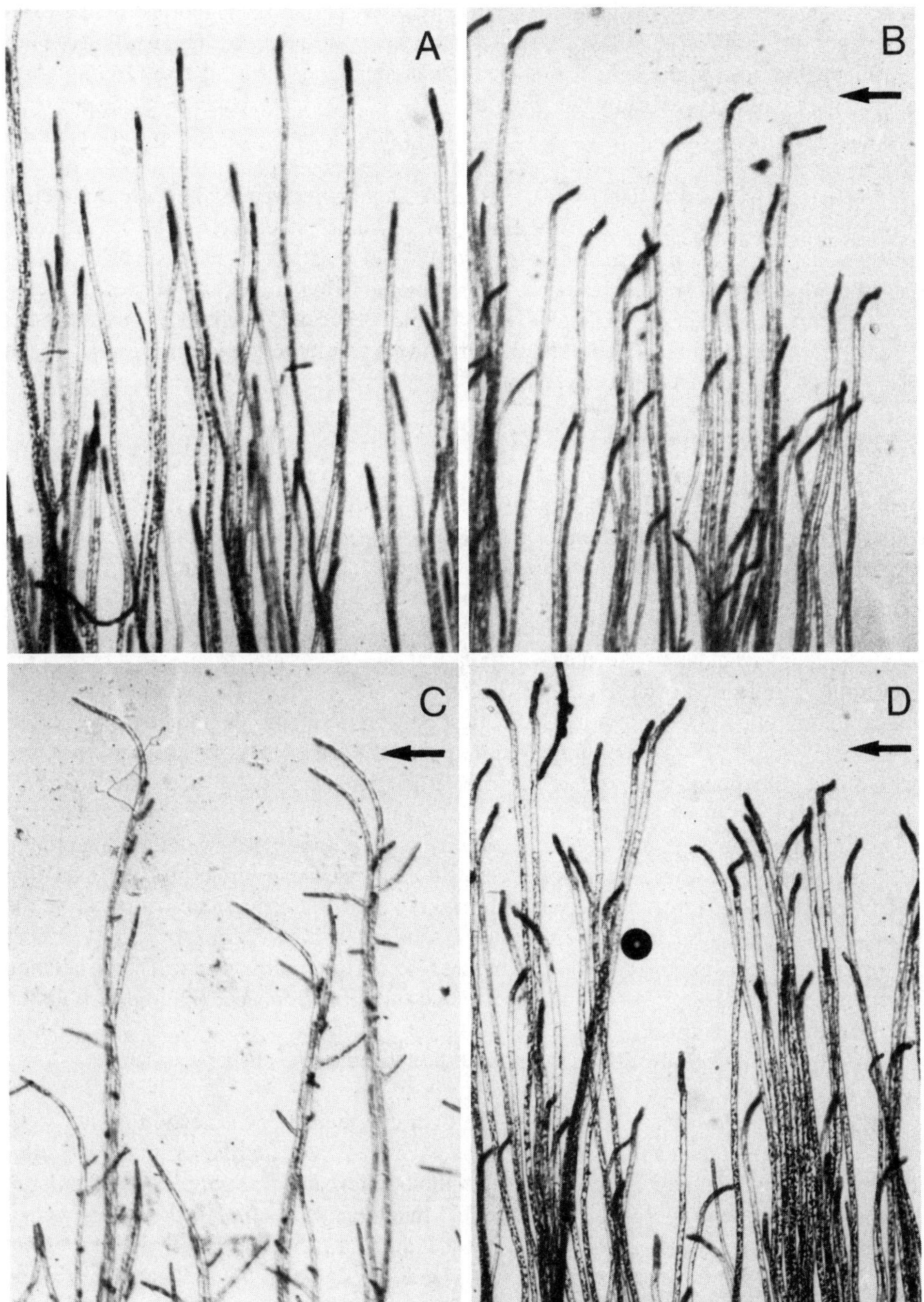

FIGURE 2. Different phototropic responses of *Ceratodon purpureus*. The light directions are marked by arrows. (A) Dark control of negative gravitropically growing protonemata filaments. (B) Positive phototrophic response induced by continuous red light (664 nm, 8 W m^{-2}) for 4 h and the beginning ''back-bending'' response after turning off the light. (C) Negative phototropic bending by ''bowing'' after 23 h blue light irradiation (420 nm, 18 W m^{-2}). (D) Phototropic response after 12 h irradiation with blue light (460 nm, 14 W m^{-2}). The protonemata show no clear-cut growth direction; the number of positive- or negative-reacting filaments is more or less equal.

aments in mosses and liverworts causes a reorientation of the main morphological axis, according to the irradiance gradient. Depending on the light intensity, the growth direction is oriented either toward the light source (i.e., positive phototropism) or away from it (i.e., negative phototropism), as shown in Figure 1. The ability to respond to changes in the irradiation conditions enables the protonemata to optimize their growth, depending on the light environment.

Protonemata exposed to linearly plane-polarized light perceive the direction of the electrical vector (**E**) of the light and grow either perpendicular or parallel to **E**. Polarotropic responses are supposed to result from the absorption of polarized light by dichroic photoreceptors, which are structurally fixed in a specific orientation at the periphery of the cells, probably near or within the plasmalemma. An example is explained in Figure 3. Investigations of tropic responses with polarized light are performed in order to get information about the orientation of the competent photoreceptors.[53]

B. RESPONSES OF SPORES

Continuous unilateral irradiation of *Funaria* spores with red light induces the direction of chloronemal outgrowth in a complex manner, depending on the light intensity range.[54] In low and medium irradiances of unpolarized light *Funaria* spores germinate at the brightest part, while with high light intensities chloronemata emerge from the darkest part. In polarized light of low irradiance the spore showed no sensitivity to the electrical vector. With increased light intensities the cells respond to vertically polarized light by a positive polarotropic outgrowth, perpendicular to the plane of polarization. In the higher intensity ranges the polarotropic germination occurred at the shaded side of the spore. The authors explain the lack of polarotropism in the low-irradiance response by a diffuse distribution of the photoreceptors within the cell. Accordingly, with increased fluence rates the receptor molecules appear structurally bound in an ordered manner at the periphery of the spore, mediating the polarotropic outgrowth.

A similar phenomenon was described for phytochrome-dependent spore germination in different fern species, where no action dichroism could be demonstrated for the induction by red light pulses.[55] However, the young fern protonema filament is known to exhibit the classical polarotropic response.[56-58] The authors theorize that phytochrome originally present in the spore and inducing the germination process is not structurally bound. The installation of membrane-associated phytochrome, responsible for polarotropism, might begin with the polar growth.

Obviously, spore germination in *Funaria*, although induced by light, does not necessarily depend on a strong absorption gradient within the spore.

Polar germination occurs well in uniform light environment (see Section II) and even in darkness. Thus, unilateral irradiation is not a prerequisite for the origin of polarity. Ultrastructural investigation of moss spore morphology revealed a slight preformed polarity in both cell content and spore wall structure.[59-62] Inhomogeneity of the spore wall structure results from varying thickness of the intine and the complementary distribution of wall material in the exine. The intine is thickest where the exine is thinnest, and this is the position where the spore coat ruptures in the beginning of germination.[19,60] Thus, germination follows a preformed polarity in uniform environment, but if the spores are exposed to an irradiation gradient the light signal will be the dominant determinative influence in the induction of polarity.

C. RESPONSES OF THE PROTONEMAL TIP CELL

Elongation in protonemata is restricted to the tip region of the apical cell of each filament. Tip growth is restricted locally to a small curved area of the apex, and by unknown mechanisms the protonemal cells are able to grow with a rather constant, specific width. Thus,

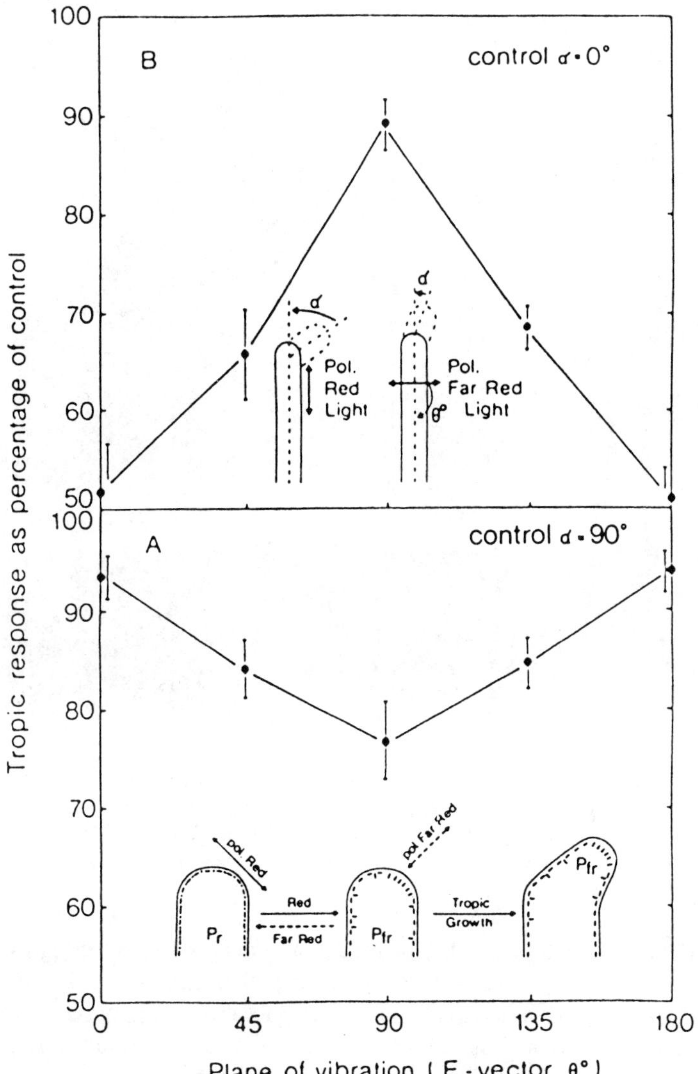

FIGURE 3. Effect of polarized red (664 nm) and far-red (730 nm) light in the phototropic response of *Ceratodon purpureus* protonemata. (A) The filaments were irradiated with polarized red light of different vibration planes for 90 min with 5 min red/5 min dark cycles. (B) The reversibility experiments were performed with red light with a vibration plane parallel to the long axis of the cell and then subsequently irradiated with far-red light of different vibration planes, also with a cycle of 5 min red/5 min far-red light.

tip growth is the most obvious expression of structural polarity, characterized by the polar distribution of Golgi vesicles.[63,64] Phototropic growth responses in protonemata start with the displacement of the limited growth region (growth center) from the apex of the filament toward one side of its apical dome, known as ''bulging'' growth in contrast to ''bowing'' growth, as shown in Figures 1A,B and 2B,C,D.[56,65] Due to this shifting of the growth center, the shape of a phototropically bending protonemal filament is an abrupt change in growth direction expressed as a sharp angle. The causal analysis of bending growth by bulging, including the displacement of the growth center, has yet to be analyzed.

An experimental advantage of mosses, compared, for example, with higher plants, is

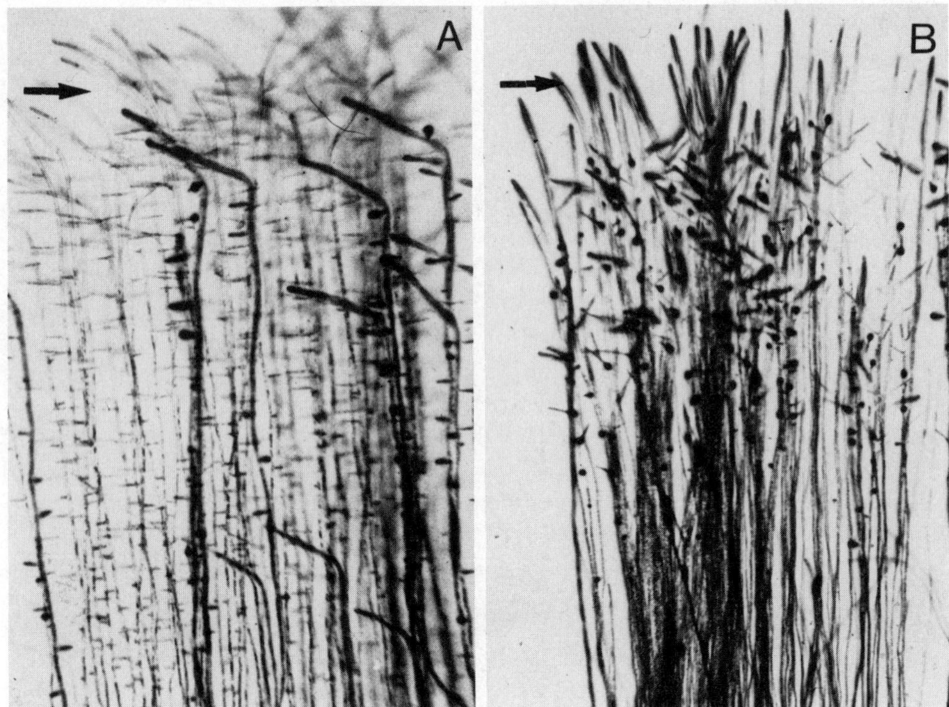

FIGURE 4. Wild-type (A) and mutant strain (B) from *Ceratodon purpureus*. The phototropically altered mutant was selected from spores treated with a mutagen.[66] The direction of illumination is shown by arrows. The protonemata were irradiated unilaterally with red light (664 nm, 15 W m^{-2}) for 12 h. The mutant is expressing a very much reduced response compared to the wild-type. The branching patterns of both wild-type and mutant are not oriented phototropically.

the rather simple methodical possibility of producing mutants and selecting morphogenetically altered strains.[17] It has also been possible to obtain various phototropically altered mutant strains from *C. purpureus*. In Figure 4 the phototropic responses of the wild-type and a mutant strain are shown. It is important to recognize that the branching pattern is not altered between wild-type and mutant strain. Those mutants may be valuable in elucidating the molecular regulation of phototropism.

Based on studies with polarized light in fern and moss protonemal tip cells, a useful model for local arrangement and orientation of phytochrome molecules has been established. This model served to explain many phytochrome-triggered reorientation phenomena in lower plants.[41,44,53,56] According to this model, the dichroic arranged photoreceptor molecules are associated with the plasma membrane in the tip region of the apical cell. In the P_r state the photoreceptor molecules are oriented with their axes of maximum absorption parallel to the cell surface. Upon conversion to the P_{fr} state the axes are aligned in a radial orientation. When irradiated with either red or far-red linearly polarized light, absorption occurs best with those photoreceptors whose axes coincide with the direction of the electrical vector (Figure 2). Depending on the symmetry of the irradiated area, differences in the absorption of polarized red light at the tip create an asymmetric distribution of P_{fr}.

Usually, interpretations of photo- and polarotropic responses in mosses and ferns indicate that the reorientation of tip growth in red light is directed toward the region of highest P_{fr} concentration. Kadota et al.[58] clearly demonstrated by microbeam experiments that the cells detect the difference in the P_{fr} amount of the extreme tip and the subapical regions of the apical dome and that these differences in the P_{fr} level regulate apical growth. Unpolarized light acts on the photoreceptor molecules in the same manner as polarized light, but since unpolarized light contains all directions of electrical vectors the gradients might be less distinct. Besides polarotropic effects, the establishment of absorption gradients in an asymmetrically irradiated cell is, of course, due to differences in the light intensity striking the irradiated side and the opposite side. It is likely that these differences decrease with increasing irradiance.

It is evident that an interpretation of phytochrome-derived photo- or polarotropic responses should involve a careful consideration of the specific absorption conditions and the resulting P_{fr} distribution. Positive phototropism of protonema tip cells is exclusively phytochrome dependent in *C. purpureus*. Protonemata of *Ceratodon* precultured in darkness for 5 days exhibit a positive phototropic response to unilateral irradiation with red light.[44,66] In contrast to ferns[57] and *F. hygrometrica* grown from spores under polarotropic conditions,[41] phototropic responses in vegetatively grown and dark-adapted moss protonema of *C. purpureus* could not be induced by brief light exposures.[44,66] In *Ceratodon,* growth toward the light source is maintained by either continuous unilateral irradiation with red light or a continuous cycle of 5 min red and 5 min dark. The end of the irradiation leads to a back-bending of protonemal filaments toward the initial growth direction in darkness, indicating that the reorientation of the morphological axis is not stable (Figure 1A). Reversibility was demonstrated by replacing the dark period with 5 min far-red light in the irradiation program, leading to the suppression of the response. The modulation of the growth direction in *Ceratodon* seems to be a consequence of the competition between P_{fr} and the mechanism which controls tip growth in darkness.

Studies with linearly polarized light confirmed for *Ceratodon* protonema tip cells the above-described model for phytochrome localization, with the absorption axes of P_r and P_{fr} lying parallel and perpendicular to the cell surface (Figure 3).[44]

A different situation was described for protonemata of *Physcomitrium turbinatum.* Nebel[39] evaluated an action spectrum for phototropic effectiveness by the null point method. A standard irradiation of 677 nm red light was given from one side of the tip cell and a second irradiation from the opposite, counteracting the phototropic response to the standard irradiation. The effectiveness of different wavelengths of polarized and unpolarized light to give a null response was recorded. Under these conditions Nebel[39] found a peak in the far-red region; i.e., simultaneous irradiation with red and far-red from opposite directions gave a positive phototropic response toward far-red light. Nevertheless, far-red alone was ineffective. From the synergistic effect of red and far-red and additional investigation with polarized light the author concluded the photoreceptors to be disc-shaped, absorbing both red and far-red light in the plane of the plasma membrane. It was supposed that the response depended on either the total amount of quanta absorbed or the consequent cycling of the photoreceptors between the P_r and P_{fr} states.

Jenkins and Cove[40] investigated phototropism and polarotropism in primary chloronemata of *Physcomitrella patens* wild-type and mutant strains; the latter differ partly from the wild-type in their response to light. For studies of polarotropism, germinating spores which had been preirradiated with white light were exposed to unilateral linearly polarized light of different qualities, and the reorientation of the emerging primary chloronemata was described.

The results confirm that tropic responses in *Physcomitrella patens* are controlled by phytochrome[17] in both wild-type and mutant strains. The wild-type chloronemata grew phototropically positive and perpendicular to **E** in red and blue light of low fluence rates, but reoriented their growth lateral to the incident light and parallel to **E** at high fluence rates. The transition from the "low-light" to the "high-light" response occurred abruptly at a certain fluence rate. The mutant strains exhibited only the "high-light" reaction, even with the lowest fluence rates. In green and far-red light the wild-type grew phototropically positive and perpendicular to **E**, whereas the mutants again tended to exhibit increasing parallel polarotropism with higher fluence rates. Thus, the mutant strains of *Physcomitrella* showed the same qualitative responses as the wild-type, but responded at much lower fluence rates with a "high-light" reaction.

Phototropism was studied by unilateral irradiation of chloronema filaments with unpolarized red and far-red light. The results confirmed the observations obtained with polarized light. Wild-type chloronemata showed positive phototropic responses in low light intensities, but grew lateral to the light direction in high irradiances. Again, the mutant strains seemed to react more sensitively to the light intensity, as they only exhibited the "high-light" response. In far-red light the protonemata exhibited a positive phototropic response. The response of the filaments in high irradiances of red light could be changed in part to positive phototropic growth by simultaneous application of far-red light. These findings indicate that tropic responses in mosses cannot be explained simply by growth toward the highest P_{fr} concentration in all species. In *Physcomitrella* the explanation might be applicable to the "low-light" response, but with high irradiances the chloronemata rather tend to grow away from the highest P_{fr} concentration. The authors opine that the responses to light are determined by the ratio P_{fr}/P_{tot}. If a certain critical value of P_{fr}/P_{tot} is exceeded under high irradiance conditions at the irradiated side, growth reorients toward regions containing less P_{fr}.

Obviously, far-red irradiation of the cells in *Physcomitrella* under the described conditions, i.e., after a prolonged preirradiation with white light, installs a P_{fr}/P_{tot} ratio optimal for positive phototropic growth. The response of *Physcomitrella* chloronemata to blue light resembles the response to red light. It is possible to speculate that the coinciding response might be due to the phytochrome absorption in the blue spectral range. Nevertheless, the involvement of a blue-absorbing pigment system besides phytochrome cannot be ruled out.

This problem is of special importance in polarotropism of the liverwort *Sphaerocarpus*. Steiner[67,68] demonstrated for *Sphaerocarpus* germ tubes a polarotropic response to blue and near UV, but no response to wavelengths above 550 nm could be detected. In plane-polarized blue light the germ tubes were oriented strictly perpendicular to **E**, but grew randomly in linearly polarized red light. By changing the direction of **E**, a polarotropic reorientation of the germ tubes occurred in a manner well known from *Dryopteris*. Red or far-red light, given either prior to, simultaneously with, or after the polarotropic induction, had no influence on the blue light response.

Thus, in *S. donnellii* a blue-light-absorbing pigment system controls the redirection of germ tube growth. From its polarotropic behavior a surface parallel (orientation) of the main absorption axis of these photoreceptors must be concluded.[66,67]

A rather different situation can be described for the blue-light-induced negative phototropic response in *C. purpureus*. We detected this response during long-term irradiation experiments with high intensities of blue light ($>10 \text{ W} \cdot \text{m}^{-2}$). After about 8 h of continuous irradiation the beginning of a negative phototropic growth response could be observed (Figures 1B and 2C). We interpret this response to be an avoidance reaction of the filaments to protect the cells against high irradiance conditions.

Furthermore, there is a distinct spectral region between 430 and 470 nm in which no preferential growth direction can be maintained by the protonema filaments, as shown in Figure 2D. Longer wavelengths lead exclusively to positive phototropism. It was not possible to obtain any negative response with red light, not even with the highest intensity which could be applied ($>60 \text{ W} \cdot \text{m}^{-2}$). For further understanding of blue and red light responses more experiments are required.

D. SIGNAL TRANSDUCTION

Regulation of physiological processes by light requires mechanisms for transduction and amplification of the light signal in order to generate an operative biochemical stimulus at the cellular level. In phytochrome-controlled responses the intracellular biochemical signal derives, in an unknown way, from the action of P_{fr}. It is one of the recently favored hypotheses of signal transduction that calcium ions (Ca^{2+}) may act as a second messenger which links the excitation of phytochrome to physiological processes. Evidence for this hypothesis in lower plants comes from phenomena such as the light-induced membrane depolarization in the alga *Nitella*[69] and the phytochrome-mediated uptake of calcium in *Mougeotia* cells.[70]

Ca^{2+} has been recognized as a major component of transcellular currents that occur in light-induced polarization of fern spores, zygotes of fucoid algae, and pollen grains, as well as in local growth of tubular cells.[71] In mosses, involvement of Ca^{2+} has been approached in studies on spore germination, induction of branching, bud formation, and phototropism of protonema tip cells. When spores of *F. hygrometrica* cultured in a medium with $Ca(NO_3)_2$ as a single salt are exposed to an external electrical field, they form rhizoids toward the positive electrode.[72] Forced Ca^{2+} entry at the point of outgrowth was thought to accompany the polarization effect. This was confirmed by germination studies in an external Ca^{2+} gradient, where outgrowth occurred at the point of highest Ca^{2+} uptake. The site of chloronema emergence, however, was scarcely affected by the electric field. Although this investigation was not performed in close connection with a light effect, it provides evidence that Ca^{2+} uptake into the spore also might be involved in the light-induced polarization process. In spores of the fern *Onoclea sensibilis,* which might be considered an analogous system, external Ca^{2+} is required for a short time after the application of a light signal to mediate the phytochrome response.[73] P_{fr} is assumed to install a Ca^{2+} transport system, and Ca^{2+} is considered a second messenger in spore germination. Nevertheless, recent studies on light-induced polarization of the *Fucus* zygote contradict the Ca^{2+} hypothesis.[74] Using the fluorescent Ca^{2+} chelator chlorotetracycline (CTC) the intracellular distribution of membrane-bound Ca^{2+} was visualized during the process of axis formation and further polar growth. The results indicate that Ca^{2+}, although essential to maintain tip growth, is not involved in the generation of the polar axis, i.e., the induction of polarity.

In the filamentous protonema phytohormones, especially cytokinins, are involved in the regulation of some light-dependent developmental events. Both the formation of caulonemal side branches and the determination of branch initials into gametophore buds in *Funaria* can be induced by the external supply of cytokinins. As the induction of side branches and buds occurs in fundamentally different concentration ranges, these two processes have to be distinguished as independently stimulated by cytokinins.[47] The involvement of Ca^{2+} in both cytokinin-regulated steps has been investigated.

In cytokinin-induced formation of branch initials, Saunders[75] measured an enhanced influx of Ca^{2+} at the future point of outgrowth immediately after the cytokinin treatment. A localized increase in Ca^{2+} concentration is required for cell division, which precedes the formation of a lateral outgrowth.[76-78] Induction of branches does not depend on high amounts of externally applied cytokinin and occurs spontaneously with addition of cytokinin under standard conditions.[79] It can therefore be considered that Ca^{2+} influx might be induced by light and regulated by the internal level of cytokinin. As the mode of light action on caulonemal branching remains unclear and phytochrome involvement has not yet been demonstrated, there is no strong basis for further discussion regarding a specific second-messenger function of Ca^{2+} in this respect.

In contrast to branching, the induction of bud formation by the application of cytokinins seems to be independent of external Ca^{2+}, although further development of existing bud initials cannot occur without Ca^{2+}.[79] There is again no evidence that Ca^{2+}-mediated, phytochrome-controlled transformation of branch initials into bud initials occurs either directly or via cytokinins.

The question of Ca^{2+} involvement in mediating light-induced polarity changes is of special interest in the phototropic responses of protonemata. With CTC, a strong tip-to-base gradient of membrane-associated Ca^{2+} was demonstrated in tip cells of *F. hygrometrica*[80]

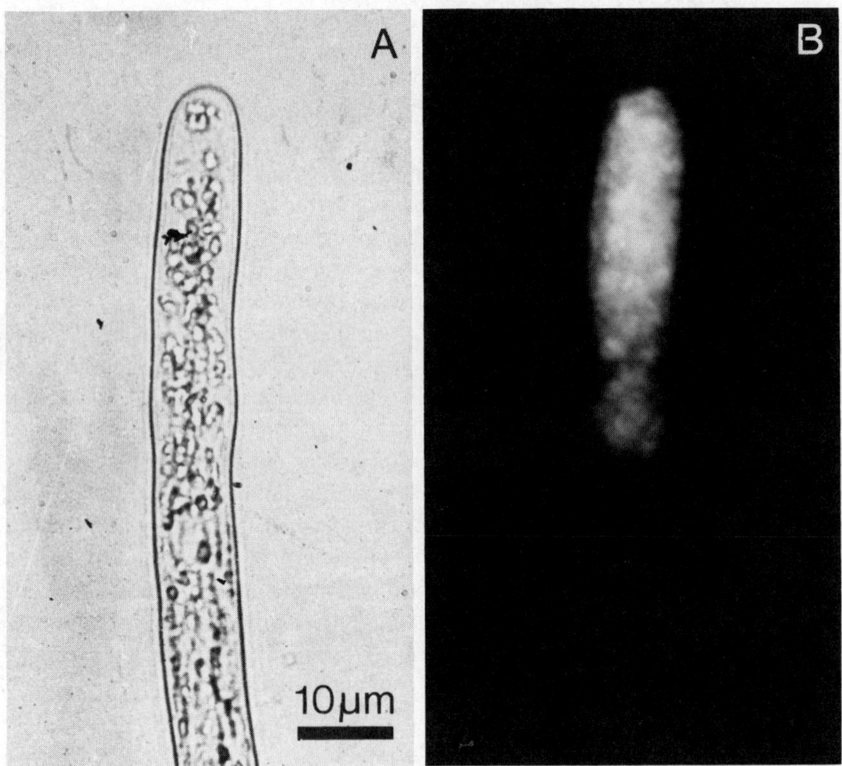

FIGURE 5. Fluorescence labeling of membrane bound Ca^{2+} in a *Ceratodon* protonema filament with chlorotetracycline.[66] (A) Protonema tip region in phase contrast microscopy, and (B) after fluorescence excitation.

and *C. purpureus*[66], as shown in Figure 5. In *Ceratodon* the Ca^{2+} gradient reorients during phototropism according to the light-dependent changes in the growth direction. With a treatment that abolishes the CTC gradient it was demonstrated that the dislocation of the growth center toward the irradiated flank of the cell is initiated independently of CTC-detectable Ca^{2+}. The expression of the newly arranged polarity, however, depends on polar growth and is inseparably associated with a high Ca^{2+} concentration at the tip.[81] The authors could demonstrate convincingly that during growth inhibition, under conditions of no detectable CTC-calcium gradient, a regulatory metabolite is synthesized by unilateral irradiation of the protonemal filaments. This unknown agonist accumulated depending on the irradiation time and became operational exclusively with P_{fr}, which is supposed to be the membrane-localized photoreceptor for the position of the newly installed growth center. Under the described condition the light signal can be stored.[81]

After the reversion of P_{fr} to P_r no response occurred, but a second P_{fr} induction became photomorphogenetically active with very small red light doses, which could never induce a phototropic response in non-preirradiated protonemal filaments. These results emphasize the synthesis and the accumulation of a specific metabolite in close cooperation with active phytochrome. This messenger is not identical to calcium.

Ca^{2+} is absolutely necessary to maintain cell division and polar growth in moss protonemata. A high concentration of Ca^{2+} is always associated with the growth region and is perhaps due to the Ca^{2+} requirement for secretory processes. However, no data support the hypothesis that Ca^{2+} might act as a second messenger. In our understanding a second messenger is a *specific* link between receptor and regulating cascade in a developmental response. This could be shown in some animal systems, but so far not in plants.

In the regulation of plant development by environmental factors, i.e., light, gravity, and phytohormones, Ca^{2+} seems to be involved in the stimulus-response coupling of many processes.[82,83] In responses to light, not only phytochrome-dependent phenomena, but also blue light effects are mediated by changes in the intracellular Ca^{2+} level.[83] Unfortunately, the strictly necessary information about modulations of cytoplasmic free Ca^{2+} in cells so far has not been obtained.[84]

Further investigations are required to elucidate transduction mechanisms, including Ca^{2+} regulation, phospholipid metabolism,[85] and alternative mechanisms.

E. RESPONSES OF MOSS PROTOPLASTS

One main problem in the elucidation of the primary biochemical or molecular reactions in plant photomorphogenesis is the processing of the light signal in a restricted number of target cells integrated in a complex tissue.[1] The investigation of total homogenates from tissues or especially from seedlings always involves the risk of artifacts or of finding no significant answer because it is masked by a nonspecific background. Tip cells of moss protonemata in some ways offer an excellent test system for studying the correlation between structure and function, but the same problem as with other tissues arises in biochemical investigations. We are measuring a singular response of the tip cell accompanied by a high background of nonresponding cells. A possible method of improving the "signal-to-noise" ratio may be the use of single tip cells derived from regenerating protoplasts.

Plant protoplasts have been described as "naked" cells because the cell wall has been removed. The isolated protoplast is a very special form of a living cell because the outer plasma membrane is fully exposed and is the only barrier between the external environment and the interior of the cell. From the earliest studies with protoplasts it became obvious that such plant material could be used for analyzing several biochemical and molecular biological problems for which cells are not suitable.[86,87] Importantly, protoplasts isolated from mosses show a very high viability and an excellent regeneration rate.[88] It could be shown that protoplasts and protonemal cells regenerated from moss protoplasts are very suitable test systems for studying photomorphogenesis.

Polar cell division in protoplasts from *Physcomitrella patens* could be induced after 1 day of regeneration by blue light and red light of relatively high light intensities (15 $W \cdot m^{-2}$).[89] The optimal light condition for *C. purpureus* was one third of that (5 $W \cdot m^{-2}$).[90] *Ceratodon* exhibited cell wall regeneration, cell division, and filamentous growth in complete darkness without any preferred polarization axis, in contrast to *Physcomitrella,* which also regenerated a new cell wall in darkness, but needed light to continue regeneration. This again emphasizes strong species-specific differentiation patterns. Furthermore, it has to be recognized that young filaments from protoplasts have much shorter cells than those of vegetatively grown protonemata.

However, the question arises whether protoplasts are comparable to the cells of the mother tissue from which they have been obtained.[91] Protoplasts and young regenerated filaments from *C. purpureus* showed remarkable differences in their photomorphogenetic responses compared with the tissue from which they had been derived.[92] Young filaments originating from protoplasts showed a pronounced positive phototropic response with blue light irradiation. This response was extremely different from that of the mother protonemal tissue, which expressed neither a negative nor a positive phototropic response with blue light illumination of comparable light intensities. The blue-light-induced mode of bending of protoplast-regenerated young protonema is a typical "bowing" response, whereas the filaments under red light illumination exhibit bending by "bulging" (Figure 6A,B; see also Figure 3A,B), typical for all red-light-induced phototropic responses in *Ceratodon.*

Using this technique it is possible to obtain large amounts of "pure" tip cell regenerates, which remarkably improve the signal-to-noise ratio in biochemical experiments. However,

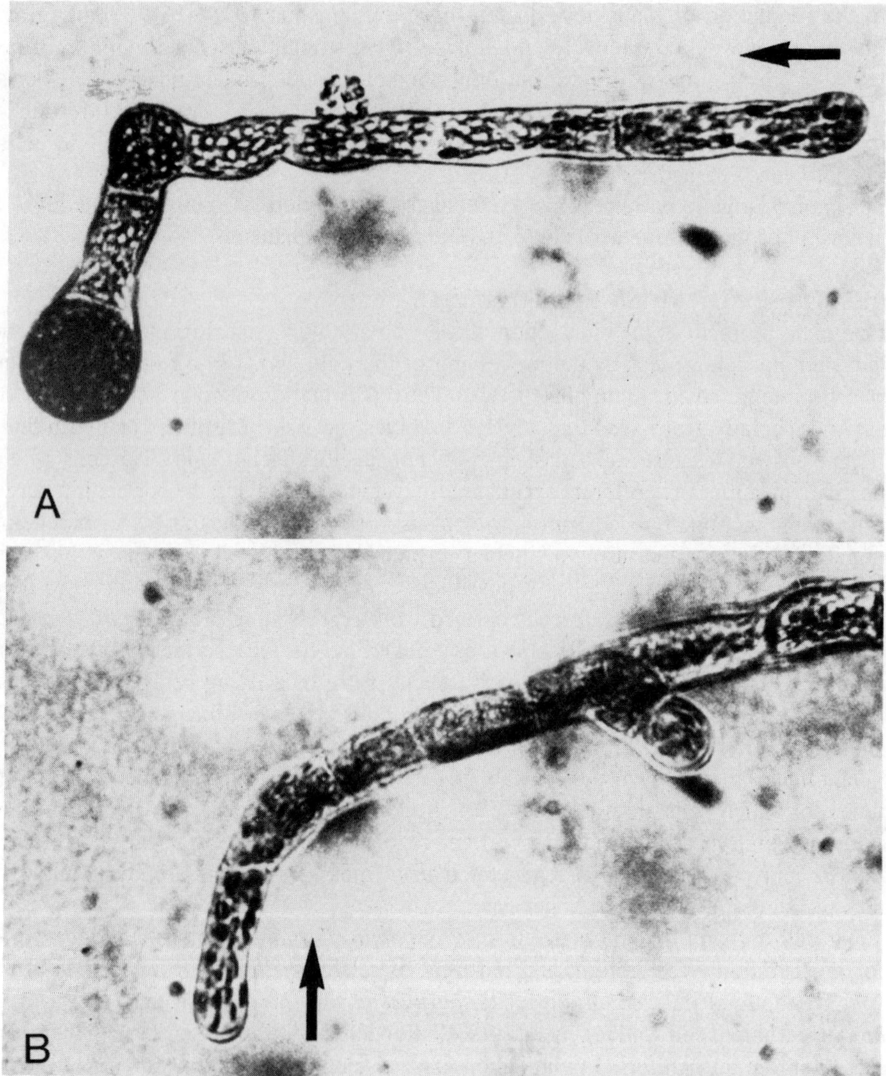

FIGURE 6. Protoplasts were isolated from 3-day-old protonemata of the moss *Ceratodon purpureus*. The protoplasts were treated as described by Wilbert[87,92] and were totally covered with semisolid agar. This procedure is necessary to maintain and fix the position of protoplasts in order to establish defined directions of irradiation. Unfortunately, photographing of the regeneration products is seriously complicated by the medium. Another culture medium with greater transparency than agar could not be used. After having formed two cells the regenerated cells were illuminated unilaterally with a light beam oriented 90° to the cell axis with red light (A) and blue light (B) for 5 and 7 days, respectively. The red-irradiated young filaments grew with a pronounced "bulging", whereas the blue-light-irradiated ones grew with "bowing". The response to blue light was remarkably slower than to red light, but in both cases a positive phototropic response was expressed. The newly regenerated cells were much shorter than the cells from normal protonemal filaments (compare with Figure 1).

even though rather high yields of regenerated cells are obtained, this method is not simple. It is more or less impossible to obtain an amount of material comparable to that normally used for biochemical experiments. However, with a reasonable technical effort one can collect amounts required for microanalytical analyses.

V. CONCLUSION

The major importance of light in controlling the growth and differentiation of mosses and liverworts has been highlighted. Many of the morphogenetic effects of light are clearly regulated by phytochrome, and classical red/far-red reversibility has been demonstrated for a variety of responses. However, there is evidence for more complicated HIR responses under continuous illumination. Further investigation of HIRs is required especially regarding blue-light-mediated responses. It is evident that much remains to be learned about many aspects of bryophyte photomorphogenesis. Perhaps of greatest concern is that so little is known about the cellular mechanisms through which light exerts its effects. The aim should be to enlighten the "black box" which still exists between the formation of P_{fr} and the signal processing until the onset of differential gene activity.

Although valuable information can still be obtained by *in vivo* observations or by using simple equipment, it will be impossible, even with these not highly differentiated gametophytes, to elucidate the molecular mechanism of light-regulated morphogenesis. Bryophytes have much to offer as experimental systems for such investigations. They resemble land plants and green algae in important features of their metabolism, but they form the only group of land plants in which a gametophyte is dominant. The gametophyte shows very good vegetative reproduction, but a perpetuating genotype. Mosses are very conservative organisms. The genetic stability, which is also expressed by tissues and cells in artificial culture, is an important feature for physiological, biochemical, and molecular biological studies. It allows consistent long-term culture of moss cells with high growth rates and good yields. The simple filamentous structure and the tip growth facilitate experiments on polarity induction and structure-to-function correlation. It is likely to be particularly advantageous to select morphogenetically altered mutants from moss spores for physiological and genetic characterization.

Cell suspension cultures have increased our knowledge about biochemical regulation in photomorphogenesis considerably.[93,94] Cell cultures of mosses should provide biochemical models which can contribute to the understanding of some key questions. How does membrane-bound phytochrome function? Are there special agonists involved in the signal transduction? What is the relationship between biochemical regulation and the appearance of a specific growth pattern? Is there an evolution of phytochrome? It is not expected that this plant system will provide complete answers to all questions, but it will help us to understand the relationship between biochemical control mechanisms and development of structural specifications.

REFERENCES

1. **Mohr, H.,** Pattern specification and realization in photomorphogenesis, in *Encyclopedia of Plant Physiology,* Vol. 16A, Shropshire, W., Jr. and Mohr, H., Eds., Springer-Verlag, Berlin, 1983, 336.
2. **Rüdiger, W.,** Phytochrome — the chromophore, in *Photomorphogenesis in Plants,* Kendrick, R. E. and Kronenberg, G. H. M., Eds., Martinus Nijhoff, Dordrecht, The Netherlands, 1986, 17.
3. **Fukshansky, L. and Schäfer, E.,** Models in photomorphogenesis, in *Encyclopedia of Plant Physiology,* Vol. 16A, Shropshire, W., Jr. and Mohr, H., Eds., Springer-Verlag, Berlin, 1983, 69.
4. **Hartmann, K. M.,** A general hypothesis to interpret "high energy phenomena" of photomorphogenesis on the basis of phytochrome, *Photochem. Photobiol.,* 5, 349, 1966.
5. **Schäfer, E.,** A new approach to explain "high irradiance response" of photomorphogenesis on the basis of phytochrome, *J. Math. Biol.,* 2, 45, 1975.
6. **Senger, H. and Schmidt, W.,** Diversity of photoreceptors, in *Photomorphogenesis in Plants,* Kendrick, R. E. and Kronenberg, G. H. M., Eds., Martinus Nijhoff, Dordrecht, The Netherlands, 1986, 137.

7. **Cosgrove, D. J.,** Photomodulation of growth, in *Photomorphogenesis in Plants,* Kendrick, R. E. and Kronenberg, G. H. M., Eds., Martinus Nijhoff, Dordrecht, The Netherlands, 1986, 341.
8. **Butler, W. L., Hendricks, S. B., and Siegelmann, H. W.,** Action spectra of phytochrome in vitro, *Photochem. Photobiol.,* 3, 521, 1964.
9. **Smith, H.,** The perception of light quality, in *Phytomorphogenesis in Plants,* Kendrick, R. E. and Kronenberg, G. H. M., Eds., Martinus Nijhoff, Dordrecht, The Netherlands, 1986, 187.
10. **Listowski, M. A.,** Über den Einfluβ verschiedenfarbigen Lichtes auf die Keimung der Sporen und Entwicklung der Protonemen einiger Moose, *Bull. Acad. Pol. Sci. Lett. Ser. B Nat.,* 6, 631, 1927.
11. **Bhatla, S. C.,** In vitro studies on growth and differentiation in the moss *Bryum argenteum* Hedw., *Beitr. Biol. Pflanz.,* 57, 157, 1983.
12. **Krupa, J.,** Studies on the physiology of germination of spores of *Funaria hygrometrica* (Sibth.), *Acta Soc. Bot. Pol.,* 33, 179, 1964.
13. **Inoue, H.,** Studies on spore germination and the earlier stages of gametophyte development in the Marchantiales, *J. Hattori Bot. Lab.,* 23, 148, 1960.
14. **Paolillo, D. J., Jr. and Kass, L. B.,** The germinability of immature spores in *Polytrichum, Bryologist,* 76, 163, 1973.
15. **Bauer, L. and Mohr, H.,** Der Nachweis des reversiblen Hellrot-Dunkelrot-Reaktionssystems bei Laubmoosen, *Planta,* 54, 68, 1959.
16. **Schild, A.,** Untersuchungen zur Sporenkeimung und Protonemaentwicklung bei dem Laubmoss *Physcomitrella patens,* Ph.D. thesis, University of Mainz, Federal Republic of Germany, Mainz, 1981.
17. **Cove, D. J., Schild, A., Ashton, N. W., and Hartmann, E.,** Genetic and physiological studies of the effect of light on the development of the moss *Physcomitrella patens, Photochem. Photobiol.,* 27, 249, 1978.
18. **Krupa, J.,** Studies on the physiology of *Funaria hygrometrica* spore germination. IV. The effect of blue and red light on the germination of spores, *Acta Soc. Bot. Pol.,* 41, 97, 1972.
19. **Valanne, N.,** The germination of moss spores and their control by light, *Ann. Bot. Fenn.,* 3, 37, 1966.
20. **Mohr, H.,** Die Beeinflussung der Keimung von Farnsporen durch Licht und andere Faktoren, *Planta,* 46, 534, 1956.
21. **Sugai, M. and Furuya, M.,** Photomorphogenesis in *Pteris vittata.* I. Phytochrome-mediated spore germination and blue light interaction, *Planta,* 8, 737, 1967.
22. **Sugai, M., Tomizawa, K., Watanabe, M., and Furuya, M.,** Action spectrum between 250 and 800 nanometers for the photoinducted inhibition of spore germination in *Pteris vittata, Plant Cell Physiol.,* 25, 205, 1984.
23. **Tomizawa, K., Sugai, M., and Manabe, K.,** Relationship between germination and P_{fr} level in spores of the fern *Lygodium japonicum, Plant Cell Physiol.,* 24, 1043, 1983.
24. **Steiner, A. M.,** Der Einfluβ von Licht und Temperatur auf die Sporenkeimung bei *Sphaerocarpus donnellii* Aust. (Hepaticae), *Z. Bot.,* 52, 245, 1964.
25. **Haupt, W.,** Effects of nutrients and light pretreatment on phytochrome-mediated fern-spore germination, *Planta,* 164, 63, 1985.
26. **Tomizawa, K., Manabe, K., and Sugai, M.,** Changes in phytochrome content during imbibition in spores of the fern *Lygodium japonicum, Plant Cell Physiol.,* 23, 1305, 1982.
27. **Fechner, A. and Schraudolf, H.,** Translation and transcription in imbibed and germinating spores of *Anemia phyllitidis* (L.) Sw., *Planta,* 161, 451, 1984.
28. **Manabe, K., Ibushi, N., Nakayama, A., Takaya, S., and Sugai, M.,** Spore germination and phytochrome biosynthesis in the fern *Lygodium japonicum* as affected by gabaculine and cycloheximide, *Physiol. Plant.,* 70, 571, 1987.
29. **Ahmad, S. M., Sultan, N., and Dager, J. C.,** Studies on the effect of light of different qualities and moisture on spore germination and gametophyte development in *Plagiochasma intermedium* L. & G., *Biologia (Lahore),* 23, 137, 1977.
30. **Szweykovska, A.,** Kinetin-induced formation of gametophores in dark cultures of *Ceratodon purpureus, J. Exp. Bot.,* 14, 137, 1963.
31. **Chopra, R. N. and Gupta, U.,** Dark-induction of buds in *Funaria hygrometrica* Hedw., *Bryologist,* 70, 102, 1967.
32. **Krupa, J.,** Studies on the physiology of germination of spores of *Funaria hygrometrica, Acta Soc. Bot. Pol.,* 36, 57, 1967.
33. **Stetler, D. A. and DeMaggio, A. E.,** Ultrastructural characteristics of spore germination in the moss *Dawsonia superba, Am. J. Bot.,* 63, 438, 1976.
34. **Mitra, G. C., Allsopp, A., and Wareing, P. F.,** The effects of light of various qualities on the development of protonema and bud formation in *Pholia nutans* (Hedw.) Lindb., *Phytomorphology,* 9, 47, 1959.
35. **Briere, C., Buis, R., and Larpent, J.-P.,** Cellular growth and cellular division in relation to age and illumination in the protonema of *Ceratodon purpureus* Brid., *Z. Pflanzenphysiol.,* 95, 315, 1979.

36. **Larpent, M. and Jaques, R.,** Role due phytochrome dans la development de protonema de *Funaria hygrometrica* Hedw., *C.R. Acad. Sci. Ser. D,* 273, 162, 1971.
37. **Jahn, H.,** Die Wirkung von Blaulicht und Kinetin auf die Protonemaentwicklung und Knospenbildung, *Flora (Jena),* 155, 10, 1964.
38. **Demkiv, O. T., Ripetsky, R. T., and Fedyk, Y. D.,** Effect of light of various wavelengths on morphology and physiology of *Funaria hygrometrica* Hedw. protonema, *Ukr. Bot. Zh.,* 28, 309, 1971.
39. **Nebel, B. J.,** Action spectra for photogrowth and phototropism in protonemata of the moss *Physcomitrium turbinatum, Planta,* 81, 287, 1968.
40. **Jenkins, G. I. and Cove, D. J.,** Phototropism and polarotropism of primary chloronemata of the moss *Physcomitrella patens:* responses of mutant strains, *Planta,* 159, 432, 1983.
41. **Bittisnich, D. and Williamson, R. E.,** Control by phytochrome of extension growth and polarotropism in chloronemata of *Funaria hygrometrica, Photochem. Photobiol.,* 42, 429, 1985.
42. **Bopp, M.,** Developmental physiology of bryophytes, in *New Manual of Bryology,* Vol. 1, Schuster, R. M., Ed., Hattori Botanical Laboratory, Nichinan, Japan, 1983, 276.
43. **Knoop, B.,** Development in bryophytes, in *The Experimental Biology of Bryophytes,* Dyer, A. F. and Duckett, J. G., Eds., Academic Press, London, 1984, 143.
44. **Hartmann, E., Klingenberg, B., and Bauer, L.,** Phytochrome-mediated phototropism in protonemata of the moss *Ceratodon purpureus* Brid., *Photochem. Photobiol.,* 38, 599, 1983.
45. **Schmiedel, G. and Schnepf, E.,** Side branch formation and orientation in the caulonema of the moss *Funaria hygrometrica:* normal development and fine structure, *Protoplasma,* 100, 367, 1979.
46. **Sievers, A. and Schnepf, E.,** Morphogenesis and polarity of tubular cells with tip growth, in *Cell Biology Monographs,* Vol. 8, Kiermayer, O., Ed., Springer-Verlag, New York, 1981, 265.
47. **Bopp, M. and Jacob, H. J.,** Cytokinin effect on branching and bud formation in *Funaria, Planta,* 169, 462, 1986.
48. **Nebel, B. J. and Naylor, A. W.,** Initiation and development of shoot-buds from protonemata in the moss *Physcomitrium turbinatum, Am. J. Bot.,* 55, 33, 1968.
49. **Kato, Y. and Watanabe, Y.,** Protonemata of aquatic moss *Amblystegium riparium* cultured in darkness and in light: growth and gametophore bud formation, *Phytomorphology,* 33, 270, 1982.
50. **Rashid, A.,** Spore germination, protonema development and bud formation in *Anoectangium thomsonii, Phytomorphology,* 20, 49, 1970.
51. **Simon, P. E. and Naef, J. B.,** Light dependency of the cytokinin-induced bud initiation in protonemata of the moss *Funaria hygrometrica, Physiol. Plant.,* 53, 13, 1981.
52. **Brandes, H.,** Der Wirkungsmechanismus des Kinetins bei der Induktion von Knospen am Protonema der Laubmoose, *Planta,* 74, 55, 1967.
53. **Hartmann, K. M., Menzel, H., and Mohr, H.,** Ein Beitrag zur Theorie der polarotropischen und phototropischen Krümmung, *Planta,* 64, 363, 1965.
54. **Jaffe, L. and Etzold, H.,** Tropic responses of *Funaria* spores to red light, *Biophys. J.,* 5, 715, 1965.
55. **Haupt, W. and Björn, L. O.,** No action dichroism for light-controlled fern-spore germination, *J. Plant Physiol.,* 129, 119, 1987.
56. **Etzold, H.,** Der Polarotropismus und Phototropismus der Chloronemen von *Dryopteris filix mas* (L.) Schott, *Planta,* 64, 254, 1965.
57. **Kadota, A., Wada, M., and Furuya, M.,** Phytochrome-mediated phototropism and different dichroic orientation of P_r and P_{fr} in protonemata of the fern *Adiantum capillus-veneris* L., *Photochem. Photobiol.,* 35, 533, 1982.
58. **Kadota, A., Wada, M., and Furuya, M.,** Phytochrome-mediated polarotropism of *Adiantum capillus-veneris* L. protonemata as analyzed by microbeam irradiation with polarized light, *Planta,* 165, 30, 1985.
59. **Brown, R. C. and Lemmon, B. E.,** Ultrastructure of sporogenesis in a moss, *Ditrichum pallidum.* III. Spore wall formation, *Am. J. Bot.,* 67, 918, 1980.
60. **Mogensen, G. S.,** The spore, in *New Manual of Bryology,* Vol. 1, Schuster, R. M., Ed., Hattori Botanical Laboratory, Nichinan, Japan, 1983, 325.
61. **Olesen, P. and Mogensen, G. S.,** Ultrastructure, histochemistry and notes on germination stages of spores in selected mosses, *Bryologist,* 81, 393, 1978.
62. **Schulz, D.,** Darstellung der submikroskopischen Strukturen lufttrockener Moossporen, *Ber. Dtsch. Bot. Ges.,* 85, 193, 1972.
63. **Schnepf, E.,** Cellular polarity, *Annu. Rev. Plant Physiol.,* 37, 23, 1986.
64. **Sievers, A. and Schnepf, E.,** Morphogenesis and polarity of tubular cells in tip growth, in *Cell Biology Monographs,* Vol. 8, Kiermayer, E., Ed., Springer-Verlag, New York, 1981, 265.
65. **Hejnowicz, Z. and Sievers, A.,** Mathematical model of geotropically bending *Chara* rhizoids, *Z. Pflanzenphysiol.,* 66, 34, 1971.
66. **Hartmann, E.,** Influence of light on phototrophic bending of moss protonemata of *Ceratodon purpureus* (Hedw.) Brid., *J. Hattori Bot. Lab.,* 55, 87, 1984.

67. **Steiner, A. M.,** Action spectrum for polarotropism of the germ tube of the liverwort *Sphaerocarpus donnellii* Aust., *Planta,* 86, 343, 1969.
68. **Steiner, A. M.,** Dose response behavior for polarotropism of the germ tube of the liverwort *Sphaerocarpus donnellii* Aust., *Planta,* 86, 334, 1969.
69. **Weisenseel, M. H. and Ruppert, H. K.,** Phytochrome and calcium ions are involved in light-induced membrane depolarization in *Nitella, Planta,* 137, 225, 1977.
70. **Dreyer, E. M. and Weisenseel, M. H.,** Phytochrome mediated uptake of calcium in *Mougeotia* cells, *Planta,* 146, 31, 1979.
71. **Weisenseel, M. H. and Kicherer, R. M.,** Ionic currents as control mechanism in cytomorphogenesis, in *Cell Biology Monographs,* Vol. 8, Kiermayer, O., Ed., Springer-Verlag, New York, 1981, 379.
72. **Chen, T.-H. and Jaffe, L. F.,** Forced calcium entry and polarized growth of *Funaria* spores, *Planta,* 144, 401, 1979.
73. **Wayne, R. and Hepler, P. K.,** The role of calcium ions in phytochrome-mediated germination of spores of *Onoclea sensibilis* L., *Planta,* 160, 12, 1984.
74. **Kropf, D. L. and Quatrano, R. S.,** Localization of membrane-associated calcium during development of fucoid algae using chlorotetracycline, *Planta,* 171, 158, 1987.
75. **Saunders, M. J.,** Cytokinin activation and retribution of plasma-membrane ion channels in *Funaria, Planta,* 167, 402, 1986.
76. **Saunders, M. J. and Hepler, P. K.,** Localization of membrane-associated calcium following cytokinin treatment in *Funaria* using chlorotetracycline, *Planta,* 152, 272, 1981.
77. **Saunders, M. J. and Hepler, P. K.,** Ca^{2+}-ionophore A23187 stimulates cytokinin-like mitosis in *Funaria, Science,* 217, 943, 1982.
78. **Saunders, M. J. and Hepler, P. K.,** Calcium antagonists and calmodulin inhibitors block cytokinin-induced bud formation in *Funaria, Dev. Biol.,* 99, 41, 1983.
79. **Markmann-Mulisch, U. and Bopp, M.,** The hormonal regulation of protonema development in mosses. IV. The role of Ca^{2+} as cytokinin effector, *J. Plant Physiol.,* 129, 155, 1987.
80. **Reiss, H.-D. and Herth, W.,** Calcium gradients in tip growing plant cells, visualized by chlorotetracycline fluorescence, *Planta,* 146, 615, 1979.
81. **Hartmann, E. and Weber, M.,** Storage of the phytochrome-mediated phototropic stimulus of moss protonemal tip cells, *Planta,* 175, 39, 1988.
82. **Roux, S. J.,** Ca^{2+} and phytochrome action in plants, *BioScience,* 34, 25, 1984.
83. **Hepler, P. K. and Wayne, R. O.,** Calcium and plant development, *Annu. Rev. Plant Physiol.,* 36, 397, 1985.
84. **Kauss, H.,** Some aspects of calcium-dependent regulation in plant metabolism, *Annu. Rev. Plant Physiol.,* 38, 47, 1987.
85. **Hartmann, E. and Pfaffmann, H.,** Involvement of phospholipid-metabolism in signal transduction and development in plants, in *Phytochrome and Plant Photomorphogenesis,* Furuya, M., Ed., Yamada Science Foundation, Okazaki, Japan, 1986, 107.
86. **Gottwald, R., Beutelmann, P., and Hartmann, E.,** Protoplasts as a tool for isolating intact chloroplasts from the moss *Leptobryum pyriforme,* in *Methods in Bryology,* Glime, J. M., Ed., Hattori Botanical Laboratory, Nichinan, Japan, 1988, 73.
87. **Hartmann, E., Wilbert, E., and Jahnen, W.,** Fluorescence labelling of moss protoplasts with calcofluor, in *Methods in Bryology,* Glime, J. M., Ed., Hattori Botanical Laboratory, Nichinan, Japan, 1988, 203.
88. **Grimsley, N. H., Ashton, N. W., and Cove, D. J.,** The production of somatic hybrids by protoplast fusion in the moss, *Physcomitrella patens, Mol. Gen. Genet.,* 154, 97, 1977.
89. **Jenkins, G. I. and Cove, D. J.,** Light requirements for regeneration of protoplasts of the moss *Physcomitrella patens, Planta,* 157, 39, 1983.
90. **Jahnen, W. and Hartmann, E.,** Microfluorometric determination of early regeneration of moss protoplasts, in *Protoplasts 1983—Poster Proceedings,* Potrykus, I., Harms, C. T., Hinnen, A., Hütter, R., King, P. J., and Shillito, R. D., Eds., Birkhäuser Verlag, Basel, 1983, 183.
91. **Boller, T. and Galbraith, D. W.,** Plant protoplasts as tools for physiological studies, in *Protoplasts 1983—Lecture Proceedings,* Potrykus, I., Harms, C. T., Hinnen, A., Hütter, R., King, P. J., and Shillito, R. D., Eds., Birkhäuser Verlag, Basel, 1983, 130.
92. **Wilbert, E.,** *Untersuchungen über Regenerationsleistungen und Polaritätsinduktion bei Protoplasten von Ceratodon purpureus* Hedw., Master's thesis, University of Mainz, Federal Republic of Germany, Mainz, 1987.
93. **Hahlbrock, K., Knobloch, K.-H., Kreuzaler, F., Potts, J. R. M., and Wellmann, E.,** Co-ordinated induction and subsequent activity changes of two groups of metabolically inter-related enzymes: light-induced synthesis of flavonoid glycosides in cell suspension cultures of *Petrosilinum hortense, Eur. J. Biochem.,* 61, 199, 1976.
94. **Schröder, J., Kreuzaler, F., Schäfer, E., and Hahlbrock, K.,** Concomitant induction of phenylalanine ammonia-lyase and flavanone synthetase mRNAs in irradiated plant cells, *J. Biol. Chem.,* 254, 57, 1979.

Chapter 4

HORMONES OF THE MOSS PROTONEMA

Martin Bopp

TABLE OF CONTENTS

I. INTRODUCTION

From the main groups of hormones in higher plants only auxins and cytokinins are documented as natural signal substances in mosses.[1,2] Not only do both hormone groups exist in mosses, but they also have basic functions in the regulation of normal development. Among the other three groups of growth regulators (gibberellins, abscisic acid, and ethylene) abscisic acid and ethylene have been identified in moss protonema (Figure 1).[3] However, it has not yet been finally clarified whether ethylene acts as a hormone to regulate differentiation or senescence, as was concluded from experiments with ethylene precursor 1-aminocyclopropane-1-carboxylic acid (ACC). Gibberellins, quite often tested as effectors for different developmental processes such as gametangial formation, have never delivered data clear enough to consider a true physiological effect.[4,5] To date there have been no reports demonstrating the occurrence of gibberellins in mosses, especially in the protonema. Our own experiments[118] failed to show detectable amounts of gibberellins in *Funaria hygrometrica* with different bioassays. To accomplish this, more detailed and sophisticated tests or direct methods (ELISA) for the determination of the endogenous content of active gibberellins are needed.

The same is true for abscisic acid (ABA), which in physiological concentrations of the micromole range inhibits growth, branching, and differentiation of moss protonema,[6,77] and it can be regarded as an antagonist to auxins and cytokinins (Figure 2). In very recent determinations ABA was found in the protonema of *F. hydrometrica* wild-type and two mutants. Therefore, it is possible that this phytohormone is involved as an internal substance in the natural process of growth regulation and differentiation.[119]

A factor with a hormone-like activity — a second messenger or mediator of hormones in animals — is cyclic adenosine 3′,5′-monophosphate (cAMP). It is well established that in higher plants cAMP plays only a minor role, if any.[7] For mosses, however, several reports deal not only with the occurrence of cAMP in the protonema,[8,9] but also with the effect of cAMP tested in a series of experiments concerning different steps of development.[10] Therefore, the possible effects of cAMP or other second messengers on moss protonema will be discussed in Section IV.

II. DEVELOPMENT OF THE PROTONEMA

In this report only the protonemata of mosses are considered because they represent a "morphogenetic system" with a very clear and characteristic stepwise differentiation. Auxins and cytokinins, and probably cAMP, show very clearly defined functions in this system. In contrast to this, the effects of hormones or hormone-like substances seem to be much less strongly associated with the growth of gametophores, sexual differentiation,[11] or the growth and development of sporophytes.[12,12a] Such effects of exogenous hormones have been published repeatedly, but in no case has there been any evidence for the natural occurrence of hormones in the gametophores or the distribution, transport, and metabolism of a hormone. Even the exact localization of the effects on apical cells, meristematic tissue, developing leaves, etc. is unknown. Quantitative effects alone, which change the size of a plant, the number of sexual organs, or the time of appearance of an organ, etc., are not sufficient to establish a certain hormone as an internal regulating substance. The research work on protonemata, however, fulfills all the criteria needed to recognize a substance as a "hormone".

In order to understand the details, a short description of protonema development in mosses is necessary. This is used as a paradigm for such research on *F. hygrometrica, Physcomitrella patens, Ceratodon purpureus,* etc.

Protonema development starts in the same way from spores, single cells, cells from

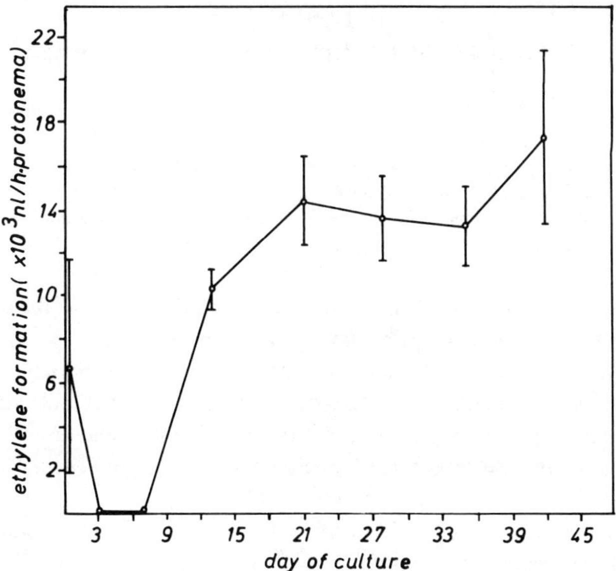

FIGURE 1. Rate of ethylene formation by the protonemata of *Funaria hygrometrica* after different days of culture.[3] Protonemata were transplanted to 25-ml vials 5 days after sowing. Ethylene was determined by gas chromatography.

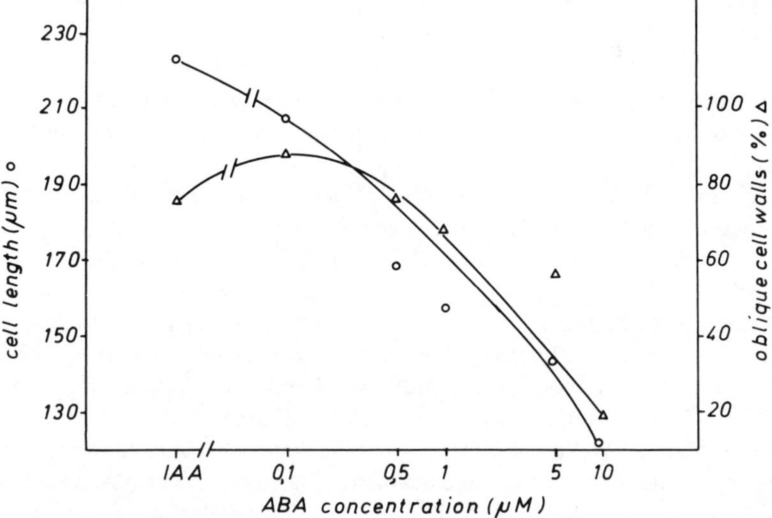

FIGURE 2. Effect of ABA on cell length and the number of oblique cell walls in the protonemata of *Funaria hygrometrica*.[6] All protonemata were treated with 1 μ*M* IAA together with the different concentrations of ABA for 2 days.

protoplasts,[13] or as regeneration products from small pieces of tissue. They usually produce a single filament from one cell. However, a few primary filaments can originate from spores, one of which is a small primary rhizoid. The others with a larger diameter and numerous chloroplasts are the chloronemata. Filaments grow exclusively at the very tip.[14] Depending on the cultural and environmental conditions the chloronema changes continuously to a second type of filament, the caulonema.[15] Both types can be distinguished by their cellular

TABLE 1
Characterization of Chloronema and Caulonema

	Chloronema	Caulonema
Morphological	Branching irregular; transverse septa	Branching regular; oblique septa
Cytological	Walls colorless; round chloroplasts	Walls brown; spindle-shaped chloroplasts
Physiological	Positive phototropic; nontarget cells for bud induction	Negative phototropic; target cells for bud induction

character (Table 1). The pronounced differences concern the position of the cell wall between two cells and the form of branching, which is very regular in the caulonema, where side branches always arise at the apical end of the cells of the main filament. They can grow either as secondary chloronemata,[2] as secondary caulonemata,[16] or as buds after the formation of a three-sided apical cell. These buds are transformed side branches[17,18] and the starting point of the gametophores. Although the branching is a very regular fact,[14,19] determination of further differentiation depends on factors which are discussed in the following pages.

Further description is limited to these three steps of development: chloronema, caulonema, and bud (gametophore) formation.

III. AUXINS

Several effects of auxins on moss development have been known for more than 30 years. These include inhibition of protonema growth, stimulation of rhizoid formation, transformation of buds to filaments, torsion of young stems, and complete suppression of leaves on gametophores.[15,20,21] Even with all these effects it was far from clear whether they could be regarded as hormone specific or were merely the manifestation of a general growth inhibition by unspecific high auxin concentrations. Not until the studies of Johri and coworkers[22-25] could it be established beyond a reasonable doubt that auxin is involved in the development of the protonema of mosses as a substance with the character of a signal or a morphogen.

Protonemata of *F. hygrometrica* were cultivated in Erlenmeyer flasks in a suspension of Knop solution without shaking. Under such conditions they remained in the chloronema stage. With the addition of low concentrations of indole-3-acetic acid (IAA) or 1-naphthaleneacetic acid (NAA) the protonemata differentiated into caulonemata.[22,23] The strength of this effect depends on the auxin concentration and on the cell density in the inoculum (Figure 3). The lower the original cell density, the lower the auxin concentration necessary. At cell densities below 0.1 mg protonema per milliliter of solution, caulonemata appear spontaneously. This density effect depends on the release of IAA oxidase from the protonema cells, which degrades IAA in the nutrient solution. Peroxidase and IAA oxidase have the highest activity in the medium when the cell density is highest,[24] whereas cyclic nucleotide phosphodiesterase, nitrate reductase, and protein kinase show the highest activity at low cell density. Because of the high auxin oxidase activity, caulonema production needs more external auxin molecules per cell in dense cultures. Nonauxins like β-NAA, cytokinins, or gibberellins do not influence caulonema differentiation.[25]

The results of the suspension culture studies were confirmed by Lehnert and Bopp[6] with experiments on a solid agar substrate at low irradiance (ca. 1 W m^{-2} = low light intensity [LLI]). Under such conditions protonemata grow only as chloronemata; however, after treatment with low concentrations of IAA, formation of oblique cell walls in the growing filaments starts immediately. The optimal effect occurs at 0.1 m*M* IAA (Figure 4).[6] Caulonema cells are the exclusive site for bud induction by cytokinins in *Funaria* and other mosses. Therefore, protonemata grown in LLI react to cytokinin only when pretreated with

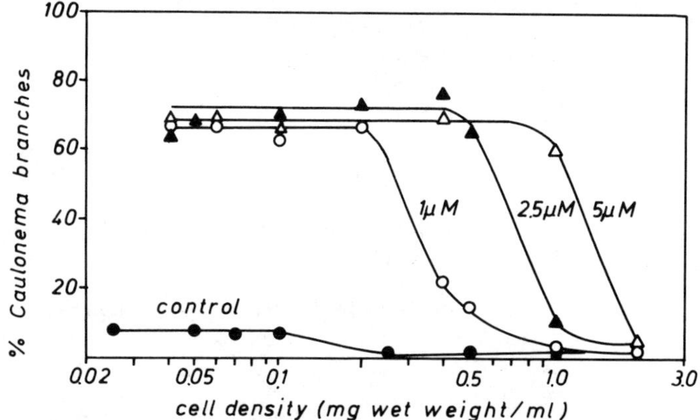

FIGURE 3. Effect of increased IAA concentrations (1 to 5 μ*M*) on the caulonema formation in suspension cultures of *Funaria hygrometrica* with different cell densities.[22] Cells grew for 5 days in the relevant conditions.

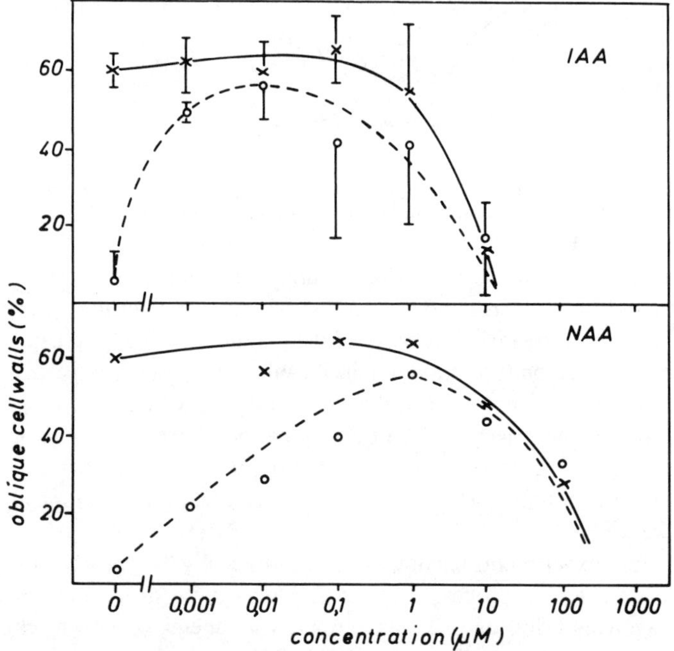

FIGURE 4. Formation of oblique cell walls as a character of caulonema in *Funaria hygrometrica*.[6] The protonemata were grown at 3000 (——) and 1000 (-----) lux for 48 h on agar substrate with different concentrations of NAA or IAA.

auxin.[6] Hence, the number of buds induced by kinetin in LLI can be used to quantify the auxin effect.[26]

Other auxins like 2,4-dichlorophenoxyacetic acid (2,4-D) and α-NAA differ only quantitatively from IAA (Figure 4). The same is true for substances found as IAA precursors in higher plants. In Figure 5, different biosynthetic pathways of IAA in higher plants are shown. Two of them are identified to be active in mosses because treatment with the relevant precursors gives results identical to those for treatment with auxin: the indole-3-pyruvic acid

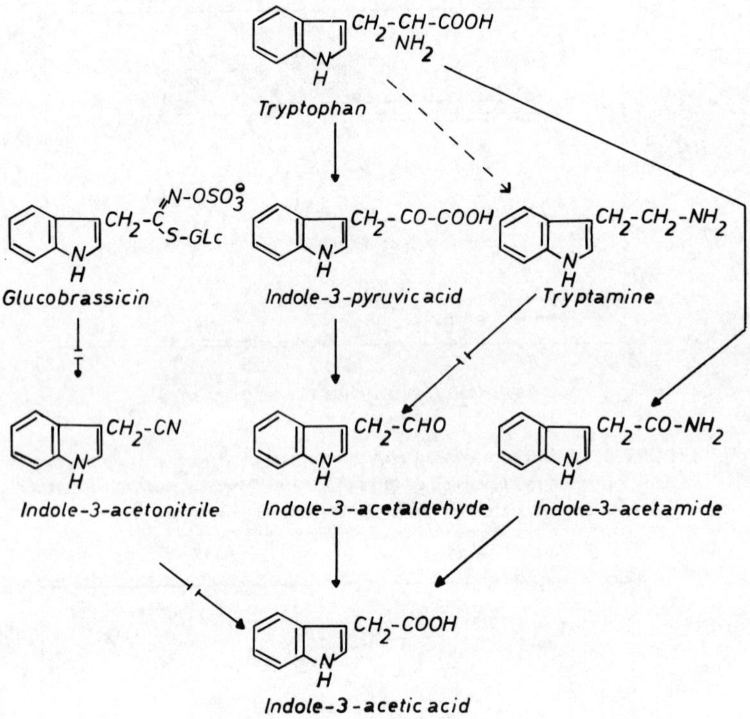

FIGURE 5. Pathways of IAA synthesis in higher plants and in moss protonemata. Pathways shown by interrupted arrows are not reported in mosses.

to indole-3-acetaldehyde pathway and, interestingly, the indo-3-acetamide pathway (which in higher plants only plays a role after infection with *Agrobacterium,* where the genes for this pathway have been identified as parts of the T region of the Ti plasmid).[27,28] Because tryptamine and indoleacetonitrile did not induce caulonemata in our experiments, we assume that these two pathways do not exist in the moss protonema. If the two other pathways are working, the enzymes necessary for the transformation of the precursors must be present, which so far has not been demonstrated directly by an enzyme preparation. In a cell-free system from *Funaria* protonema, IAA was synthesized from ^{3}H-tryptophan via the indole-pyruvate pathway.[29]

All the previous experiments cannot prove unequivocally the participation of endogenous auxins in the differentiation process, even if they give quite strong evidence. Experiments with auxin antagonists bring us one step further. Caulonema reverts to chloronema when treated with (parachlorophenoxy)isobutyric acid (PCIB), an antagonist to auxin.[30,31] In the presence of PCIB, cell diameters become smaller, the cross walls in the new, growing filaments are transverse, and all effects demonstrate a reversion from caulonema to chloronema growth (Figure 6), which corresponds to an antagonistic reaction to auxin. Furthermore, PCIB and auxin both inhibit growth of the protonema when applied in high concentrations, but the inhibition by IAA can be reduced or nearly nullified with PCIB.[32] This means that PCIB can reduce the effect of endogenous IAA as well as exogenous auxin. However, the previous results can be explained only if endogenous auxin is actively involved in the chloronema-caulonema transition.

A more convincing demonstration of the participation of endogenous auxin comes from experiments with mutants. Among a series of X-ray-induced mutants of *F. hygrometrica*[33] Hatanaka-Ernst[32] isolated one (Fu 219) which consisted only of chloronemata. This mutant could be normalized in the presence of 0.5 μM IAA. However, at that time it was not

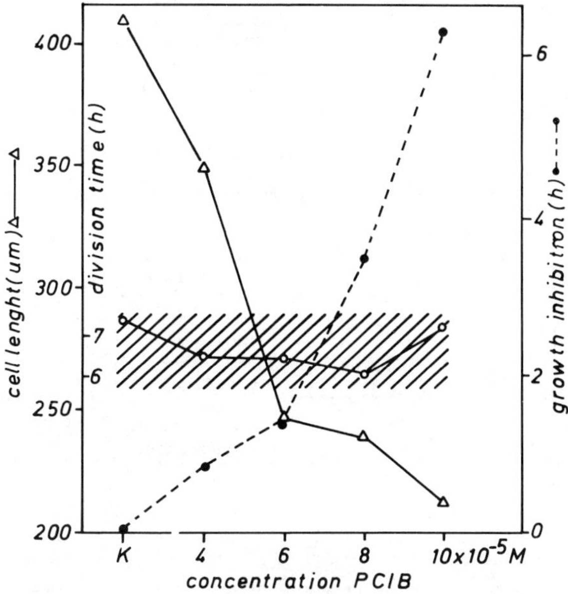

FIGURE 6. *Funaria hygrometrica* protonemata treated with (parachlorophenoxy)isobutyric acid (PCIB). Correlation between cell length and duration of the growth stops after beginning of the treatment. The division rate is not changed by PCIB.

possible to analyze the mutant in more detail. What remains difficult to understand is that the ability to react with auxin appears only after a culture time of 18 (!) months on Knop agar. Buds appeared within 3 weeks when transferred to an auxin-containing substrate, which makes it very difficult to determine the type of deficiency. Therefore, mutants have to be found which show the genetic change immediately after treatment. Since 1977 Cove and co-workers[2,35] have produced numerous mutants of the moss *P. patens* by the mutagenesis method, basically developed by Adelberg et al.[34] In this method a spore suspension is treated during germination with *N*-methyl-*N′*-nitro-*N*-nitrosoguanidine (NTG). From the surviving spores (about 10%) mutants are selected by different selection methods.[35] Among those mutants, several were changed in respect to their reaction to auxins and/or cytokinins (Figure 7).[36,37] From the mutants at least nine categories could be identified. The first group (categories 1 to 3) was defective in auxin uptake or sensitivity.[2] So the category 1 mutants (NAR 112), which grow only as chloronema, do not change growth form or growth rate when treated with NAA concentrations between 0.5 and 12.5 μM. This mutant is unable to form buds. The same insensitivity to external auxin was found for category 2 mutants, but they can produce caulonemata spontaneously. Finally, the category 3 mutants look like the wild-type, but whereas wild-type protonema are inhibited in growth by high auxin concentrations,[20] the mutant is not altered by 12.5 μM NAA or 50 μM 2,4-D.

The response of the category 4 mutant may serve as an example for categories 4 to 7. It resembles Fu 219 of *Funaria*. Small protonemata consist mainly of chloronemata; only a few caulonema-like filaments are formed at the periphery. Mutants of this type are unaffected by exogenous cytokinin alone. However, quite normal caulonema production is restored by treatment with α-NAA (0.125 to 0.250 μM).[37] Also, normal bud formation occurs after auxin treatment. The mutant becomes sensitive to exogenous cytokinin. The reaction is the same as for the wild-type strain of *Funaria* cultivated in LLI.

To explain the behavior of this category of mutants Cove and Ashton[2] suggest that a block in the synthetic pathway prevents the synthesis of endogenous auxin, which is progressively less strong in the mutants with higher numbers.[5-7]

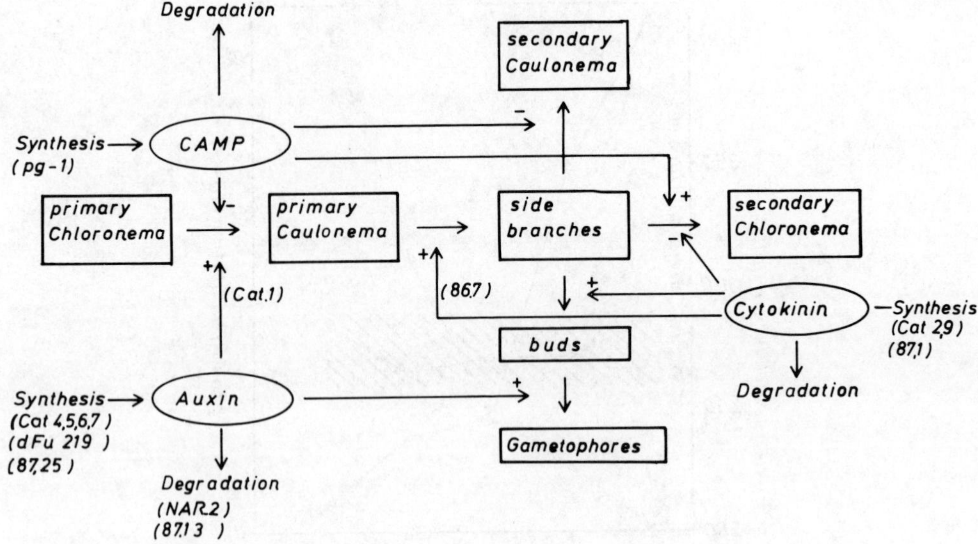

FIGURE 7. Scheme of the development of the protonema with the points of origin of the different mutants.[2] (Adapted from Cove, D. J. and Ashton, N. W., in *The Experimental Biology of Bryophytes,* Dyer, A. F. and Duckett, J. G., Eds., Academic Press, Orlando, FL, 1984, 183. With permission.)

TABLE 2
Bud Formation in the Wild-Type and the Mutant NAR-2 of *Funaria hygrometrica*[1]

Treatment	Optimal concentration(μM)	Number of buds formed in 10 μM kinetin: Wild-type (800 lux)	Number of buds formed in 10 μM kinetin: Mutant (3000 lux)
Control	0	1	0
IAA	10	~120	>500
Indole-3-acetaldehyde	100	~120	>500
Indole-3-pyruvic acid	10	~120	>500
Tryptophan	1000	~150	>500
Indole-3-acetamide	100	>280	>700
Tryptamine	100	0	~30

Note: In 3000 lux all protonemata of the wild-type produce high numbers of buds with kinetin. After a pretreatment with IAA or its precursors for 3 days the protonemata are subjected to 10 μM kinetin together with the tested substances.

In contrast to the hypothesis of Cove and Ashton,[2] experiments with "auxin-resistant" mutants of *F. hygrometrica* (NAR) indicate another way to reduce the endogenous auxin content. The *Funaria* mutants were selected in the same way as the *Physcomitrella* mutants, on a substrate with a high concentration of NAA (5 to 50 μM). The most clearly defined mutant, NAR-2, behaves as a type 4 mutant of *P. patens*; NAR-4 and -5 resemble the type 5 to 7 mutants.[2] The development of NAR-2 can be normalized by a supplement of about 0.1 to 1 μM IAA, α-NAA, or 2,4-D. The precursors indicated in Figure 5 induce the formation of caulonemata, which have the ability to form buds after kinetin treatment (Table 2). Only tryptamine has no significant effect, similar to observations in the wild-type under LLI conditions.[38] Therefore, a block in the biosynthetic pathway between tryptophan and

TABLE 3
Total Activity of IAA Oxidase (ΔA 260 nm × 30 min^{-1} × g^{-1} fw) (Σ of the Soluble and Particulate Fractions) of the Wild-Type and NAR-2 Protonema at Various Ages of the Protonema[1]

Days of growth	Wild-type	NAR-2
7	1.642 ± 0.05	5.294 ± 0.28
10	2.508 ± 0.03	5.16 ± 0.11
13	2.015 ± 0.15	4.103 ± 0.13

Note: The means and standard errors are from four replicates. Cultures raised on minimal medium without auxin on agar.

IAA could not have caused the auxin deficiency. A reduced sensitivity in the mutants can be excluded by the experiments because the effective concentrations are the same for the mutant NAR-2 and the wild-type in LLI. These are the reasons for assuming that the auxin requirement is due to an enhanced degradation of auxin in the protonema as was demonstrated recently by Atzorn et al.[121] using an auxin ELISA test.

A comparison of the activity of auxin oxidase in the soluble and particulate protein fractions of the wild-type and the NAR-2, calculated per gram fresh weight, gives up to 20 times higher values in the mutant than in the wild-type (Table 3). In contrast to auxin oxidase, peroxidase activity in the mutant is much lower than in the wild-type,[38] and the hypothesis can be put forward that only one enzyme, a peroxidase isoenzyme acting either as peroxidase or IAA oxidase, is mutated to a higher activity of IAA oxidase. The effect of the two enzymes was shown earlier by the experiments of Sharma et al.[24] where enzymes delivered in the substrate changed the differentiation of the suspension culture.[23,40] Thus, the hypothesis is supported that auxin degradation by auxin oxidase causes the auxin requirement in NAR-2.

In more recent experiments mutants were induced by UV treatment of protoplasts.[120] With such treatment around 10% of the protoplasts survive and can be cultivated on agar. They regenerate cells which give rise to new protonemata. After several days morphologically characterized mutants can be selected. Auxin-insensitive and auxin-sensitive types were found among such selected new forms. One mutant strain, growing only as chloronemata on the Knop agar surface, was not repaired by tryptophan treatment; thus, in this case, the auxin synthesis pathway may have been interrupted somewhere between tryptophan and IAA. Other mutants behave like NAR-2. They very quickly degrade auxin synthesized from exogenous tryptophan.[121]

For a considerable period nothing was known about the internal auxins in mosses.[2] However, in the last few years several reports have demonstrated unequivocally that IAA is a natural constituent of the moss protonema. Ashton et al.[39] found 2.1 ng g^{-1} fresh weight (fw) or 75 ng g^{-1} dry weight (dw) IAA in *P. patens* using gas chromatography and ion-monitory mass spectrometry. Jayaswal and Johri[29] detected between 1.9 and 5.0 ng g^{-1} protonema cells (fw) with an indole-γ-pyrone fluorometric assay.

Both results are in good agreement with respect to the total amount. They show that IAA is present in a concentration sufficient to induce the transition from chloronema to caulonema. From protonemata of *F. hygrometrica* the IAA content was analyzed with an enzyme immunoassay (ELISA).[1,121] With this method the concentration for the wild-type grown in high light intensity (HLI) conditions was about 0.4 nmol g^{-1} fw (or 70 ng g^{-1} fw), which may be about ten times more than reported earlier. In LLI about 0.15 nmol g^{-1}

IAA was detected (depending on the age of the protonema), and the mutant NAR-2 contained 0.12 nmol g^{-1} fw in HLI. The differences in quantity should be viewed with caution, since amounts of auxin can vary according to the method of extraction and the sensitivity of the test. For example, Markmann-Mulisch[40] found between 1.1 and 18.5 ng g^{-1} dw of IAA in the protonema of *Polytrichum formosum* depending on the extraction method.

More important than the absolute amount is the relationship between the values of IAA in the protonemata grown in different conditions, which allow caulonema formation or not, and the values in relevant mutants. In LLI the IAA content in wild-type protonema is about one third of that in HLI, and the mutant NAR-2 contains an amount similar to the LLI wild-type.

Thus, it seems clear that endogenous IAA in the range of 10^{-9} mol/g fw is responsible for the transition from chloronema to caulonema. All conditions which lower the endogenous auxin content, such as PCIB treatment, reduction of the light intensity, and suspension culture, cause the failure of caulonema formation. Mutants not able to form caulonemata spontaneously either have a block in the biosynthetic pathway of IAA or have a rate of IAA degradation higher than the possible synthesis.

Besides the amount of hormone, the distribution of the substance within the plant plays an equally important role. The mode of transport is the crucial prerequisite of a coordinated development, as demonstrated by the following observations: when the apical cell of a caulonema is killed, all cells within the filament lose their caulonemal character;[16,41,42] the same is found when filaments are isolated from the whole protonema, but because the removal of the apical cell alone has this effect a signal from that cell must maintain the caulonema status in all cells of the filament. From the previous experiments we can conclude that auxin represents this signal. A polar auxin transport from the tip to the base of a filament was observed in whole protonemata[48] and in isolated rhizoids fed ^{14}C-IAA,[49] whereas other substances like amino acids or purines are transported from the base of a filament to the apical region.[42,55] The transport mechanism includes a pH-dependent passive auxin influx[50] and a 2,3,5-triiodobenzoic acid (TIBA)-inhibited active efflux.[50,51] Therefore, isolated protoplasts accumulate more auxin when incubated with TIBA than without.[52] Details of the transport seem to be different from those described in higher plants; thus, in basal cells more auxin is accumulated than in the apical cell, which creates a gradient along the filament. However, it remains to be proved whether this is the cause for the polar transport as such.

At the moment nothing is known about the kind of action of auxin in moss cells, but we can elucidate a few crucial points to see what auxin really does. Because moss protonemata have tip growth without any elongation in the older cell walls, the normal effect of auxin on wall extensibility may play no role in moss protonemata. One exception is the formation of tmemata[3,122] in which very short cells elongate along the whole cell.

However, a comparison of the apical growth rates of chloronemata and caulonemata shows that caulonema filaments always grow faster than those of the chloronema.[16] The growing region of the caulonema contains a characteristic "tip body"[56] in which vesicles transport wall material to the growing area. These tip bodies are responsible for the high growth rate. Such a distinct apical zone is lacking in the slow-growing chloronema cells.

In general, the side branches start as chloronemata and change after a while to caulonemata. This is shown by the growth rate (Figure 8). With the elongation of the side branches, the growth rate increases up to the point when the caulonema character is reached. The treatment with PCIB reduces the growth rate together with the change from caulonema to chloronema (Figure 6).[31] Therefore, a possible first effect of auxin can be the stimulation of growth rate, associated with the transition from chloronema to caulonema.[16]

At the molecular level the first step of auxin action appears to be its binding to a membrane-bound receptor,[53] but the nature of the auxin receptor in mosses is completely

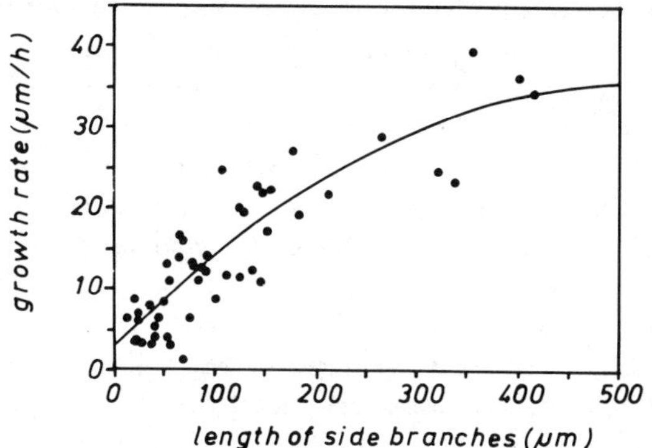

FIGURE 8. Correlation between the length of side branches growth rate and of caulonema filaments of *Funaria hygrometrica*.[16]

unknown. In the only investigation to date, Lomax-Reichert et al.[54] found that auxin binding was barely detectable in total membrane preparations of auxin-resistant and wild-type *Physcomitrella*. Presumably, more data about the molecular level of auxin action which can be applied to mosses as well will be available shortly.[123]

The scheme worked out for *Funaria* and *Physcomitrella*, and certainly for many other mosses with a similar type of development, may not be valid for all moss protonemata in exactly the same fashion. Exceptions to these rules can be found and are reported repeatedly. They concern different species[43-45] or different culture conditions[46,47] where caulonema formation is either completely missing or not necessary for bud formation. Clearly, in this case IAA may play a minor role in the differentiation processes.

IV. CYCLIC AMP

Only a limited number of reports deal with the occurrence and the effect of cAMP in the moss protonemata.[10] The main results were obtained with suspension cultures of *F. hygrometrica*. As mentioned before, in a suspension culture with an inoculum of <50 μg ml^{-1} plant material, the protonema grows only in the form of chloronema filaments. Caulonema formation requires the supplementation of IAA depending on concentration and cell density (Figure 3). This auxin effect can be antagonized by a simultaneous application of cAMP at concentrations higher than 0.1 μM.[8] Some other derivatives of adenine can also antagonize the IAA-elicited response, whereas pyrimidines are ineffective.[57] In the moss *Pylaisiella*, which grows without caulonemata in suspension cultures, an antagonistic effect of auxin and cAMP is seen in the elongation of chloronema filaments.[58]

The specificity of the cAMP-auxin antagonism on protonema differentiation is supported by experiments with a spontaneous mutant of *F. hygrometrica*, pg-1. It produces about 65% caulonema filaments in the suspension without auxin. However, cAMP changes the relationship between caulonema and chloronema and enhances the initiation of chloronema filaments so that the caulonema portion is reduced to 35%.[57] It seems that the mutant has limited ability to synthesize endogenous cAMP.

As evidence of the endogenous cAMP content, Handa and Johri[9] isolated a factor which could not be distinguished from authentic ^{3}H-cAMP by the following three criteria: copurifying in a solvent system, stimulation of protein kinase activity, and degradation by phosphodiesterase. The quantitative determination of the factor identified in these tests allows a comparison of the cAMP content in the wild-type and in the mutant. In the wild-type cultures with auxin consisting of 65 to 75% caulonemata the content corresponded completely with that in pg-1, whereas in a control culture between four and seven times more cAMP was present. This is consistent with the assumption that cAMP is an internal regulatory substance in *Funaria*. In recent experiments it was shown that, in *Funaria*, cAMP-dependent protein kinase may inhibit the cytokinin-induced response, which results in the greening of the cells and the formation of chloronemata according to the effect of exogenous cAMP in the previous experiments.[59,60] Kaul and Sachar,[61] however, were unable to detect incorporation of ^{14}C-adenine in cAMP, and they concluded that the cAMP content of the moss must be too low to be detected.[10] Also, a few other results are not fully consistent with the action of cAMP as a second messenger in mosses.[2,62]

As for all true endogenous signal substances, metabolism — synthesis and degradation — is an important factor. Therefore, the importance of cAMP can be supported if the enzymes necessary for metabolism can be found in the protonema. In a survey no adenylate cyclase (the enzyme which synthesizes cAMP) was found in higher plants, but the activity was high in algae and in the protonema of *F. hygrometrica*. The level of activity was on the same order of magnitude as in slime molds, which need cAMP for cell aggregation.[63] Adenylate cyclase activity was also localized in the shoot apices of the moss *Bryum argenteum*.[116]

The enzyme which regulates the total amount of cAMP by degradation is cyclic nucleotide phosphodiesterase (PDE). It was first detected in *Sphagnum* among the mosses,[117] and later it could be demonstrated in the suspension cultures of *Funaria*. The level depended on the cell density and, surprisingly, was highest at low cell densities, in contrast to auxin oxidase.[24] However, this result fits well with the caulonema-chloronema distribution. High activities of PDE reduce the internal cAMP content and, therefore, favor the formation of caulonemata.

Finally, substances which inhibit PDE should have an effect similar to exogenous cAMP because they reduce internal cAMP degradation. This was the case using theophylline, aminophylline, and the synthetic PDE inhibitor ICI #58301. All substances enhanced the formation of chloronemata in the presence of IAA,[8,57] which means that the relation between IAA and cAMP is changed to favor cAMP. The effect on the mutant pg-1 is similar and in agreement with the assumed defective cAMP production in this mutant.

Other substances acting as so-called "second messengers" will be mentioned in the discussion on the scheme of cytokinin reactions (Section V).

V. CYTOKININS

Soon after the identification of kinetin as a cell division factor[64] it was found that the formation of buds was drastically enhanced in the protonema of the moss *Tortella caespitosa* when the protonema reached the "ripeness to buds"[21] (later defined as the "critical size").[65] The effect of kinetin on bud induction was confirmed for many moss species with and without a typical caulonema. All synthetic and natural substances with the characteristics of a cytokinin (adenine derivatives with an N^6-substituted side chain with five or more carbon atoms) transform cells from filamentous growth to bud growth with a three-sided apical cell (Table 2).[12] In some mosses cytokinin ribosides are less effective than the free base.[66,67]

The transformation process studied in *F. hygrometrica* starts shortly after the application of exogenous cytokinins. After 3 h the growth of side branches of the caulonema stops if they are shorter than 80 μm; then the apical area of the filament becomes dome-shaped and the deposition of new wall material moves from the very tip region to the side of the cell, forming a round "bud initial".[17,18,68] In most cases, cytokinins induce "moruloid" buds as a consequence of the relatively high cytokinin concentrations used.[69]

Under standard conditions buds are formed only on caulonema cells. Mutants which need auxin for caulonema formation, therefore, react to cytokinin after auxin treatment,[36] like mosses grown under conditions which do not allow spontaneous caulonema differen-

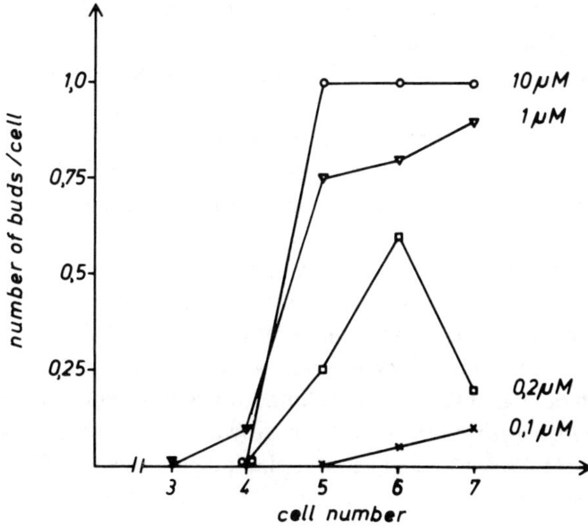

FIGURE 9. Bud induction in *Funaria hygrometrica* at different concentrations of kinetin (0.1 to 10 μ*M*). Distribution of the buds on the caulonema cells. Cell number beginning at the tip cell = 1. For each point 20 cells are counted.

tiation.[6,25,70] This means that auxin creates the target cells for cytokinin. Because auxin always has to act first,[6] this type of interaction is called "sequential interaction".[26] It is a question of definition whether the caulonema cell itself or the side branch is the true target cell.

Experiments where cytokinins can induce buds on very young protonemata, or in low concentrations directly on chloronema cells,[45,46,71-73] are not contradictory to this concept. They only demonstrate that a strong morphological definition of the "target" cells is questionable if it should be generalized.

Nevertheless, in "normal" protonema of *Funaria* and many other mosses the number of buds depends on the number of caulonema cells which are able to react[65,74] and on the cytokinin concentration.[75] Not only the number of buds but also the position along the caulonema filament is always determined (Figure 9). At low concentrations only the most sensitive cells are able to form a bud.[75] This effect facilitates the use of bud formation as a quick and sensitive bioassay for cytokinins.[76,77] In general, concentrations lower than 0.1 μ*M* are ineffective for bud induction.

High external osmotic pressure applied together with cytokinin suppresses the regular formation of a bud. Instead, spiral filaments are formed which give rise to buds when the osmotic pressure is reduced; this occurs even in filaments longer than 80 μm. It shows that plasmolysis only affects the expression of the developmental program and not its initiation by cytokinin.[78]

Besides bud formation, cytokinins have another quite important effect; they stimulate cell division in protonema.[79,79a] Comparable to this is the induction of branching of caulonema cells. Under standard conditions one of the characteristic events in the caulonema is the very regular spontaneous formation of side branches. If this is not the case, the cells form side branches under the influence of exogenous cytokinins. Saunders and Hepler[80-82] used this effect to study the molecular mechanism of cytokinin action. When branching takes place spontaneously it is expected that the cytokinin requirement for this process is saturated by the internal cytokinin level, which should be much lower than for bud formation.

Protonemata cultivated without phosphate produce caulonemata free of side branches.[16]

The branching can be induced with cytokinins in the picomolar concentration range, whereas buds require micromolar concentrations in the medium.[83] Therefore, the two cytokinin-stimulated processes, cell division and bud formation, can be separated.

In order to obtain a first hint whether endogenous cytokinins are involved in both processes, experiments with cytokinin antagonists can be used. However, only one report showed that the effect of exogenous cytokinins can be reduced by a simultaneous application of substances acting as anticytokinins,[84] perhaps because they can block an endogenous receptor. However, such receptors have not been identified in mosses so far.[85]

A better way to study the activity of endogenous cytokinins is their direct identification. The first reports of endogenous cytokinins dealt with abnormally growing mosses, for which it was assumed that the endogenous cytokinin level was quite high.[86] Beutelmann and Bauer[87] have identified a substance called "bryokinin" by gas chromatography and mass spectrometry as N^6-Δ^2 isopentenyl adenine in the medium on which a callus of the hybrid *F. hygrometrica* × *Physcomitrium pyriforme* had been grown. The concentration of this cytokinin released into the substrate was in the millimolar range. The so-called OVE mutants of *Physocomitrella patens* were another source of cytokinin. They spontaneously produce 20 to 50 times more buds than the wild-type on a minimal medium and resemble, therefore, the wild-type protonemata grown in a high exogenous cytokinin concentration.[2] These overproducers (category 9 of the *Physcomitrella* mutants) contain an enhanced endogenous cytokinin concentration in comparison to the wild-type. The OVE mutants belong to at least three complementation groups.[88] This means that the mutation of three different genes can lead to the same phenotype, and it remains to be seen whether this involves the rate of synthesis or the rate of degradation of the cytokinins which are altered. Also, in *Funaria* one OVE mutant was found which has not yet been analyzed in detail.[120]

Cytokinin was determined in the substrate on which the overproducers of *Physocomitrella patens* had been grown. Isopentenyl adenine could be identified as the main component after supplying ^{14}C-adenine in the substrate. Zeatin was found in a concentration 12 to 22 times lower.[86,89,90] In a preliminary note the overproduction was regarded more as a consequence of higher synthesis than of reduced degradation.[2] This was confirmed by studies on the metabolism of cytokinin in OVE mutants and the wild-types. In both no clear differences in cytokinin breakdown were found. To come to a final conclusion, investigation of metabolism using low concentrations of tritiated cytokinin is needed because the very high exogenous concentrations normally used alter the phenotype of the wild-type in the direction of the OVE mutant. This also may change the internal metabolism in the same way.[91]

With the highly sensitive radioimmunoassay or enzyme immunoassay, cytokinin also can be detected in the protonema of the wild strains of *Physcomitrella* and *F. hygrometrica*.[1,101,119] In the wild strain, as well, the content of isopentenyl adenine is always much higher than that of zeatin, independent of the age of the protonema or the species.

This observation leads to the speculation that the synthesis of cytokinin in the moss protonema starts with the transfer of isopentenyl to AMP by isopentenyl transferase, an enzyme identified as a gene product of the T region of *Agrobacterium*,[28,92,93] so that isopentenyl adenine (IPA), as the first active cytokinin, remains the main component. In higher plants, in general, zeatin (being the hydroxylation product of IPA) is present in higher concentrations than IPA. The process of hydroxylation may be less effective in mosses so that IPA remains the main cytokinin.

In a very recent paper the distribution of IPA within the protonema cells of OVE mutants of *Physcomitrella* was studied using a monoclonal antibody with high affinity to IPA. After fixation and embedding of the material at low temperature the position was visualized by an indirect immunogold labeling. Only the cell wall of a part of all cells was labeled.[94] Perhaps this detected only the cytokinin which left the cell and was, therefore, present for a short time in the wall. In the substrate cytokinins can always be found in high amounts.[86,87,91]

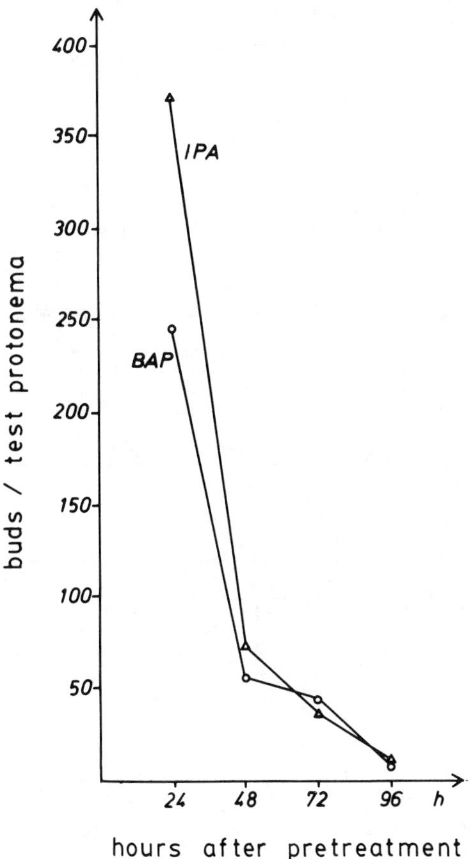

FIGURE 10. Bud induction in *Funaria hygrometrica* by substances released from BAP- or IPA-pretreated protonema. The protonemata were cultivated for 24 h on BAP- or IPA-containing substrate and then transferred to a cellophane layer covering a test protonema. This was changed every 24 h and the number of buds induced at this test protonema was counted. The figure shows that most of the buds inducing substances (cytokinins) are released during the first 24 h.

If the *Funaria* protonema is saturated with IPA or benzyladenine it releases part of the cytokinin into the substrate, which enables the induction of bud formation in a test protonema (Figure 10).[95]

For cytokinin to induce bud formation it must be present for several days, until a definite stage of the young bud has been reached. If cytokinin is removed by washing before this stage all the buds revert to the protonema stage.[96] A photolabile cytokinin which is destroyed by ultraviolet (UV) light loses its activity if the protonema is exposed to UV light. Also, in this case the induced buds are reversed to the protonema stage.[97] However, if the buds are induced by kinetin and kinetin is removed after 24 or 48 h of treatment (which is possible if the protonema grows on a cellophane surface)[98] all the buds also revert to protonema. This is not the case with other cytokinins, especially the endogenous IPA.[95]

Therefore, the catabolism of the different cytokinins involves different pathways in moss protonema. Kinetin is metabolized very quickly to adenine and adenine derivatives by a direct cleavage of the side chain.[55,99] Essentially the same results were obtained with ben-

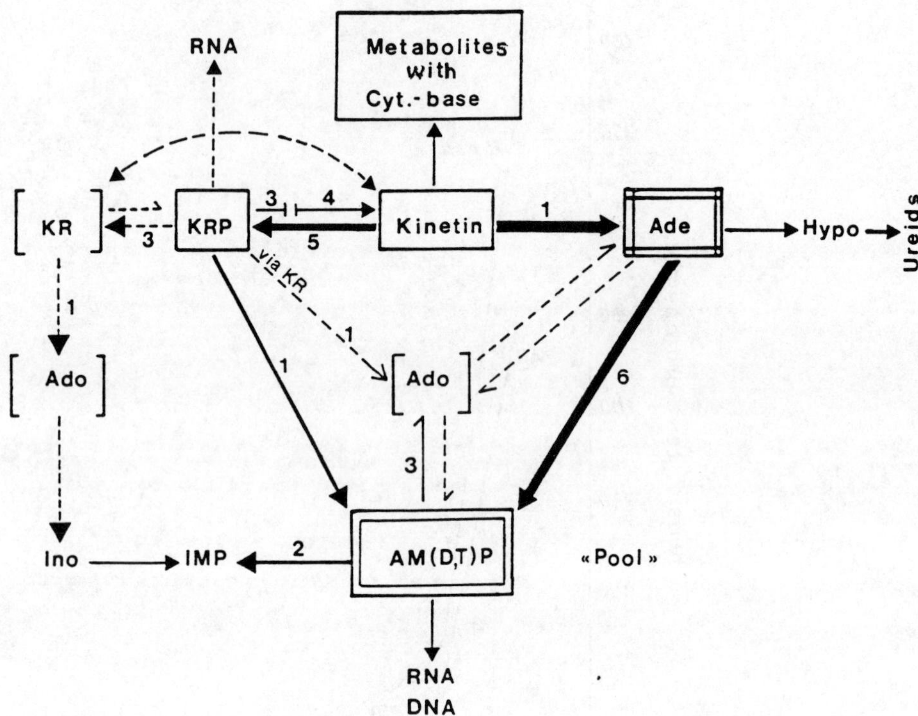

FIGURE 11. Scheme of the kinetin metabolism in *Funaria hygrometrica*.[101] Enzymes involved in the schemata of Figures 11 and 12: (1) kinetin oxidase, (2) deaminase, (3) 5′-nucleotidase, (4) nucleosidase, (5,6) purine phosphoryltransferase. Hypo = hypoxanthine (6-hydroxypurine); KRP = kinetin ribotide; KR = kinetin riboside; Ado = adenosine; Ade = adenine; Ino = inosine (6-hydroxypurine riboside); IMP = inosine monophosphate; BAP = benzylaminopurine; BAMP = benzylaminopurine ribotide; BAPR = benzylaminopurine riboside.

zylaminopurine (BAP), zeatin, or IPA in *Physcomitrella*.[91] At least in *Funaria* quantitative differences exist between kinetin and the other cytokinins.[75,100]

To study this problem the enzyme which removes the side chain from kinetin, kinetin oxidase (cytokinin oxidase), was extracted and analyzed *in vitro*. In Figure 11 it is shown that the enzyme is very active on kinetin. After 24 h about 35 to 40% of the initial radioactivity of ^{14}C-labeled kinetin could be detected in adenine and 10% in hypoxanthine, whereas BAP (Figure 12) was practically not degraded to adenine, but was very strongly metabolized to the benzyladenine ribotide and riboside, from which the benzyl group could then be removed.[101,102]

The different activities of kinetin oxidase *in vitro* explain why kinetin is metabolized much quicker than the other natural or artificial cytokinins. These results agree very well with the reaction of the protonema on external application of cytokinins.

The last question concerns the primary effect of cytokinins in mosses. This problem was studied by Saunders and Hepler,[80-82,105] who observed that under certain conditions exogenous cytokinin was needed for the branching of caulonema cells, which is a prerequisite for bud induction.[17] An electron microscope study of the cytokinin-induced side branch has shown that the origin of a filament is the first step and that the transition to a bud is the second step.[18] This confirms that formation of side branch and bud are two independent steps. The two processes, however, were not separated in the relevant reports, which makes some results difficult to understand. Nevertheless, they give a new insight into the action of cytokinins.

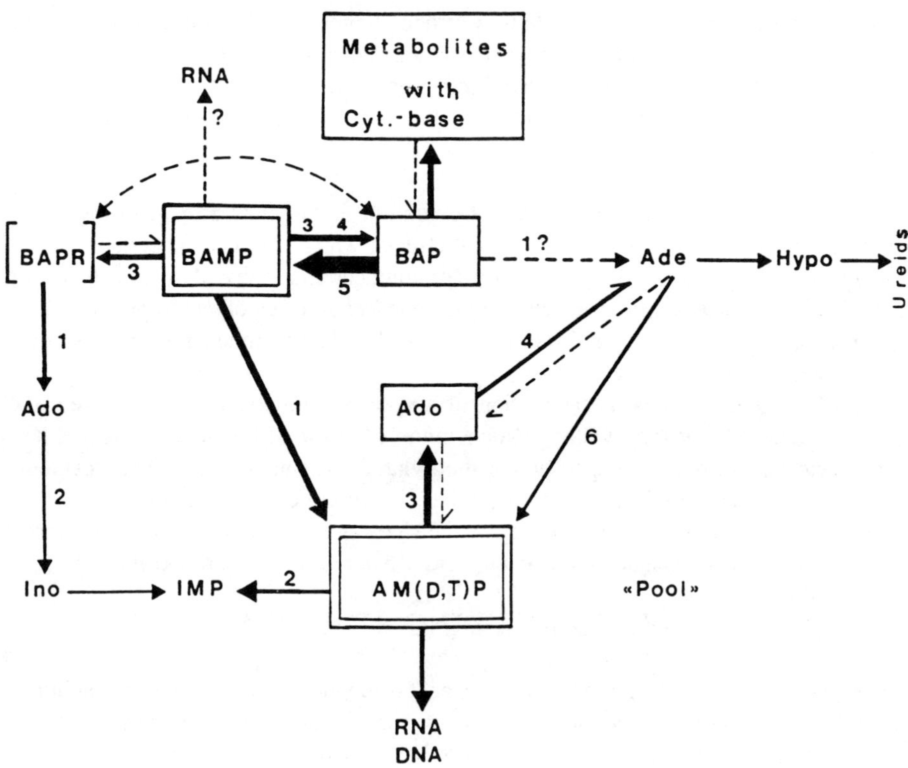

FIGURE 12. Scheme of the BAP metabolism in *Funaria hygrometrica*.[101] See caption of Figure 11 for explanation of abbreviations.

The formation of side branches can be regarded as a specific asymmetrical (unequal) cell division.[80-82] There is strong evidence that cytokinins stimulate this asymmetrical division in *F. hygrometrica* by increasing the free Ca^{2+} content in the cytoplasm. The relevance of Ca^{2+} in the cytokinin effect can be accepted if the following three criteria are fulfilled: (1) a rise in Ca^{2+} occurs during the process in question, (2) an artifically enhanced Ca^{2+} content should trigger the process, and (3) a blockage of Ca^{2+} increase must inhibit the process.[103] All three points could be confirmed in *Funaria*: intracellular Ca^{2+} is enhanced after cytokinin treatment, as shown by chlorotetracycline (CTC) fluorescence;[80] cell division is induced by the Ca^{2+} ionophore A23187 under the influence of external Ca^{2+};[81] and Ca^{2+} antagonists like La^{3+}, D600, or Verapamil prevent cell division and bud formation in the presence of cytokinin.[82] Furthermore, inhibition of calmodulin, which is necessary for Ca^{2+} action, blocks bud initiation. From these experiments it can be concluded that extracellular Ca^{2+}, uptake of Ca^{2+}, and calmodulin are essential for the effect of cytokinin on asymmetrical cell division.[82] This means that the primary effect of cytokinin involves the opening of Ca^{2+} channels so that the Ca^{2+} content is changed in a temporal and spatial manner within the target cells.

However, it seems that this model applies only to side branch formation. A detailed analysis of the transition to buds in *Funaria* in Ca^{2+}-free medium has shown that reduction in the number of buds alone does not give correct data[72,82] because in most cases the factors reduce the number of reacting cells and, therefore, no buds can be formed.[40,104] In a Ca^{2+}-free medium or when Ca^{2+} antagonists are used, the formation of round or dome-shaped cells can occur even when no Ca^{2+} can be detected in the cells with CTC fluorescence. All further cell divisions, however, are suppressed,[104] which is in agreement with a general inhibition of cell division.

In more recent studies with a vibrating probe, Saunders[105,106] found that in untreated caulonema cells a maximal inward current was located at the nuclear region. This current increased along the whole cell after cytokinin treatment and was finally localized at the presumptive division site. It is very probable that Ca^{2+} is a component of this current. In the initiation of side branch formation a new orientation of the elements of the cytoskeleton is involved.[60]

The Ca^{2+} activity, as a mediator of the cytokinin effect, may be part of the phosphatidylinositol trisphosphate system of hormonal effects, which results at least in the liberation of endogenous Ca^{2+} pools.[7] In preliminary communications it was described that inositol trisphosphate (IP_3) increases immediately after the application of cytokinin to protonemata[107,108] due to the activation of phospholipase C, as was also demonstrated for an application of IAA to vesicles from carrot cells.[115]

The IP_3 system includes the activation of a specific protein kinase. However, such a protein kinase which phosphorylates proteins involved in the differentiation process has not yet been determined, nor were specific phosphorylated proteins detected. The "caulonema-specific proteins" separated by gradient gel microelectrophoresis certainly were not this type of protein;[109-111] it seems that they belong to the ribosomes. In two-dimensional microelectrophoresis, proteins of isolated ribosomes were separated at the same position.[112-114]

VI. CONCLUDING REMARKS

The preceding account has shown that the hormonal system of mosses includes the sequential interaction of auxin and cytokinin as a main component and perhaps cAMP as an antagonistic system. All these components have been characterized very clearly. The moss protonema may, therefore, play a very important role in further investigations on the primary action of the hormones at the molecular level. However, much more sophisticated research methods are required to reach this very high aim. When such chain reactions are demonstrated in mosses, conclusions for higher plants can also be made even if not all steps and parts of the hormonal system are fully identical in both plant groups. It is of further importance in the progress of our knowledge that detailed analyses of the hormonal system of moss protonemata have shown small differences in synthesis, degradation, and transport, etc. These differences can be understood as steps on the way to the highly functional hormone system in higher plants.

ACKNOWLEDGMENTS

I want to thank Mrs. Sis very much for checking the English, Mrs. B. Nickel for typing the manuscript, and Mrs. U. Keßler for the art work. My own research included in this paper was supported over many years by the Deutsche Forschungsgemeinschaft, which is gratefully acknowledged.

REFERENCES

1. **Bopp, M. and Bhatla, S. C.,** Hormonal regulation of development in mosses, in *Hormonal Regulation of Plant Growth and Development,* Vol. 2, Purohit, S. S., Ed., Agro-Botanical Publishers, Old Ginanni, Bikaner, India, 1985, 65.
2. **Cove, D. J. and Ashton, N. W.,** The hormonal regulation of gametophytic development in bryophytes, in *The Experimental Biology of Bryophytes,* Dyer, A. F. and Duckett, J. G., Eds., Academic Press, London, 1984, 177.

3. **Rohwer, F. and Bopp, M.,** Ethylene synthesis in moss protonema, *J. Plant Physiol.*, 117, 331, 1985.
4. **Chopra, R. N. and Bhatla, S. C.,** Regulation of gametangial formation in bryophytes, *Bot. Rev.*, 49, 29, 1983.
5. **Chopra, R. N. and Kumra, P. K.,** Hormonal regulation of growth and antheridial production in three mosses grown *in vitro, J. Bryol.*, 12, 49, 1983.
6. **Lehnert, B. and Bopp, M.,** The hormonal regulation of protonema development in mosses. I. Auxin-cytokinin interaction, *Z. Pflanzenphysiol.*, 110, 379, 1983.
7. **Poovaiah, B. W. and Reddy, A. S. N.,** Calcium messenger system in plants, *Crit. Rev. Plant Sci.*, 6, 47, 1987.
8. **Handa, A. K. and Johri, M. M.,** Cell differentiation by 3′,5′-cyclic AMP in a lower plant, *Nature (London),* 259, 480, 1976.
9. **Handa, A. K. and Johri, M. M.,** Cyclic adenosine 3′,5′-monophosphate in moss protonema, *Plant Physiol.*, 59, 490, 1977.
10. **Bhatla, S. C. and Chopra, R. N.,** Biological significance of cyclic adenosine 3′,5′-monophosphate in mosses, *Physiol. Plant,* 57, 383, 1983.
11. **Bopp, M. and Bhatla, S. C.,** Physiology of sexual reproduction in mosses, in *Sexual Differentiation in the Plant Kingdom,* Durand, R. and Durand, B., Eds., CRC Press, Boca Raton, FL, 1990, in press.
12. **Bopp, M.,** Developmental physiology of bryophytes, in *New Manual of Bryology,* Vol. 1, Schuster, R. M., Ed., Hattori Botanical Laboratory, Nichinan, Miyazaki, Japan, 1983, 276.
12a. **Menon, M. K. C. and Lal, M.,** Problems of development in mosses and moss-allies, *Proc. Indian Acad. Sci. Sect. B,* 47, 115, 1981.
13. **Mejia, A., Spangenberg, G., Koop, H. U., and Bopp, M.,** Microculture and electrofusion of defined protoplasts of the moss *Funaria hygrometrica, Bot. Acta,* 101, 166, 1988.
14. **Schmiedel, G. and Schnepf, E.,** Side branch formation and orientation in the caulonema of the moss, *Funaria hygrometrica:* Normal development and fine structure, *Protoplasma,* 100, 367, 1979.
15. **Kofler, L.,** Contribution a l'etude biologique des mousses cultivees *in vitro:* germination des spores, croissance et dévеloppement du protonéma chez *Funaria hygrometrica, Rev. Bryol. Lichenol.*, 28, 1, 1959.
16. **Knoop, B.,** Development in bryophytes, in *The Experimental Biology of Bryophytes,* Dyer, A. F. and Duckett, J. G., Eds., Academic Press, London, 1984, 143.
17. **Bopp, M.,** The hormonal regulation of protonema development in mosses. II. The first steps of cytokinin action, *Z. Pflanzenphysiol.*, 113, 435, 1984.
18. **Conrad, P. A., Steucek, G. L., and Hepler, P. K.,** Bud formation in *Funaria:* organelle redistribution following cytokinin treatment, *Protoplasma,* 131, 211, 1986.
19. **Schmiedel, G. and Schnepf, E.,** Side branch formation and orientation in the caulonema of the moss *Funaria hygrometrica:* experiments with inhibitors and with centrifugation, *Protoplasma,* 101, 47, 1979.
20. **Bopp, M.,** Die Wirkung von Heteroauxin auf Protonemawachstum und Knospenbildung von *Funaria hygrometrica, Z. Bot.*, 41, 1, 1953.
21. **Gorton, B. S. and Eakin, R. E.,** Development of gametophyte in the moss *Tortella caespitosa, Bot. Gaz. (Chicago),* 119, 31, 1957.
22. **Johri, M. M.,** Differentiation of caulonema cells by auxin in suspension cultures of *Funaria hygrometrica,* in *Plant Growth Substances,* Hirokawa Scientific, Tokyo, 1974, 925.
23. **Johri, M. M. and Desai, S.,** Auxin regulation of caulonema formation in moss protonema, *Nature (London),* 245, 223, 1973.
24. **Sharma, S., Jayaswal, R. W., and Johri, M. M.,** Cell-density-dependent changes in the metabolism of chloronema cell cultures, *Plant Physiol.*, 64, 154, 1979.
25. **Johri, M. M.,** Regulation of morphogenesis. I. Regulation of cell differentiation and morphogenesis in lower plants, in *Frontiers of Plant Tissue Culture,* Thorpe, T. A., Ed., Int. Assoc. Plan Tissue Cultures, University of Calgary, Calgary, Alberta, 1978, 27.
26. **Bopp, M.,** Probleme der Interaktion zwischen Phytohormonen, *Ber. Dtsch. Bot. Ges.*, 92, 323, 1979.
27. **Schröder, J., Buchmann, I., and Schröder, G.,** Enzymes of auxin and cytokinin biosynthesis encoded in Ti plasmids, in *Plant Growth Substances 1985,* Bopp, M., Ed., Springer-Verlag, Berlin, 1986, 177.
28. **Morris, R. O., Powell, G. K., Beaty, J. S., Durley, R. C., Hommes, N. G., Lica, L., and MacDonald, E. M. S.,** Cytokinin biosynthetic genes and enzymes from *Agrobacterium tumefaciens* and other plant-associated prokaryotes, in *Plant Growth Substances 1985,* Bopp, M., Ed., Springer-Verlag, Berlin, 1986, 185.
29. **Jayaswal, R. K. and Johri, M. M.,** Occurrence and biosynthesis of auxin in protonema of the moss *Funaria hygrometrica, Phytochemistry,* 24, 1211, 1985.
30. **Sood, S. and Hackenberg, D.,** Interaction of auxin, antiauxin and cytokinin in relation to the formation of buds in moss protonema, *Z. Pflanzenphysiol.*, 91, 385, 1979.
31. **Bopp, M.,** The hormonal regulation of morphogenesis in mosses, in *Proceedings in Life Sciences,* Skoog, F., Ed., Springer-Verlag, Berlin, 1980, 351.

32. **Hatanaka-Ernst, M.,** Entwicklungsphysiologische Untersuchungen an strahleninduzierten Protonema-mutanten von *Funaria hygrometrica* Sibth., *Z. Pflanzenphysiol.,* 55, 259, 1966.
33. **Oehlkers, F.,** Entwicklungsgeschichte röntgeninduzierter Mutanten von *Funaria hygrometrica, Z. Vererbungsl.,* 96, 234, 1965.
34. **Adelberg, E. A., Mandel, M., and Chen, G. C. C.,** Optimal conditions for mutagenesis by *N*-methyl-*N'*-nitro-*N*-nitrosoguanosidine in *Escherichia coli* K12, *Biochem. Biophys. Res. Commun.,* 18, 788, 1965.
35. **Ashton, N. W. and Cove, D. J.,** The isolation and preliminary characterization of auxotrophic and analogue resistant mutants of the moss *Physcomitrella patens, Mol. Gen. Genet.,* 154, 87, 1977.
36. **Cove, D. J., Ashton, N. W., Featherstone, D. E., and Wang, T. L.,** The use of mutant strains in the study of hormone action and metabolism in the moss *Physcomitrella patens,* in *Proc. 4th John Innes Symp. 1979,* John Innes Charity, Norwich, England, 1980, 231.
37. **Ashton, N. W., Grimsley, N. H., and Cove, D. J.,** The isolation and physiological analysis of mutants of the moss *Physcomitrella patens,* using auxin and cytokinin resistant mutants, *Planta,* 144, 437, 1979.
38. **Bhatla, S. C. and Bopp, M.,** The hormonal regulation of protonema development in mosses. III. Auxin-resistant mutants of the moss *Funaria hygrometrica* Hedw., *J. Plant Physiol.,* 120, 233, 1985.
39. **Ashton, N. W., Schultze, A., Hall, P., and Bandurski, R. S.,** Identification and estimation of indole-3-acetic acid in gametophytes of the moss *Physcomitrella patens, Planta,* 164, 142, 1985.
40. **Markmann-Mulisch, M.,** Untersuchungen über die Rolle des Calciums bei der cytokininabhängigen Knospeninduktion und Entwicklung am Protonema der Laubmoose *Funaria hygrometrica* und *Polytrichum formosum,* Ph.D. Dissertation, Universität Heidelberg, Heidelberg, Federal Republic of Germany, 1985.
41. **Knoop, B.,** Untersuchungen zum Regenerationsmechanismus bei *Funaria hygrometrica* Sibth. III. Auslösung durch Inhibitoren und Unterdrückung der apikalen Dominanz, *Z. Pflanzenphysiol.,* 77, 350, 1976.
42. **Bopp, M. and Knoop, B.,** Regulation de la differenciation chez le protonema des mousses, *Soc. Bot. Fr.,* 121, 145, 1974.
43. **Spiess, L. D., Lippincott, B. B., and Lippincott, J. A.,** Effect of hormones and vitamin B_{12} on gametophore development in the moss *Pylaisiella selwynii, Am. J. Bot.,* 60, 708, 1973.
44. **Lal, M.,** In vitro production of apogamous sporogonia in *Physcomitrium coorgense* Broth., *Phytomorphology,* 11, 263, 1961.
45. **Sood, S. and Chopra, R. N.,** A record preponement of bud induction in the moss *Entodon myurus, Z. Pflanzenphysiol.,* 60, 390, 1973.
46. **Reski, R. and Abel, W. O.,** Induction of budding on chloronemata and caulonemata of the moss *Physcomitrella patens,* using isopentenyladenine, *Planta,* 165, 354, 1985.
47. **Klein, B. and Bopp, M.,** Effect of activated charcoal in agar on the culture of lower plants, *Nature (London),* 230, 474, 1971.
48. **Larpent-Gourgaud, M.,** Problemes poses par les echanges intercellulaires dans le protonema des bryophytes: mise en evidence des phenomenes de transport, *Colloq. Bryol. Soc. Bot. Fr.,* 121, 161, 1974.
49. **Rose, S. and Bopp, M.,** Uptake and polar transport of indoleacetic acid in moss rhizoids, *Physiol. Plant.,* 58, 57, 1983.
50. **Rose, S., Rubery, P. H., and Bopp, M.,** The mechanism of auxin uptake and accumulation in moss protonemata, *Physiol. Plant.,* 58, 52, 1983.
51. **Rose, S., Eberhardt, I., and Bopp, M.,** Temperature-dependent auxin efflux from moss protonema, *Z. Pflanzenphysiol.,* 109, 243, 1983.
52. **Bopp, M. and Geier, U.,** Protoplasts and transport, in *Methods in Bryology,* Glime, J., Ed., Hattori Botanical Laboratory, Nichinan, Miyazaki, Japan, 1988, 87.
53. **Venis, M.,** *Hormone Binding Sites in Plants,* Longman, New York, 1985.
54. **Lomax-Reichert, T., Ashton, N. W., and Ray, P. M.,** Naphthaleneacetic acid and fusicoccin binding to membrane fractions of the moss *Physcomitrella patens,* in *Plasmalemma and Tonoplast: Their Functions in the Plant Cell,* Marme, D., Marre, E., and Hertel, R., Eds., Elsevier, Amsterdam, 1982, 303.
55. **Erichsen, U., Knoop, B., and Bopp, M.,** Uptake, transport and metabolism of cytokinin in moss protonema, *Plant Cell Physiol.,* 19, 839, 1978.
56. **Schmiedel, G. and Schnepf, E.,** Polarity and growth of caulonema tip cells of the moss *Funaria hygrometrica, Planta,* 147, 405, 1980.
57. **Handa, A. K. and Johri, M. M.,** Involvement of cyclic adenosine 3',5'-monophosphate in chloronema differentiation in protonema cultures of *Funaria hygrometrica, Planta,* 114, 317, 1979.
58. **Spiess, L. D.,** Antagonism of cytokinin-induced callus in *Pylaisiella selwynii* by nucleosides and cyclic nucleotides, *Bryologist,* 82, 47, 1979.
59. **Saunders, M. J. and Boullion, K. J.,** Calcium regulation of cytokinesis in *Funaria, Plant Physiol.,* 80, 60, 1986.
60. **Saunders, M. J.,** unpublished data.
61. **Kaul, R. and Sachar, R. C.,** On the presence of adenosine 3',5'-cyclic monophosphate in moss *(Funaria hygrometrica), Biochem. Biophys. Res. Commun.,* 104, 126, 1982.

62. **Schneider, J., Szweykowska, A., and Spychala, M.,** Evidence against mediation of cAMP in bud-inducing effect of cytokinin in moss protonema, *Acta Soc. Bot. Pol.*, 44, 588, 1975.
63. **Hintermann, R. and Parish, R. W.,** Determination of adenylate cyclase activity in variety of organisms: evidence against the occurrence of the enzyme in higher plants, *Planta,* 146, 459, 1979.
64. **Skoog, F. and Miller, C. O.,** Chemical regulation of growth and organ formation in plant tissues cultured *in vitro, Soc. Exp. Biol. Symp.*, 11, 118, 1957.
65. **Bopp, M. and Brandes, H.,** Versuche zur Analyse der Protonemaentwicklung der Laubmoose. II. Uber den Zusammenhang zwischen Protonemadifferenzierung und Kinetinwirkung bei der Bildung von Moosknospen, *Planta,* 62, 116, 1964.
66. **Whitacker, B. D. and Kende, H.,** Bud formation in *Funaria hygrometrica*. A comparison of the activities of three cytokinins with their ribosides, *Planta,* 121, 93, 1974.
67. **Spiess, L. D.,** Comparative activity of isomers of zeatin and ribosylzeatin on *Funaria hygrometrica, Plant Physiol.*, 55, 583, 1975.
68. **Bopp, M. and Fell, J.,** Manifestation der Cytokinin abhängigen Morphogenese bei der Induktion von Moosknospen, *Z. Pflanzenphysiol.*, 79, 81, 1976.
69. **Szweykowska, A. and Maćkowiak, T.,** On the development of gametophores in *Funaria hygrometrica* and *Ceratodon purpureus* in liquid cultures, *Acta Soc. Bot. Pol.*, 31, 296, 1962.
70. **Johri, M. M.,** The protonema of *Funaria hygrometrica* as a system for studying cell differentiation, in *Form Structure and Function in Plants,* Mohan Ram, H. Y., Shah, J. J., and Shah, C. K., Eds., Sarita Prakashan, Meerut, India, 1975, 116.
71. **Spiess, L. D., Lippincott, B. B., and Lippincott, J. A.,** Influence of octopine, calcium and compounds that affect calcium transport on zeatin-induced bud formation by *Pylaisiella selwynii, Am. J. Bot.*, 71, 1416, 1984.
72. **Spiess, L. D., Lippincott, B. B., and Lippincott, J. A.,** Influence of certain plant growth regulators and crown gall related substances on bud formation and gametophore development of the moss *Pylaisiella selwynii, Am. J. Bot.*, 59, 233, 1972.
73. **Nehlsen, W.,** A new method for examining induction of buds by cytokinin, *Am. J. Bot.*, 66, 601, 1979.
74. **Jahn, H.,** Der Einfluβ von Kinetin auf die Anlage der Stämmchen von *Funaria hygrometrica* Sibth., *Flora,* 154, 568, 1964.
75. **Bopp, M., Gerhäuser, D., and Keβler, U.,** On the hormonal system of mosses, in *Plant Growth Substances 1985,* Bopp, M., Ed., Springer-Verlag, Berlin, 1986, 263.
76. **Hahn, H. and Bopp, M.,** A cytokinin test with high specificity, *Planta,* 83, 115, 1968.
77. **Valadon, L. R. G. and Mummery, R. S.,** Quantitative relationship between various growth substances and bud production in *Funaria hygrometrica*. A bioassay for abscisic acid, *Physiol. Plant.*, 24, 232, 1971.
78. **Schnepf, E., Deichgräber, G., and Bopp, M.,** Growth, cell wall formation and differentiation in the protonema of the moss, *Funaria hygrometrica:* effects of plasmolysis on the developmental program and its expression, *Protoplasma,* 133, 50, 1986.
79. **Szweykowska, A., Dornowska, E., Cybulska, A., and Wasiek, G.,** The cell division response to cytokinins in isolated cell cultures of the protonema of *Funaria hygrometrica* and its comparison with the bud induction response, *Biochem. Physiol. Pflanz.*, 162, 514, 1971.
79a. **Szweykowska, A., Korez, J., Jaskiewicz-Mroczkowska, B., and Metelska, M.,** The effect of various cytokinins and other factors on the protonemal cell division and the induction of gametophores in *Ceratodon purpureus, Acta Soc. Bot. Pol.*, 41, 403, 1972.
80. **Saunders, M. J. and Hepler, P. K.,** Localization of membrane-associated calcium following cytokinin treatment in *Funaria hygrometrica* using chlorotetracycline, *Planta,* 152, 272, 1981.
81. **Saunders, M. J. and Hepler, P. K.,** Calcium ionophore A23182 stimulates cytokinin-like mitosis in *Funaria, Science,* 217, 943, 1982.
82. **Saunders, M. J. and Hepler, P. K.,** Calcium antagonists and calmodulin inhibitor block cytokinin-induced bud formation in *Funaria, Dev. Biol.*, 99, 41, 1983.
83. **Bopp, M. and Jacob, H. J.,** Cytokinin effect on branching and bud formation in *Funaria, Planta,* 169, 462, 1986.
84. **Bopp, M.,** Action mechanism of cytokinins in mosses (a model for the effects of cytokinins), in *Plant Growth Substances 1973,* Proc. 8th Int. Conf. Plant Growth Substances, Hirokawa Publishing Co., Tokyo, 1974, 934.
85. **Gardner, G., Sussman, M. R., and Kende, H.,** In vitro cytokinin binding to the particulate cell fraction from the protonema of *Funaria hygrometrica, Planta,* 143, 67, 1978.
86. **Wang, T. L., Cove, D. J., Beutelmann, P., and Hartmann, E.,** Isopentenyladenine from mutants of the moss, *Physcomitrella patens, Phytochemistry,* 19, 1103, 1980.
87. **Beutelmann, P. and Bauer, L.,** Purification and identification of a cytokinin from moss callus cells, *Planta,* 133, 215, 1978.

88. **Grimsley, N. H., Featherstone, D. R., Courtice, G. R. M., Ashton, N. W., and Cove, D. J.,** Somatic hybridization following protoplast fusion as a tool for the genetic analysis of development in the moss *Physcomitrella patens, Symp. Biol. Hung.*, 22, 363, 1980.
89. **Wang, T. L., Horgan, R., and Cove, D. J.,** Cytokinins from the moss *Physcomitrella patens, Plant Physiol.*, 68, 735, 1981.
90. **Wang, T. L., Beutelmann, P., and Cove, D. J.,** Cytokinin biosynthesis in mutants of the moss, *Physcomitrella patens, Plant Physiol.*, 68, 739, 1981.
91. **Wang, T. L., Futers, T. S., McGeary, F., and Cove, D. J.,** Moss mutants and the analysis of cytokinin metabolism, in *The Biosynthesis and Metabolism of Plant Hormones,* Crozier, A. and Hillmann, J. R., Eds., Cambridge University Press, London, 1984, 135.
92. **Akiyoshi, D. E., Klee, H., Amasino, R. M., Nester, E. W., and Gordon, M. P.,** T-DNA of *Agrobacterium tumefaciens* encodes an enzyme of cytokinin biosynthsis, *Proc. Natl. Acad. Sci. U.S.A.*, 81, 5994, 1984.
93. **Akiyoshi, D. E., Regier, D. A., Jen, G., and Gordon, M. P.,** Cloning and nucleotide sequence of the tzs gene from *Agrobacterium tumefaciens* strain T37, *Nucleic Acids Res.*, 13, 2773, 1985.
94. **Eberle, J., Wang, T. L., Cook, S., Wells, B., and Weiler, E. W.,** Immunoassay and ultrastructural localization of isopentenyladenine and related cytokinins using monoclonal antibodies, *Planta,* 172, 289, 1987.
95. **Bopp, M.,** How can external hormones regulate the morphogenesis of mosses?, *J. Hattori Bot. Lab.*, 53, 159, 1982.
96. **Brandes, H. and Kende, H.,** Studies on cytokinin-controlled bud formation in moss protonema, *Plant Physiol.*, 43, 827, 1968.
97. **Sussman, M. R. and Kende, H.,** The synthesis and biological properties of 8-azido-N^6-benzyladenine, a potential photoaffinity reagent for cytokinin, *Planta,* 137, 91, 1972.
98. **Bopp, M., Jahn, H., and Klein, B.,** Eine einfache Methode, das Substrat während der Entwicklung von Moosprotonemen zu wechseln, *Rev. Bryol. Lichenol.*, 33, 219, 1964.
99. **Bopp, M. and Erichsen, U.,** Metabolism of the cytokinins in mosses, in *Metabolism and Molecular Activities of Cytokinins,* Guern, J. and Peaud-Lenoel, C., Eds., Springer-Verlag, Berlin, 1981, 105.
100. **Bopp, M. and Gerhäuser, D.,** Localization of cytokinin responses in the moss *Funaria hygrometrica, Biol. Plant.*, 27, 265, 1985.
101. **Gerhäuser, D.,** Cytokinin Turnover in Moss Protonemata {*Funaria hygrometrica* (L.) Sibth.}, Ph.D. dissertation, Universität Heidelberg, Heidelberg, Federal Republic of Germany, 1987.
102. **Gerhäuser, D. and Bopp, M.,** Cyclokininoxydases in mosses, *J. Plant Physiol.*, 135, 680, 1990.
103. **Jaffe, L. F.,** Calcium explosions as triggers of development, *Ann. N.Y. Acad. Sci.*, 339, 86, 1980.
104. **Markmann-Mulisch, U. and Bopp, M.,** The hormonal regulation of protonema development in mosses. IV. The role of Ca^{2+} as cytokinin effector, *J. Plant Physiol.*, 129, 155, 1987.
105. **Saunders, M. J.,** Cytokinin activation and redistribution of plasma membrane ion channels in *Funaria:* a vibrating-microelectrode and cytoskeletal inhibitor study, *Planta,* 167, 402, 1986.
106. **Saunders, M. J.,** Correlation of electrical current influx with nuclear position and division in *Funaria* caulonema tip cells, *Protoplasma,* 132, 32, 1986.
107. **Tabbot, J. and Saunders, M. J.,** Fluctuation of IP_3 concentration during cytokinin-induced bud formation in *Funaria, Plant Physiol.*, 80, 113a, 1986.
108. **Conrad, P. A. and Hepler, P. K.,** The PI cycle and cytokinin-induced bud formation in *Funaria, Plant Physiol.*, 80, 60a, 1986.
109. **Erichsen, J., Knoop, B., and Bopp, M.,** On the action mechanism of cytokinins in mosses: caulonema specific proteins, *Planta,* 135, 161, 1977.
110. **Bopp, M., Erichsen, U., Nessel, M., and Knoop, B.,** Connection between the synthesis of differentiation-specific proteins and the capacity of cells to respond to cytokinin in the moss *Funaria, Physiol. Plant.*, 42, 73, 1978.
111. **Sood, S., Brenner, K., and Bopp, M.,** The occurrence of caulonema-specific proteins in different moss species, *Planta,* 138, 299, 1978.
112. **Bellte, W.,** Charakterisierung eines mit der Differenzierung zusammenhängenden Proteinmusters bei *Funaria hygrometrica,* Ph.D. Dissertation, Universität Heidelberg, Heidelberg, Federal Republic of Germany, 1980.
113. **Murach, K.-F., Beltle, W., and Bopp, M.,** Crude-tissue-extract analysis by micro-electrophoresis in cylindrical continuous polyacrylamide gradient gels, *Electrophoresis,* 3, 337, 1982.
114. **Beltle, W., Zimmerman, S., Bopp, M., and Murach, K.-F.,** Two-dimensional polyacrylamide gel electrophoresis in micro-gels: gradient gel electrophoresis followed by dodecyl sulfate gel electrophoresis, *Electrophoresis,* 4, 143, 1983.
115. **Zbell, B. and Walter, C.,** About the search for the molecular action of high-affinity auxin-binding sites on membrane-localized rapid phosphoinositide metabolism in plant cells, in *Plant Hormone Receptors,* Klämbt, D., Ed., Springer-Verlag, Berlin, 1987, 141.

116. **Bhatla, S. C. and Chopra, R. N.,** Subcellular localization of adenylate cyclase in the shoot apices of *Bryum argenteum* Hedw., *Ann. Bot.*, 54, 195, 1984.
117. **Amrhein, N.,** Cyclic nucleotide phosphodiesterases in plants, *Z. Pflanzenphysiol.*, 72, 249, 1974.
118. **Rademacher, W. and Bopp, M.,** unpublished data.
119. **Atzorn, R. and Bopp, M.,** unpublished data.
120. **Bopp, M. and Keβler, U.,** unpublished data.
121. **Atzorn, R., Bopp, M., and Merdes, U.,** The physiological role of indole acetic acid in the moss *Funaria hygrometrica* Hedw. II. Mutants of *Funaria hygrometrica* which exhibit enhanced catabolism of indol-3-acetic acid, *J. Plant Physiol.*, 135, 526, 1989.
122. **Bopp, M., Quiades, H., and Schnepf, E.,** Formation and disintegration of tmema cells in moss protonema, in preparation, 1990.
123. **Zbell, B., Schwendemann, I., and Bopp, M.,** High affinity GTP-binding on microsomal membranes prepared from moss protonema of *Funaria hygrometrica, J. Plant Physiol.*, 134, 639, 1989.

Chapter 5

GROWTH REGULATING SUBSTANCES IN MOSSES

Satish C. Bhatla and Sadhana Dhingra-Babbar

TABLE OF CONTENTS

I. INTRODUCTION

The development of the moss protonema and subsequent differentiation of buds are regulated by a variety of environmental factors such as light intensity, photoperiod, temperature, and humidity. Under experimental conditions, the moss protonema behaves abnormally when subjected to certain "stress conditions" such as low light intensity and submerged culture conditions. At times the development suppressed by such stress factors can be normalized by subjecting the cultures to specific hormone treatments.[1,2] One such example of hormone-mediated stress response is observed in *Funaria hygrometrica*. The protonema of *Funaria* raised in liquid basal medium under stationary culture conditions does not exhibit a normal differentiation pattern and remains in the chloronema state. Addition of indole-3-acetic acid (IAA) or α-naphthaleneacetic acid (α-NAA) to the culture medium leads to caulonema formation.[3] Similarly, the protonema grown on semisolid basal medium under low light intensity (800 lux) remains in the chloronema state and exhibits further differentiation only when treated with physiological concentrations of IAA.[4]

Auxins are not the only active components involved in protonema differentiation. By now it is well established that exogenously supplied cytokinins induce and promote the formation of buds in a wide variety of mosses.[5] In *Funaria* cytokinins act only when chloronema has differentiated into caulonema. Thus, in this moss auxins and cytokinins act sequentially on protonema differentiation and bud formation.[2] When chloronema cultures grown under low light intensity (800 lux) are subjected to kinetin treatment and transferred to kinetin-free medium, buds do not appear. However, when auxin treatment is followed by cytokinin, or if both the hormones are supplied together, a large number of buds are induced.[4] This demonstrates the sequential interaction of the two hormones in regulating protonema development.

Studies have further shown that auxins interact antagonistically with a third kind of growth regulator, cyclic adenosine 3′,5′-monophosphate (cAMP), in regulating protonema differentiation.[6,7] cAMP, added in concentrations of 10 to 100 n*M* in liquid cultures, maintains the protonema of *F. hygrometrica* in the chloronema state. Furthermore, in a caulonema-overproducing mutant (pg-1) of *Funaria*, application of 100 n*M* cAMP leads to a reduction of caulonema formation.[6,7]

On subjecting the moss protonema to specific hormone treatment the results obtained lead to the speculation that endogenous hormones are involved in protonemal growth and development. The above observations are further supported by the use of inhibitors/activators of hormone biosynthesis and degradation. A detailed account of these investigations is given in an earlier review[2] and also in Chapter 4 of this volume.[8]

In the present review we aim to provide information on the naturally occurring growth-regulating substances in mosses. The biological relevance of the physiological reactions of exogenously supplied growth hormones in moss development can be confirmed only if evidence is available for the presence of these substances in the moss system. Detection of the enzymes involved in the biosynthesis and degradation of the naturally occurring growth hormones is all the more necessary, along with evidence for the uptake and transport system of these substances.

II. AUXINS

A. DETECTION AND ESTIMATION

Rapid advances in the use of techniques such as high performance liquid chromatography (HPLC), capillary gas chromatography (GC), and mass spectrometry (MS) have resulted in the development of new methods for the separation, identification, and quantitative assay of plant hormones. HPLC is employed for the purification of plant hormones from tissue extracts. MS can provide information regarding the structures and amounts of phytohormones present. Recently there has been increasing use of immunological techniques for plant hormone analysis. Plants present unusual difficulties in the application of immunological methods, since the hormones are present in micromolar amounts in a background of millimolar quantities of phenylpropanes and tannins, which can interact with antibodies. Radioimmunoassays (RIAs) are the most sensitive techniques available for the identification and quantification of phytohormones.

Application of these techniques for the detection and estimation of IAA in mosses has been made very recently. In the protonema of *F. hygrometrica, Physcomitrella patens,* and *Polytrichum formosum,* presence of endogenous auxins has been demonstrated using different techniques. The total amount of the hormone determined in the tissue varies depending on the extraction procedure and the method of determination (Table 1). Using the technique of gas chromatography-selected ion monitoring-mass spectrometry (GC-MS) coupled with an isotope dilution assay with 4,5,6,7-tetradeutero-indole-3-acetic acid as the internal standard, IAA has been estimated to be present in aseptically cultured gametophytes of wild-type *Physcomitrella patens* (Hedw.) B.S.G. at a level of 0.075 $\mu g\ g^{-1}$ dry weight (dw) or 2.1 ng g^{-1} fresh weight (fw).[9] The estimated low levels of IAA seem to be explained partly because only free IAA was assayed. In many higher plant tissues most of the IAA can be present as esterified IAA or peptidyl IAA.[10] Moreover, the low level of IAA in the gametophytes also may be explained by the fact that endogenous IAA of *P. patens* can diffuse or be transported out of the tissue into the growth medium.[11-13]

Detection of IAA in the wild-type *Physcomitrella patens* does not show that IAA is the only or even the major auxin in this plant which regulates development. However, clarification of this point should be possible by assaying for IAA in a selection of auxin-sensitive, developmentally abnormal mutants. Several mutants without spontaneous caulonema formation have been isolated from *Physcomitrella* and *Funaria.* Experiments with "auxin-resistant" mutants of *F. hygrometrica* (naphthaleneacetic acid resistant [NAR]) indicate the specific involvement of IAA in protonema differentiation. Five mutants were selected on a substrate with a high concentration of NAA (50 to 500 μM).[1,14] The most clearly defined mutant, NAR-2, behaves as a type 4 mutant of *P. patens.*[13,15] It remains in the chloronema state when grown on basal medium. Normal development can be restored by supplementing the medium with 1 μM IAA, β-NAA, or 2,4-dichlorophenoxyacetic acid (2,4-D), and it leads to the formation of caulonema filaments.[1,16] Using the RIA technique the IAA contents of the wild-type and the mutant NAR-2 *Funaria* have been compared.[2] Protonema of wild-type *Funaria* grown in high light intensity (HLI, 3000 lux) contained about 0.4 nmol IAA g^{-1} fw or 70 ng g^{-1} fw. In low light intensity (LLI, 800 lux) the IAA content of the wild-type protonema was about 0.15 mol g^{-1} fw, and the mutant NAR-2 in HLI contained 0.12 nmol g^{-1} fw. It is clear from these observations that in LLI the IAA content of the wild-type is about one third of that grown in HLI. At HLI the mutant contains the same amount as the wild-type grown in LLI. Thus, it seems clear that endogenous auxin (in the range of 10^{-19} mol/g fw) is responsible for the transition from chloronema to caulonema.[16] Any situation that lowers the auxin content (LLI suspension culture, and mutants with blocked biosynthetic pathways or with high rates of degradation) must result in the suppression of caulonema formation.

A comparison of the free auxin levels in the chloronema cells of the wild-type and pg-1 mutant of *Funaria* reveals that the level of auxin in the pg-1 mutant is slightly lower than that of chloronema cells in the wild-type.[17] The pg-1 is a leaky chloronema-repressed mutant which predominantly produces caulonema filaments (>70%) in the absence of endogenous auxin.[6,7] Based on an indole-α-pyrone fluorometric assay, the level of putative IAA has been estimated to be 5.0 and 1.9 μg/kg in caulonema and chloronema cells, respectively.[17] In chloronema cells the auxin activity, based on the oat coleoptile bioassay, is three- to

TABLE 1
Growth Regulating Substances in Mosses

Plant/Tissue	Technique	Quantity	Ref.
	Indole-3-acetic acid (IAA)		
Physcomitrella patens	GC-MS coupled with isotope dilution method	0.075 μg g^{-1} dw (= 2.1 ng g^{-1} fw)	9
Polytrichum formosum	RIA	0.61—10.60 pmol mg^{-1} dw	19
Funaria hygrometrica	RIA	0.15 nmol g^{-1} fw (WT grown in LLI)	2
		0.4 nmol g^{-1} fw (WT grown in HLI)	
		0.12 nmol g^{-1} fw (mutant NAR-2 grown in HLI)	
	Indole-α-pyrone fluorometric assay	5 μg kg^{-1} fw (caulonema)	17
		1.9 μg kg^{-1} fw (chloronema)	
	N^6-(Δ^2-Isopentenyl) (i^6Ade, IPA, or iP)		
Callus from the hybrid *F. hygrometrica* × *Physcomitrium pyriforme*	GC-MS	ca. 10^{-6} *M*	39, 40
F. hygrometrica	RIA	—	2
Physcomitrella patens	GC-MS:HPLC	—	35—37
	ELISA	—	45
	Zeatin [6-(4-hydroxy-3-methylbut-2-enylamino)purine]		
Physcomitrella patens (OVE 78)	HPLC; GC-MS	—	37, 44
F. hygrometrica	RIA	—	2
	Zeatin glucoside		
Physcomitrella patens (OVE mutants)	RIA	—	44
	Cyclic adenosine 3′,5′-monophosphate (cAMP)		
F. hygrometrica	PK stimulation assay	31 pmol g^{-1} fw or 1.14 pmol mg^{-1} protein (chloronema)	55
			55
	Gilman assay	2—7 times higher than the values obtained with PK assay (chloronema)	55
	Ethylene		
F. hygrometrica	GLC	1 nl h^{-1} g^{-1} protein	80
	Acetylcholine		
Callus tissue from the hybrid *Funaria hygrometrica* × *Physcomitrium pyriforme*	TLC fraction followed by bioassay	—	82
	GLC	—	83

Note: Abbreviations: GLC — gas-liquid chromatography; GC-MS — gas chromatography-mass spectrometry; PK — protein kinase; RIA — radioimmunoassay; ELISA — enzyme-linked immunosorbent assay; TLC — thin-layer chromatography; WT — wild type; NAR-2 — naphthaleneacetic acid-resistant mutant of *Funaria hygrometrica*; LLI — low light intensity (800 lux); HLI — high light intensity (3000 lux); dw — dry weight; fw — fresh weight; —, quantity not determined.

fourfold higher than that determined by the fluorometric assay; the presence of another acidic auxin in addition to IAA is indicated. The solvent system used by the investigator did not separate this unidentified auxin from IAA. Alternatively, it is conceivable that a substance moving along with IAA quenches fluorescence in the fluorometric assay. No such substance has been detected in the caulonema cells from the mutant pg-1.

Using the RIA technique[18] Markmann-Mulisch[19] determined free IAA levels in the protonema of *Polytrichum formosum*. The amount of IAA varied between 0.61 and 10.60 pmol/mg dw of protonema, depending on the method employed for raising the cultures. The protonema contained higher amounts of free IAA when grown in weak light.

B. *IN VITRO* AND *IN VIVO* SYNTHESIS

Initial evidence for the IAA biosynthetic capacity of mosses is based on the observed effects of the IAA precursors on caulonema formation. Among higher plants four different pathways of IAA biosynthesis are well documented.[20] They involve indole-3-acetonitrile, tryptamine, indole-3-pyruvic acid, and indole-3-acetamide, respectively, as intermediates. The starting point for all pathways is the amino acid tryptophan. Experiments on *F. hygrometrica* have demonstrated that tryptophan can replace IAA in creating target sites which exhibit bud induction in response to subsequent cytokinin treatment.[21] Indole-3-acetaldehyde and indole-3-pyruvic acid (IPA) evoke a similar morphogenetic response, as do IAA and tryptophan, leading to the formation of caulonema filaments. However, tryptamine and indole-3-acetonitrile (which represent the other two pathways) do not induce caulonema formation. This can be explained by the absence of the enzymes amino oxidase and nitrilase, required for the conversion of these precursors to IAA.[4]

Recent investigations on the chloronema cultures of *F. hygrometrica* have shown that the relative uptake of ^{3}H-tryptophan at low cell density (0.025 mg/ml) is ca. tenfold greater than that at high cell density (2.5 mg/ml).[17] ^{3}H radioactivity cochromatographed with IPA and IAA, and no radioactivity was found in tryptamine, confirming earlier observations that the tryptamine pathway of IAA biosynthesis does not exist in the moss tissue. It is thus clear that moss tissues can synthesize IAA and that the tryptophan is metabolized into IAA via the transamination pathway. Greater accumulation of radioactivity (per milligram protein) in the form of IPA and IAA in the cells at low density is due to higher uptake of tryptophan by low-density cultures. In a cell-free system, as well, ^{3}H-tryptophan is metabolized into ^{3}H-IAA, which cochromatographs with the authentic ^{14}C-IAA.

Spectrophotometric demonstration of the enzyme tryptophan synthetase (L-serine hydrolyase [adding indole], EC 4.2.1.20) has recently been made from the extracts of protonema of the wild-type and the mutant NAR-2 of *F. hygrometrica*.[98] It is the terminal enzyme in the pathway of tryptophan biosynthesis, and it catalyzes the following three reactions:

1. Indoleglycerol phosphate + L-serine $\xrightarrow{\text{pyridoxal phosphate}}$ L-tryptophan + D-glyceraldehyde-3 phosphate

2. Indole + L-serine $\xrightarrow{\text{pyridoxal phosphate}}$ L-tryptophan

3. Indoleglycerol phosphate $\longrightarrow$ indole + D-glyceraldehyde-3-phosphate

The presence of tryptophan synthetase activity gives us positive proof for the IAA biosynthesizing system in the moss tissue.

C. DEGRADATION AND IAA OXIDASE ACTIVITY

Metabolic degradation of auxins in the moss protonema became evident by the demonstration of IAA oxidase activity in the protonema suspension cultures of *F. hygrometrica*.[22] The enzyme activity was found to be proportional to the cell density. A comparison of the

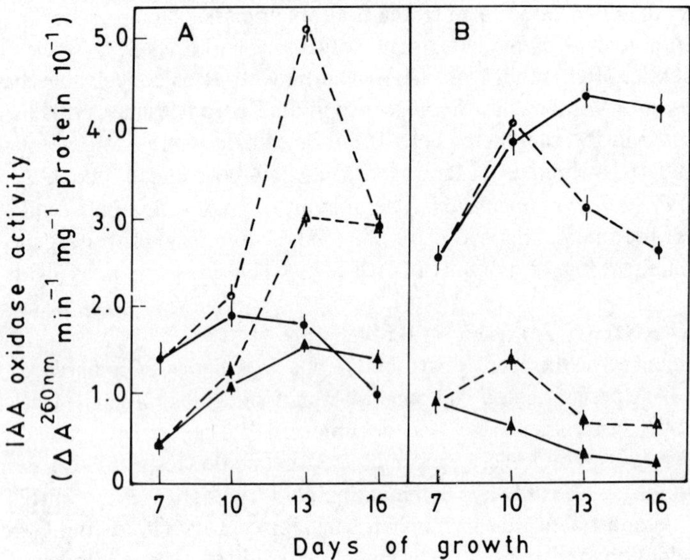

FIGURE 1. Changes in the IAA oxidase activity in the soluble and particulate fractions of the wild-type and mutant (NAR-2) protonema of *Funaria hygrometrica* at various stages of growth. Wild-type with (▲-----▲) and without (▲——▲) 10 μM IAA. NAR-2 protonema with (●-----●) and without (●——●) 10 μM IAA. (A) Soluble fraction; (B) particulate fraction. Mean and standard error from four replicates. (From Bhatla, S. C. and Bopp, M., *J. Plant Physiol.*, 120, 233, 1985. With permission.)

IAA oxidase activity in the soluble and particulate fractions of the protonema extracts from the wild-type and the mutant (NAR-2) *Funaria* reveals that the enzyme activity is about three times higher in the mutant as compared to that in the wild-type protonema of the same age (7 days old).[1] Up to this stage the wild-type protonema is in the chloronema state. The activity changes during later stages of growth, but the mutant always has a significantly higher level of enzyme activity. The specific activity of IAA oxidase (in terms of milligrams protein) in the soluble fractions of both the wild-type and mutant protonema increases during the initial phase of growth (Figure 1). In the particulate fraction the enzyme activity decreases in the wild-type and increases significantly in the mutant. With the treatment of 10 μM IAA, IAA oxidase activity is stimulated in both the fractions of the wild-type and the mutant, accompanying the formation of caulonema.

These observations indicate that the lowering of the auxin level in the mutant can be due to increased enzymatic degradation of the hormone. Peroxidase activity is relatively lower in the mutant as compared to the wild-type (Figure 2), indicating that the differences observed in the enzyme activities cannot be a consequence of variations in the protease activities. Instead, they represent true biochemical variations parallel to the morphogenetic changes. An increase in IAA oxidase activity in the soluble fraction of the wild-type protonema during caulonema formation further indicates that the endogenous auxin concentration necessary to maintain the caulonema state may be less than that required for its initiation, but a certain minimal amount of the hormone is always required.[15,21] The observation that auxin application results in an increase in IAA oxidase activity is in conformity with earlier reports on tissues of higher plants.[23-25]

D. UPTAKE AND TRANSPORT

The accumulation of a substance within the cell is the combined effect of its influx and efflux. In the case of single-stranded filaments of the moss protonema we can assume that,

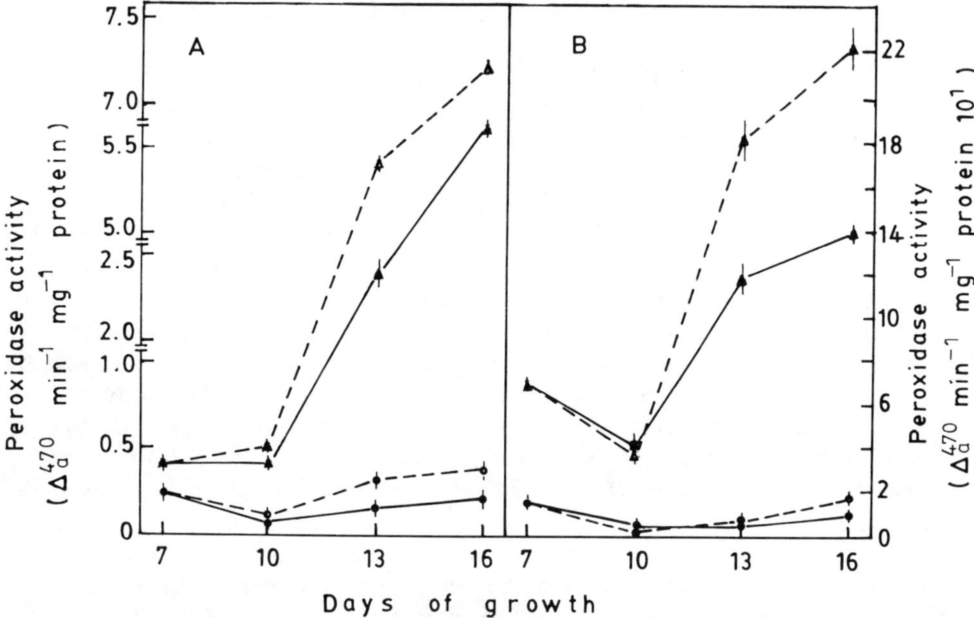

FIGURE 2. Changes in the activity of peroxidase in the soluble and particulate fractions of wild-type and mutant (NAR-2) protonema of *Funaria hygrometrica* at various stages of growth. Wild-type with (▲-----▲) and without (▲——▲) 10 μM IAA. NAR-2 protonema with (●-----●) and without (●——●) 10 μ*M* IAA. (A) Soluble fraction; (B) particulate fraction. Mean and standard error from four replicates. (From Bhatla, S. C. and Bopp, M., *J. Plant Physiol.*, 120, 233, 1985. With permission.)

because of their very narrow contact with the surrounding medium, exchange of substances between the cells and the medium is the most common mode of influx and efflux. However, a symplastic exchange of substances between adjacent cells also must take place because tip cells of the caulonema are required to maintain the filaments in a stable state of differentiation. Plasmodesmata exist between neighboring cells as symplastic connections.[26]

If the tip cell is destroyed or the rate of apical cell growth is reduced, caulonema filaments dedifferentiate into chloronema.[27] Therefore, one expects a transport of signals or signal substances from the tip cell to the base. In *F. hygrometrica* a strict polar transport of ^{14}C-IAA has been observed from the tip to the base of rhizoids.[28] Only about 10% of the radioactivity is transported in the opposite direction. This is in contrast to the direction of transport of nucleosides, sugars, amino acids, and kinetin, which are translocated mainly from the basal to the apical part of the filament.[29-31] The polar transport of auxins can occur partly through the medium as an apoplastic component and partly as a symplastic flux.[28] The equilibrium between influx and efflux of IAA is reached after about 1 h without further accumulation.[32]

The efflux of auxins is active and is regulated by carriers driven by chemiosmotic components. The efflux of IAA from the cell wall and that from protoplasts can be separated by a clear dependence on temperature.[33] In isolated protoplasts auxin accumulation, in comparison with the surrounding medium, reaches 50- to 70-fold after 30 min, whereas in intact cells, with cell wall and vacuoles, it is not more than around 10-fold.[2]

Auxin uptake has a passive and an active component. The passive accumulation of IAA in the protonema cells is pH dependent, with a strong increase between pH 4.5 and 4.7 (pK value for IAA = 4.7). This means that the pH-dependent uptake is due to the dissociation of the IAA molecule and is not due to a pH gradient between the cell and the surrounding medium. The active part of the uptake has been demonstrated by saturation experiments in which the labeled auxin is supplied along with nonradioactive IAA.[34]

III. CYTOKININS

A. DETECTION AND SUBCELLULAR LOCALIZATION

Despite the use of moss tissue in numerous studies on the physiology and biochemistry of cytokinins, it was not possible to detect the production of cytokinins by normal moss tissue until recently. The wild-type strains of mosses have been observed to produce very low levels of endogenous cytokinins.[35-37] This may be the reason that early reports on the endogenous detection of cytokinins in mosses mainly concerned the abnormal protonema.

The first evidence for the presence of cytokinins in bryophyte tissue came from the work of Bauer,[38] who isolated a substance from callus derived from the sporophyte tissue of the moss hybrid *F. hygrometrica* × *Physcomitrium pyriforme*. This substance possesses properties akin to kinetin, and Bauer therefore termed it "bryokinin". Subsequent work by Beutelmann[39,40] led to the characterization of "bryokinin", using GC-MS, as N^6-(Δ^2-isopentenyl) adenine (iP or i^6Ade), occurring at a concentration of ca. 10^{-6} *M*. Another substance whose properties resemble an adenine derivative[41] has been isolated from several mosses. This compound, factor H, has been shown to increase the number of buds and to shorten the time required for bud production.[29,42] In the cultures of cytokinin-overproducing (OVE) mutants of *Physcomitrella patens* fed with ^{14}C-adenine, labeled cytokinin has been detected in the growth medium and also as the natural product of cells.[35-37] In these mutants, labeled i^6Ade appears to be the major cytokinin in the moss tissue (Figure 3). It is also released from the protonema into the medium (Figure 4). Besides the main component (i^6Ade), zeatin is also present in small quantities, its concentration being 12 to 22 times less than that of i^6Ade. A comparison of the cytokinins detected by GC-MS and by RIA shows that the two techniques are in reasonable agreement. Furthermore, RIA has revealed the presence of a third cytokinin in moss culture medium.[44] This probably represents a zeatin glucoside. Glucosides do, in fact, cross-react in the RIA for zeatin. Mosses, therefore, may produce a third cytokinin as well. RIA has further shown that the protonema of *F. hygrometrica* also contains i^6Ade as the main cytokinin, along with zeatin at a concentration about ten times less than that in the former[2] (Table 1).

Although much is known about the gross amounts of growth regulators in plants, there is scant knowledge of their tissue distribution and subcellular localization. Plant hormones exist in only small amounts in most tissues, which makes their localization even less feasible. These problems have been overcome recently by the use of monoclonal antibodies (mABs) with high specificity and by the choice of tissue, e.g., tissues with relatively high concentrations of the hormone. Tissue of the moss *Physcomitrella patens* offers a good choice for immunocytochemical localization of cytokinins because it is well characterized[11,12] and because mutants (OVEs) exist which overproduce cytokinins.[36,37]

Eberle et al.[45] raised mABs with specificity for isopentenyladenosine (IPAdo) to minimize cross-reactivity with unknown antigens in the tissue of *Physcomitrella patens*. These mABs are suitable for detection of femtomole amounts of IPAdo and isopentenyladenine (IPA) when used in an enzyme-linked immunosorbent assay (ELISA). Tissue of *P. patens* (OVE 201), when immunostained using indirect immunogold labeling with anti-IPmAB as the primary antibody and then viewed in a transmission electron microscope, showed that the gold particles are confined almost exclusively to the area of the cell wall, with little labeling of the cytoplasm and only background labeling in the vacuole or chloroplasts. Wild-type moss tissue, when immunolabeled in the same way, showed no appreciable labeling with IPmAB or in the presence of the IPA competitor. Labeling in the OVE protonema was not found in each cell or on each grid. It is possible that cytokinin production and secretion is distributed unevenly among the protonemal cells, thus maintaining a high apoplastic IPA concentration in the cell wall during tissue preparation. This appears to be a strong possibility because the moss mutant (OVE) exports large quantities of cytokinins in the medium and

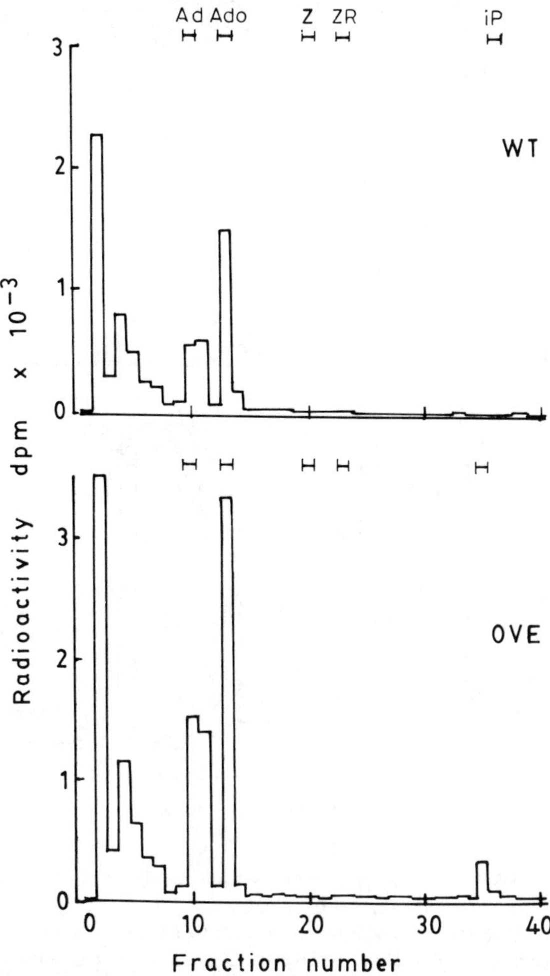

FIGURE 3. HPLC profiles of [^{14}C]Ade acidic metabolites from the tissue of *Physcomitrella patens* (wild-type and OVE 78). Moss tissue, after incubation with [^{14}C]Ade, was extracted, and the extract was subjected to ion exchange chromatography. The acidic wash was treated with alkaline phosphatase and then partitioned into butanol. The butanol phase was analyzed by HPLC on Hypersil® ODS with a 40-min gradient of 25% acetonitrile; 1-min samples were collected, dried, and measured by liquid scintillation counting. Horizontal bars indicate the elution positions of authentic compounds. Abbreviations: ZR — zeatin riboside; rest as for Figure 4. (From Wang, T. L., Futers, T. S., McGeary, F., and Cove, D. J., in *The Biosynthesis and Metabolism of Plant Hormones*, Crozier, A. and Hillman, J. R., Eds., Cambridge University Press, London, 1984, 135. With permission.)

they can be washed out of cells readily using continuous medium replacement.[96] The wall, therefore, may contain the highest concentration of the hormone.

B. BIOSYNTHESIS

Biosynthesis of free cytokinins in plant cells appears to represent a minute pathway of metabolism of the ubiquitous compound adenine. Moreover, the experimental difficulties created by this situation have greatly hindered progress toward elucidation of this pathway.

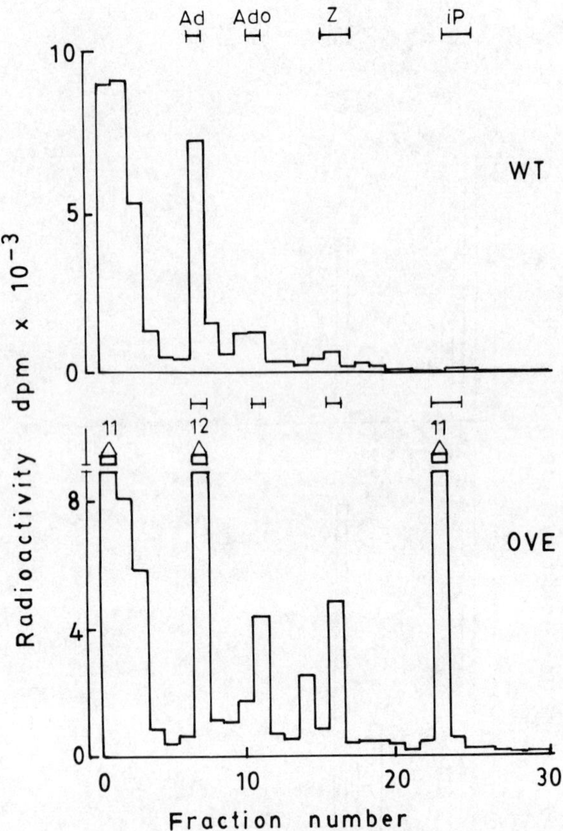

FIGURE 4. HPLC profiles of [^{14}C]-Ade metabolites from moss culture medium. Aliquots of culture medium from wild-type and OVE 78 *Physcomitrella patens* incubated with [^{14}C]-Ade for 48 h were analyzed by HPLC on Hypersil ODS using a 30-min gradient of water to 30% acetonitrile; 1-min samples were collected, dried, and measured by liquid scintillation counting. Horizontal bars indicate the elution positions of authentic compounds. Abbreviations: Ad — adenine; Ado — adenosine; Z — zeatin; iP — N^6-(Δ^2-isopentenyl)adenine. (From Wang, T. L., Futers, T. S., McGeary, F., and Cove, D. J., in *The Biosynthesis and Metabolism of Plant Hormones*, Crozier, A. and Hillman, J. R., Eds., Cambridge University Press, London, 1984, 135. With permission.)

Unlike investigations on other hormones, studies on cytokinin biosynthesis are more complicated because cytokinins exist both in a free form and as a part of another molecule — transfer RNA (tRNA). Numerous isolations of cytokinins from tRNA[46,47] following the discovery of IPAdo in the tRNA of brewer's yeast[48] led to the suggestion that tRNA may be a source of free cytokinins in plant tissues.[49] However, evidence exists both for and against this assumption.

Early work on the demonstration of biosynthesis of free cytokinins was mainly based on the cochromatography of labeled metabolites with authentic standards. Since cytokinins normally occur in very low levels in plant tissues and form a minor component of the total production from their precursors, it is technically difficult to detect radioactively labeled cytokinins by providing labeled Ade or mevalonic acid to the tissue. Thus, tissues which contain relatively high quantities of cytokinins should be an aid to *in vivo* labeling experiments. Further aid to studies on cytokinin biosynthesis can be obtained by comparing tissue variants (e.g., normal and tumor tissues, hormone-dependent and hormone-independent tissues). From this point of view *Physcomitrella patens* has proved invaluable. Genetically defined mutants (OVE) of *Physcomitrella* are available which overproduce cytokinins as compared to the wild-type tissue.

Three OVE mutants of the moss *Physcomitrella patens* have been shown to convert [8-^{14}C]-adenine (Ad) to N^6-[^{14}C]-(Δ^2-isopentenyl)adenine (i^{6}Ade), the presence of which was confirmed by thin layer chromatography (TLC), HPLC, and recrystallization to constant specific radioactivity.[36,37] Moss tissue grown in liquid culture rapidly takes up exogenous [8-^{14}C]-Ad. Samples of the culture medium show that radioactivity present in the medium falls over a 48-h incubation period to reach equilibrium at approximately 5% of the original value. Within the incubation period (48 h) Ad passes into the tissue and its metabolites are released into the medium. Fewer metabolites appear in the wild-type culture medium over the same experimental period. The uptake of [^{14}C]-Ad and its incorporation into cytokinins appear to depend on the amount of Ad available.

HPLC analysis of the moss (OVE mutant) culture medium by reverse-phase chromatography shows radioactivity cochromatographing with Ad and i^{6}Ade (Figure 4).[36,37,44] The radioactivity cochromatographing with i^{6}Ade is absent from the wild-type culture medium. TLC shows the same difference. Two-dimensional TLC of moss tissue extracts indicates several radioactive compounds in each extract. However, analysis of all OVE tissue extracts (but not wild-type) shows a spot which cochromatographs with i^{6}Ade. On the basis of these data, OVE tissue has been found to rapidly incorporate [^{14}C]-Ad into [^{14}C]-i^{6}Ade, which appears in the culture medium as well as in the tissue. In addition, the tissue of the OVE mutants contains [^{14}C]-i^{6}AMP [N^6-(Δ^2-isopentenyl)adenosine monophosphate], but the culture medium does not. Therefore, it seems possible that the synthesis of the cytokinin may occur via the nucleotide, but that i^{6}Ade is the major product in the reaction and that it passes into the culture medium. No i^{6}Ado [N^6-(Δ^2-isopentenyl)adenosine] or i^{6}Ade has been detected in either culture medium or tissue of the wild-type, although OVE mosses are known to contain the latter.

C. METABOLISM

Cytokinins not only trigger the initiation of buds; their presence is required throughout the process of bud development. Therefore, changes in the internal content of cytokinins by metabolism are of great importance. When a cytokinin base is taken up by a tissue, it normally will be broken down to Ad by side chain cleavage or modified via glycosylation or amino acylation. The formation of conjugates is a process which the cytokinins have in common with other plant hormones, herbicides, and pesticides.

Bopp and Erichsen[50] have demonstrated the formation of adenine and nucleotides from labeled kinetin. Both chloronema and caulonema filaments of *F. hygrometrica* show similar metabolic patterns; most of the label accumulates at the tip of the protonemal filament.[51] When either *Physcomitrella patens* wild-type or OVE tissue is fed with [^{14}C]-labeled benzylaminopurine (BAP), zeatin, or inorganic phosphate (iP) (ca. 10 μM), essentially the same results are obtained — the major metabolite being Ad,[44] in addition to nucleotides. No clear differences could be demonstrated in the cytokinin breakdown pattern of OVE and wild-type tissues.

Kinetin is very unstable in moss tissue.[52] After a short incubation period (1 h), products of side chain cleavage appear. Bopp and Gerhäuser[52] have tested the cleavage process *in vivo* and *in vitro* (Figure 5). They have extracted and analyzed "cytokinin oxidase" using the method of Kaminek[99] and studied the metabolism of [^{14}C]-kinetin and [^{3}H]-2iP *in vitro*.[8,43,52,53] The results have been compared with *in vivo* distribution of labeled substances after feeding the moss tissue for 24 h.

In vivo about 35 to 40% of the initial activity of kinetin could be detected in adenine

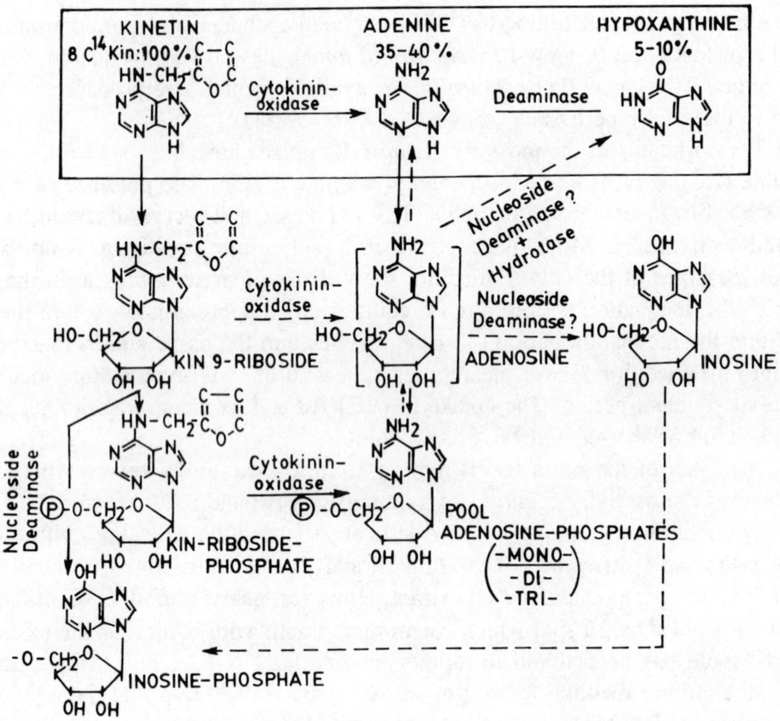

FIGURE 5. *In vitro* (box) and *in vivo* metabolism of kinetin in *Funaria hygrometrica*. [8-^{14}C]-kinetin (N^6-furfurylamino [8-^{14}C]-purine, 925 MBq in mol^{-1}) was either added to an enzyme preparation or fed to intact moss protonema for 24 h. Substances indicated in the scheme were labeled after 24 h. (From Bopp, M., Gerhäuser, D., and Kessler, U., in *Plant Growth Substances*, Bopp, M., Ed., Springer-Verlag, Berlin, 1986, 263. With permission.)

and about 5 to 10% in hypoxanthine. It appears that in addition to cytokinin oxidase a deaminase must be present in the extract. *In vivo* labeled products have been identified, as shown in Figure 5. If [^{3}H]-isopentenyladenosine is used for labeling *in vitro*, much less adenine is found, but about 30% isopentenyladenine, 15% inosine, and 10% hypoxanthine could be detected (Figure 6). A constant but low level of 2iP always remains in the cells, which is in agreement with the physiological experiments using 2iP treatment. The enzyme cytokinin oxidase extracted from the moss tissue is not completely identical to the enzyme from higher plants because 2iP is the better substrate for the latter whereas kinetin is better for the enzyme from the moss tissue.[43,53]

D. UPTAKE

Moss tissue differs from many other plant tissues by the fact that its uptake of cytokinins is slow. Figure 7 shows the uptake of zeatin and 2iP by both wild-type and OVE tissues of *Physcomitrella patens* as compared to the uptake of Ad by the same tissues.[44] Over a period of 48 h no more than 40% of the cytokinin has been removed from the medium. This is not due to an export of labeled metabolites formed from the cytokinin taken up by the tissue. Even after 14 days it has been observed that 40% of cytokinin supplied to the moss tissue remains in the culture medium. Furthermore, within 48 h the cytokinin supplied expresses its biological activity and induces bud formation on the protonema. Thus, one consequence of this poor uptake of cytokinin is that the dose-response figures, normally quoted for the biological activity of cytokinins with respect to moss tissue, could be overestimated and inaccurate.

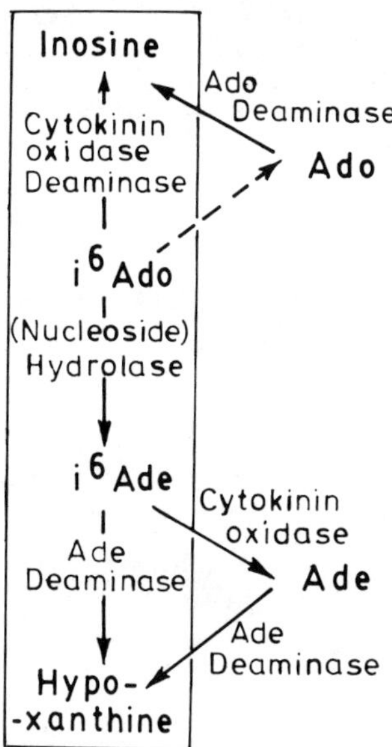

FIGURE 6. *In vitro* (box) metabolism of isopentenyladenosine (i^{6}Ado) in *Funaria hygrometrica*. N^6-(Δ^2-isopentenyl)adenosine (^{3}H-dialcohol) was added to an enzyme preparation for up to 24 h. The substances in the box were found in the reaction mixture after 16 to 24 h. The substances outside the box can be found after 12 to 24 h of *in vivo* treatment of intact moss protonema. Enzymes were not determined. (From Bopp, M., Gerhäuser, D., and Kessler, U., in *Plant Growth Substances*, Bopp, M., Ed., Springer-Verlag, Berlin, 1986, 263. With permission.)

IV. CYCLIC ADENOSINE 3′,5′-MONOPHOSPHATE

A. OCCURRENCE

Investigations have been carried out in the recent past to demonstrate the presence of cAMP and its related enzymes in the moss tissue.[54] A factor indistinguishable from cAMP has been isolated from the suspension cultures of chloronema cells of *F. hygrometrica* using the protein kinase stimulation method and Gilman assay.[55] This factor stimulates the activity of protein kinase from rabbit skeletal muscles and is chromatographically identical to 3′,5′-cAMP. Its ability to stimulate protein kinase activity is completely abolished by cyclic 3′,5′-nucleotide phosphodiesterase [(cN)PDE], the rate of inactivation being similar to that of authentic cAMP. The protein kinase (PK) assay for cAMP is highly specific and sensitive.[56] The intracellular level of cAMP determined by this method in the chloronemal cells of *Funaria* is 31 pmol g^{-1} fw (1.14 pmol mg^{-1} protein),[55] and this agrees closely with the values reported for a variety of animal tissues in the absence of stimulation by an exogenous hormone.[57] Levels of cAMP determined by the Gilman assay in the chloronema cultures of *Funaria* are two to seven times higher than those determined by the PK assay (Table 1).[55] These discrepancies are attributed to the possible presence of additional cyclic nucleotides or to interference by salts or other impurities in the sample. Thus, more rigorous purification steps are required to determine cAMP in the moss system if the Gilman assay is to be employed.

In another attempt to characterize the putative cAMP in *F. hygrometrica*, large amounts

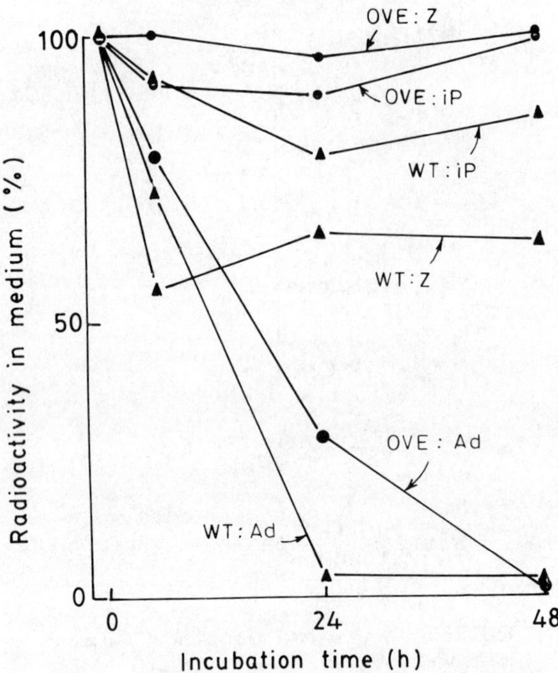

FIGURE 7. Uptake of adenine and cytokinins by the tissue of *Physcomitrella patens* (wild type and OVE 78). Tissues were incubated with 0.5 μCi [^{14}C]-cytokinin, and 10 μl aliquots of culture medium taken for liquid scintillation counting. Dry weights of tissue used (mg) — OVE 78/iP, 3.1 ± 0.3; OVE 78/Z, 9 ± 2; WT/iP, 5 ± 1; WT/Z, 13.5 ± 3.5. Abbreviations: Z — zeatin; iP — isopentenyl adenine; WT — wild type; OVE — cytokinin-overproducing mutant. (From Wang, T. L., Futers, T. S., McGeary, F., and Cove, D. J., in *The Biosynthesis and Metabolism of Plant Hormones*, Crozier, A. and Hillman, J. R., Eds., Cambridge University Press, London, 1984, 135. With permission.)

of [^{14}C]-adenine (555 to 740 kBq per batch) were fed to the cultures consisting of protonema and the gametophytic tissue, with the intention of labeling the endogenous pool of adenosine triphosphate (ATP).[58] If the system has the enzymatic potential to metabolize the labeled ATP to labeled cAMP, then it should be possible to collect enough of the putative cAMP by cochromatography with authentic cAMP. The comigrating labeled product was not hydrolyzed into 5′-AMP by (cN)PDE treatment. Furthermore, unlike the authentic cAMP, putative [^{14}C]-cAMP could not be converted to cyclic inosine monophosphate (cIMP) upon deamination with nitrous acid. Similar results were obtained with *Dicranella coarctata* by the use of labeling techniques. On the basis of these findings it is argued that cAMP, if present at all, exists in very low levels in the moss system, since the methods employed are sensitive enough to detect 1 pmol of authentic cAMP in the samples.

B. SUBCELLULAR LOCALIZATION OF ADENYLATE CYCLASE

Adenylate cyclase, the enzyme responsible for the biosynthesis of cAMP, has been very well characterized in animal systems. Its activity has been observed in algae and in the protonema of *F. hygrometrica*.[59]

ATP serves as a natural substrate for adenylate cyclase and other ATPases. Specificity for adenylate cyclase may be enhanced in the experimental system by the use of a substrate such as 5′-adenylyl imidodiphosphate (AMP-PNP), which is hydrolyzed by adenylate cyclase

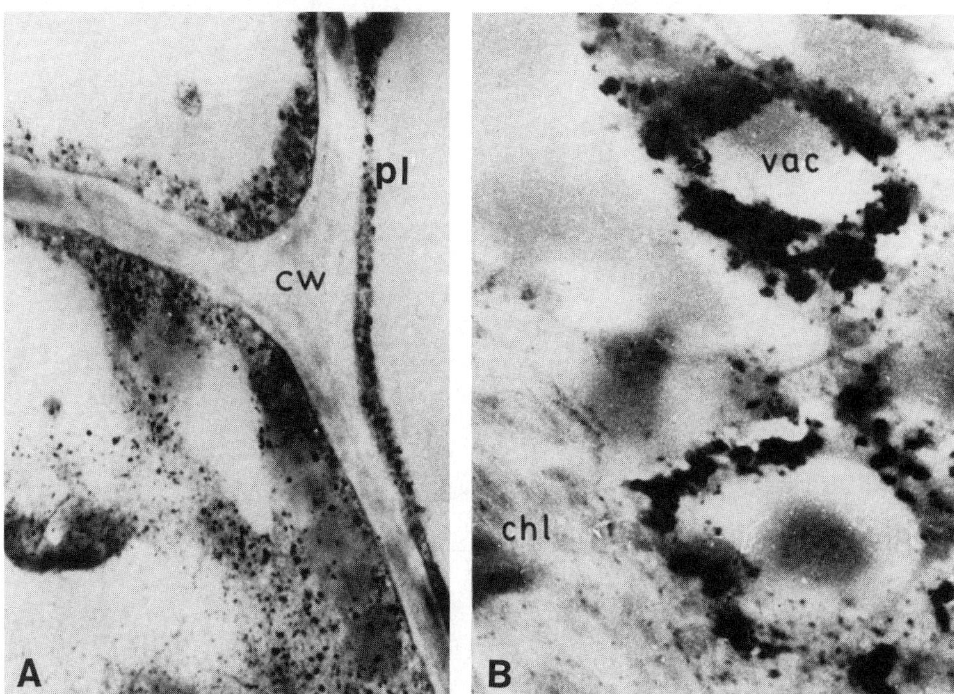

FIGURE 8. Adenylate cyclase activity in a male shoot apex of *Bryum argenteum* showing heavy staining along the (A) plasmalemma and (B) tonoplast. Sections were incubated in the presence of AMP-PNP and lightly stained with uranyl acetate. (Magnifications: A × 5,000; B × 12,000.) Abbreviations: cw — cell wall; pl — plasmalemma; vac — vacuole; chl — chloroplast. (From Bhatla, S. C. and Chopra, R. N., *Ann. Bot. (London)*, 54, 195, 1984. With permission.)

and not by other ATPases.[60] AMP-PNP is partially degraded by the enzyme to give rise to cAMP and imidodiphosphate (PNP). AMP-PNP has been used to detect adenylate cyclase activity in a variety of animal tissues by electron-histochemical means, since PNP precipitates with lead and forms an electron-dense marker.[60] Among higher plants, adenylate cyclase activity has been demonstrated by this technique in maize root tips,[61] pea hypocotyls,[62] and root nodules of *Alnus glutinosa*.[63] Subcellular localization of the cAMP-biosynthesizing enzyme has also been achieved in the shoot apices of the moss *Bryum argenteum* Hedw., using adenylyl imidodiphosphate as the substrate.[64] Considerable enzyme activity has been localized on the plasma membrane, tonoplast, and the membranous system of developing plastids (Figures 8 and 9). No activity appears to be associated with other subcellular membranes. Similar observations have been made in *Physcomitrium cyathicarpum*.[85] The observed activity of adenylate cyclase on the plastids in *Bryum* is in agreement with earlier detection of activity in the isolated intact chloroplasts of *Phaseolus vulgaris*.[65,66]

C. PURIFICATION AND CHARACTERIZATION OF CYCLIC NUCLEOTIDE PHOSPHODIESTERASE

Studies on the fate of cAMP in the chloronemal cells of *Funaria* indicate that only one tenth of the radioactive cAMP taken up by the cells remains as cAMP after 30 min of incubation and that the rest is degraded rapidly.[67] Initially the major product is adenosine, but later ADP and ATP are predominant. About 10 to 25% of the total radioactivity appears in 5′-AMP, whereas very little or none is detected in 3′-AMP.

The enzyme responsible for the degradation of cAMP has been characterized in animals,

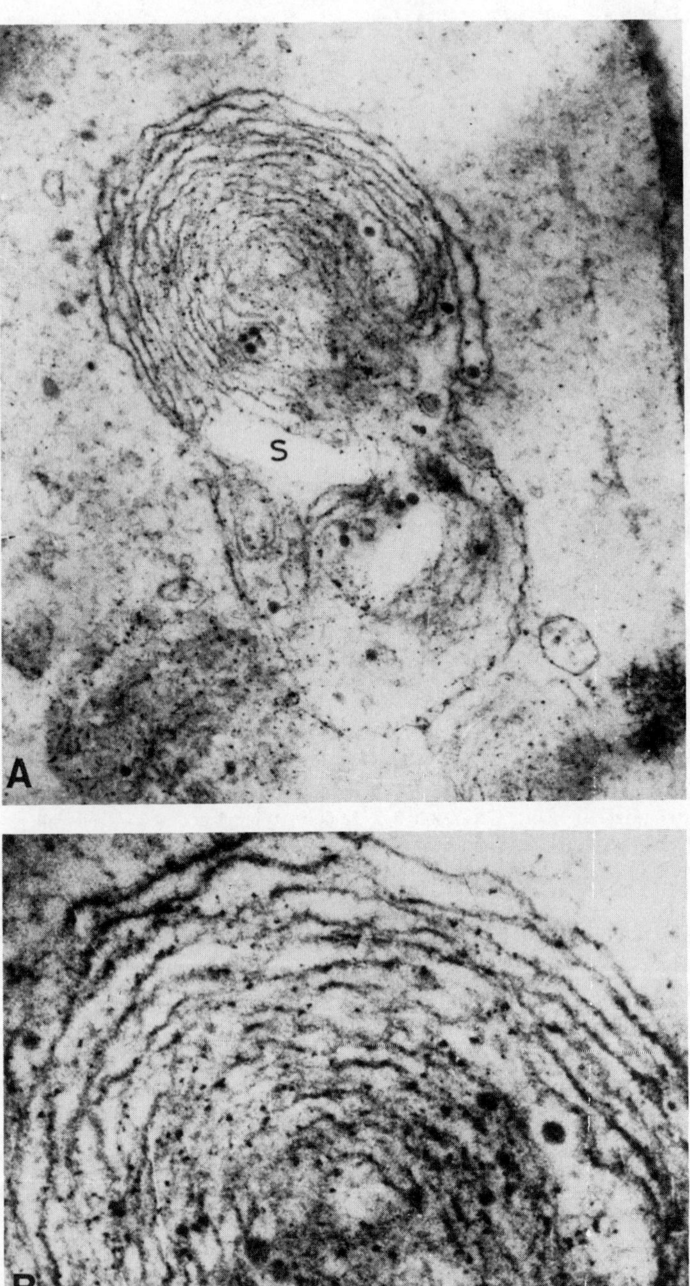

FIGURE 9. (A) Adenylate cyclase activity in a male shoot apex of *Bryum argenteum* showing heavy staining along the membranous system of a developing plastid. Sections were incubated in the presence of AMP-PNP and lightly stained with uranyl acetate. (Magnification × 6000.) (B) Enlarged view of a portion of A. (Magnification × 12,000.) Abbreviation: s — starch grain. (From Bhatla, S. C. and Chopra, R. N., *Ann. Bot. (London)*, 54, 195, 1984. With permission.)

and its activity has also been demonstrated in bacteria, cellular slime molds, yeast, and other fungi.[70] Among the autotrophic plants (other than bryophytes), phosphodiesterase activity has been demonstrated in a green alga (*Chlamydomonas reinhardtii*)[68] and in some higher plants.[69,70] Among bryophytes, activity of (cN)PDE has been observed in some taxa.[97] An extensive survey of phosphodiesterase activity in a variety of plants revealed that the properties of PDE obtained from a species of *Sphagnum* are similar to those of the enzyme present in animals and microorganisms.[97] When cAMP is incubated with extracts from this plant it is hydrolyzed into 5′-AMP, and the activity is higher at alkaline pH (8.5) than at acidic pH (5.4).

A cell-free chloronema homogenate of *F. hygrometrica*, upon differential centrifugation, shows the presence of about 30% of the total PDE activity in the 750 g pellet (particulate fraction) and about 40 to 50% of the activity in the 37,000 g supernatant (soluble fraction).[71,72] Similar results have been obtained with soybean suspension cultures.[73] The presence of (cN)PDE activity in both the particulate and the soluble fractions indicates the possible spatial distribution of the enzyme. Further work is required to determine whether (cN)PDE adheres to organelles or is present within them.

Analysis of the properties of the 86-fold-purified enzyme (obtained by DEAE-cellulose column chromatography) from the extract of chloronema cells of *Funaria* shows that PDE exists in multiple forms.[72,95] Three components can be distinguished:

1. A low-affinity component (K_m = 1.38 ± 0.37 m*M*) using 3′,5′-cAMP as substrate
2. A high-affinity component (K_m = 8.7 ± 2.8 μ*M*) for 3′,5′-cAMP
3. A component using 2′,3′-cAMP as substrate (K_m = 0.58 ± 0.11 m*M*)

The low-K_m component has a molecular weight of 85,000; its activity is zinc dependent and optimal at pH 5.5 to 6.5. It is insensitive to methylxanthines, but is stimulated by imidazole. It catalyzes the formation of both 3′-AMP and 5′-AMP from 3′,5′-cAMP. Further purification of this component of (cN)PDE, achieved by subjecting it to affinity column chromatography on a cAMP-agarose column, yields two peaks of enzyme activity.[72] The peak I enzyme has a pH optimum at 6.0, is insensitive to methylxanthines, and continues to be zinc dependent for its activity. It seems to produce both 3′-AMP and 5′-AMP as the degradation products. The peak II phosphodiesterase catalyzes the formation of only 5′-AMP from 3′,5′-cAMP, and its pH optimum is 7.5 to 8.0. Its activity is completely inhibited by methylxanthines and EDTA. The effect of EDTA can be antagonized almost totally by calcium and partially by magnesium and zinc. Thus, peak II phosphodiesterase activity (in the extract from chloronema cells of *Funaria*) has all the properties of PDE from animals, whereas the peak I activity largely resembles the nonspecific phosphodiesterase from higher plants.[70] So far, *F. hygrometrica* is the only photosynthetic plant in which both cAMP-specific and cAMP-nonspecific PDEs have been demonstrated.

D. UPTAKE AND LEACHING

In *Funaria* the uptake of [^{3}H]-cAMP exhibits a distinct dependence on the density of the culture.[67] Irrespective of cAMP concentration, cells at lower density (1 to 2 mg/ml) accumulate four to six times more radioactivity than cells at high cell density (10 mg/ml). However, at a given cell density the amount of cAMP taken up is directly proportional to the concentration of cAMP in the medium. Washing of chloronema cell suspensions of *Funaria* leads to a significant decrease in cAMP content.[55] Studies have indeed demonstrated the presence of cAMP in the culture medium.[74]

V. ETHYLENE

Involvement of ethylene in the regulation of transition from chloronema to caulonema in mosses has been established by the use of its precursor, 1-aminocyclopropane-1-carboxylic acid (ACC). ACC content and ethylene production in moss protonema coincide with progressing differentiation of chloronema to caulonema. Since caulonema differentiation in *Funaria* is regulated by auxins,[3,4,21] a correlation between auxin effect and the enhancement of ethylene formation may be more than a coincidence. Another effect of ethylene is its influence on senescence in higher plants.[75] Protonemata older than 25 days are characterized by the formation of so-called ''tmemata'' (empty cells formed by unequal division of cells, which separate the filaments). ACC can stimulate tmemata formation in younger cells.[53] Therefore, ethylene also appears to act as a kind of senescence hormone in the moss protonema.

A. DETERMINATION

So far ethylene production among mosses has been studied only in the protonema of *F. hygrometrica*.[80] Very young protonemata (5 days after sowing) produce a considerable amount of ethylene (as determined by GC) immediately after transplantation onto fresh medium. The production stops thereafter. The first outburst of ethylene in the very young protonema may be attributed to the production of ''wound ethylene'', caused by injuries to the filaments during transfer to fresh medium. A second increase occurs within ca. 21 days after transplantation and remains constant during further development. The ethylene produced during this period was around 1 nl C_2H_2 h^{-1} g^{-1} protein, the same order of magnitude as in higher plants (Table 1).

B. PRECURSORS OF ETHYLENE

In higher plants ethylene is synthesized from methionine via *S*-adenosylmethionine and ACC.[76] It appears that microorganisms can also use other substrates besides methionine as precursors for ethylene formation. Ethylene production has also been detected in several ferns.[77-79]

To obtain information about the precursors for ethylene synthesis in the protonema of *F. hygrometrica*, possible precursors (methionine, ACC, oxoglutarate, and glutamate) were provided in the nutrient medium individually.[80] Ethylene production by the moss protonemata raised on these media was significantly increased only by ACC. Methionine and glutamate had no effect, whereas ethylene production was slightly decreased with oxoglutamate. The effect of ACC together with observations on the ACC concentration in the tissue indicate that ethylene synthesis in the moss protonema occurs via the methionine-ACC pathway.

The protonema of *F. hygrometrica* produces more ethylene on IAA-containing medium.[80] When protonemata cultivated for 3 weeks are transplanted onto fresh medium containing IAA (10^{-6} *M*), the ACC content of the tissue analyzed 24 h later is nearly ten times higher (1.36 n*M* ACC mg^{-1} protein) as compared to the protonemata transplanted onto IAA-free medium (0.16 n*M* ACC mg^{-1} protein), indicating the similarity of ethylene synthesis in higher plants and the protonema of *Funaria*.

VI. ACETYLCHOLINE

A. DETERMINATION

Acetylcholine has been detected in a number of higher and lower plants.[81,86] Among mosses, so far only callus tissue regenerated from the seta of the sporophyte of the hybrid *F. hygrometrica* × *Physcomitrium pyriforme* has been used for the determination of this substance. The first evidence for the presence of acetylcholine in this tissue was obtained indirectly by means of pharmacological experiments on the heart of the frog (*Rana temporaria*) using a TLC fraction of the moss tissue extract cochromatographing with acetylcholine.[82] The heart reacts to the active fraction in the same way as it does to acetylcholine.

The bioassay procedure adopted by Hartmann[82] for the determination of acetylcholine is very sensitive but lacks specificity. The gas-liquid chromatography technique used by the same worker later,[83] however, gave conclusive proof for the presence of acetylcholine in the moss tissue (Table 1). The authenticity of the compound in the moss callus was shown by experiments as follows:

1. The retention time relative to propionylcholine (internal standard) was identical for acetylcholine from callus extract and authentic acetylcholine, and the addition of known amounts of acetylcholine to the extract resulted in a nearly quantitative recovery (88.7%) of the added acetylcholine.
2. In a sample of the tissue extract incubated with acetylcholinesterase (acetylcholine hydrolase, EC 3.1.1.7) no acetylcholine was detected.

B. PHYTOCHROME REGULATION

The physiological significance of acetylcholine in plants is not quite clear. It has been suggested that in the moss callus acetylcholine concentrations are related to phytochrome-mediated processes.[82] Hartmann and Kilbinger[83] determined acetylcholine content in the moss callus irradiated for 10 days with different radiation programs. The moss callus subjected to red light treatment contained 56 times as much acetylcholine per gram fresh weight as did the callus irradiated with red/far-red light. The acetylcholine content of the callus grown under standard white light conditions was only 27% of the amount found after red light irradiation. No acetylcholine was detected in the moss callus grown in the dark.

These investigations prove that acetylcholine concentration is regulated by phytochrome. The effect of red light on acetylcholine concentration could be reversed by exposure to far-red light. Jaffe and co-workers[81,84] suggested that acetylcholine is a highly specific hormone linked to the primary action of the far-red-absorbing form of phytochrome holochrome (P_{fr}), whereas others believe that acetylcholine interferes nonspecifically with a phytochrome-regulated process.[87] The experiments of Hartmann and Kilbinger[83] show that very little acetylcholine is formed with short-term irradiations. If the synthesis of acetylcholine is a long-term effect and depends on the duration of irradiation, it is difficult to explain the phytochrome-mediated fast responses.

VII. OTHER GROWTH REGULATORS

To date there are no confirmed reports on the existence of gibberellins in mosses. Abscisic acid has been detected recently in the protonema of *F. hygrometrica*.[100] In addition to the identified growth substances discussed so far, the moss protonema also contains and/or delivers to the substrate on which it grows some factors which play an important role in the normal growth and developmental processes. However, the chemical identities of these substances remain to be established. A very effective growth-regulating component is factor H (H = Hemmung). It was isolated from the culture medium of *F. hygrometrica* grown in liquid.[88,89] It can also be detected in the tissue extract of *Funaria* and has been found in several other mosses as well.[90] The main effect of factor H is a clear inhibition of caulonema growth, accompanied by a simultaneous stimulation of bud formation. Although this substance was highly purified[91] and physicochemically characterized, its definite chemical nature still remains unclear. Its properties, however, resemble an adenine derivative.[41] A substance quite similar to factor H, perhaps identical to it, has been found in the growth medium of *Bryum klinggraeffii*, in which it stimulates the formation of gemmae and inhibits overall growth.[92]

In the protonema of the moss *Ceratodon purpureus*, factors for branching (ramification [RF]) and cell division (CDF) were found.[93,94] The production of RF is light dependent, and it is not translocated to unilluminated parts of the protonema, whereas CDF can be translocated in both the directions in the filament. These factors appear identical to the factor H and F (F = Förderung) of *Funaria*.[42]

ACKNOWLEDGMENTS

Some of the unpublished observations cited in this review are from investigations undertaken by S.C.B. using a research grant from Stiftung VW, West Germany. The co-author (S.D.-B.) is grateful to C.S.I.R., New Delhi, for the award of Pool Officership.

REFERENCES

1. **Bhatla, S. C. and Bopp, M.,** The hormonal regulation of protonema development in mosses. III. Auxin-resistant mutants of the moss *Funaria hygrometrica* Hedw., *J. Plant Physiol.,* 120, 233, 1985.
2. **Bopp, M. and Bhatla, S. C.,** Hormonal regulation of development in mosses, in *Hormonal Regulation of Plant Growth and Development,* Vol. 2, Purohit, S. S., Ed., Agro-Botanical Publishers, Bikaner, India, 1985, 65.
3. **Johri, M. M. and Desai, S.,** Auxin regulation of caulonema formation in moss protonema, *Nature (London),* 245, 223, 1973.
4. **Lehnert, B. and Bopp, M.,** The hormonal regulation of protonema development in mosses. I. Auxin-cytokinin interaction, *Z. Pflanzenphysiol.,* 110, 379, 1983.
5. **Chopra, R. N.,** Some aspects of morphogenesis in bryophytes, in *Recent Advances in Cryptogamic Botany,* Bharadwaj, D. C., Ed., Palaeobotanical Institute, Lucknow, India, 1981, 190.
6. **Handa, A. K. and Johri, M. M.,** Cell differentiation by 3′,5′-cyclic AMP in a lower plant, *Nature (London),* 259, 480, 1976.
7. **Handa, A. K. and Johri, M. M.,** Involvement of cyclic adenosine 3′,5′-monophosphate in chloronema differentiation in protonema cultures of *Funaria hygrometrica, Planta,* 114, 317, 1979.
8. **Bopp, M.,** Hormones of the moss protonema, in *Bryophyte Development: Physiology and Biochemistry,* Chopra, R. N. and Bhatla, S. C., Eds., CRC Press, Boca Raton, FL, 1990, chap. 4.
9. **Ashton, N. W., Schulze, A., Hall, P., and Bandurski, R. S.,** Estimation of indole-3-acetic acid in gametophytes of the moss *Physcomitrella patens, Planta,* 164, 142, 1985.
10. **Bandurski, R. S. and Schulze, A.,** Concentration of indole-3-acetic acid and its derivatives in plants, *Plant Physiol.,* 60, 211, 1977.
11. **Ashton, N. W., Grimsley, N. H., and Cove, D. J.,** The isolation and physiological analysis of mutants of the moss *Physcomitrella patens*, using auxin and cytokinin resistant mutants, *Planta,* 144, 437, 1979.
12. **Ashton, N. W., Grimsley, N. H., and Cove, D. J.,** Analysis of gametophytic development in the moss, *Physcomitrella patens*, using auxin and cytokinin resistant mutants, *Planta,* 144, 427, 1979.
13. **Cove, D. J., Ashton, N. W., Featherstone, D. E., and Wang, T. L.,** The use of mutant strains in the study of hormone action and metabolism in the moss *Physcomitrella patens*, in *Proc. 4th John Innes Symp.*, Davies, D. R. and Hopwood, D. A., Eds., John Innes Charity, Norwich, England, 1980, 231.
14. **Klingbeil, I.,** Untersuchungen zum Auxinhaushalt an Mutanten von *Funaria hygrometrica,* Staatsexamensarbeit, Universität Heidelberg, Heidelberg, Federal Republic of Germany, 1982.
15. **Cove, D. J. and Ashton, N. W.,** The hormonal regulation of gametophytic development in bryophytes, in *The Experimental Biology of Bryophytes,* Dyer, A. F. and Duckett, J. G., Eds., Academic Press, London, 1984, 177.
16. **Bopp, M.,** Auxin in mosses: effects and occurrence, *Bryol. Times,* 44, 1, 1987.
17. **Jayaswal, R. K. and Johri, M. M.,** Occurrence and biosynthesis of auxin in protonema of the moss *Funaria hygrometrica, Phytochemistry,* 24, 1211, 1985.
18. **Weiler, E. W. and Ziegler, H.,** Determination of phytohormones in phloem exudate from tall species by radioimmunoassay, *Planta,* 152, 168, 1981.
19. **Markmann-Mulisch, U.,** Untersuchungen über die Rolle des Calciums bei der Cytokinin-Abhängigen Knospeninduktion und Entwicklung am Protonema der Laubmoose *Funaria hygrometrica* und *Polytrichum formosum,* Doktorarbeit, Universität Heidelberg, Heidelberg, Federal Republic of Germany, 1985.
20. **Sembdner, G., Gross, D., Liebisch, H.-W., and Schneider, G.,** Biosynthesis and metabolism of plant hormones, in *Encyclopedia of Plant Physiology,* Vol. 9, Macmillan, J., Ed., Springer-Verlag, Berlin, 1980, 281.

21. **Sood, S. and Hackenberg, D.,** Interaction of auxin, antiauxin and cytokinin in relation to the formation of buds in moss protonema, *Z. Pflanzenphysiol.*, 91, 385, 1979.
22. **Sharma, S., Jayaswal, R. K., and Johri, M. M.,** Cell-density-dependent changes in the metabolism of chloronema cell cultures. I. Relationship between cell density and enzymic activities, *Plant Physiol.*, 64, 154, 1979.
23. **Lee, T. T.,** Promotion of indoleacetic acid oxidase isoenzymes in tobacco callus cultures by indoleacetic acid, *Plant Physiol.*, 48, 56, 1971.
24. **Lee, T. T.,** Changes in indoleacetic acid oxidase isoenzymes in tobacco tissues after treatment with 2,4-dichlorophenoxyacetic acid, *Plant Physiol.*, 49, 957, 1972.
25. **Waldrum, J. D. and Davies, E.,** Subcellular localization of IAA oxidase in peas, *Plant Physiol.*, 68, 1303, 1981.
26. **Idzikowska, K. and Szweykowska, A.,** The ultrastructural aspects of the cytokinin-induced bud formation in *Ceratodon purpureus, Protoplasma,* 94, 41, 1978.
27. **Knoop, B.,** Development in bryophytes, in *The Experimental Biology of Bryophytes,* Dyer, A. F. and Duckett, J. G., Eds., Academic Press, London, 1984, 143.
28. **Rose, S. and Bopp, M.,** Uptake and polar transport of indoleacetic acid in moss rhizoids, *Physiol. Plant.*, 58, 57, 1983.
29. **Bopp, M. and Knoop, B.,** Regulation de la différentiation chez le protonéma des mousses, *Colloq. Bryol. Soc. Bot. Fr.*, 121, 145, 1974.
30. **Larpent-Gourgaud, M.,** Problemes poses par les echanges intercellulaires dans le protonema des bryophytes: mise en evidence des phenomenes de transport, *Colloq. Bryol. Soc. Bot. Fr.*, 121, 161, 1974.
31. **Overlach, U.,** Untersuchungen über den Stofftransport im Caulonema von *Funaria hygrometrica,* Diplomarbeit, Universität Heidelberg, Heidelberg, Federal Republic of Germany, 1973.
32. **Rubery, P. H. and Sheldrake, A. R.,** Carrier-mediated auxin transport, *Planta,* 118, 101, 1974.
33. **Rose, S., Eberhardt, I., and Bopp, M.,** Temperature dependent auxin efflux from moss protonema, *Z. Pflanzenphysiol.*, 109, 243, 1983.
34. **Rose, S., Rubery, P. H., and Bopp, M.,** The mechanism of auxin uptake and accumulation in moss protonemata, *Physiol. Plant.*, 58, 52, 1983.
35. **Wang, T. L., Cove, D. J., Beutelmann, P., and Hartmann, E.,** Isopentenyladenine from mutants of the moss, *Physcomitrella patens, Phytochemistry,* 19, 1103, 1980.
36. **Wang, T. L., Beutelmann, P., and Cove, D. J.,** Cytokinin biosynthesis in mutants of the moss *Physcomitrella patens, Plant Physiol.*, 68, 739, 1981.
37. **Wang, T. L., Horgan, R., and Cove, D. J.,** Cytokinins from the moss, *Physcomitrella patens, Plant Physiol.*, 68, 735, 1981.
38. **Bauer, L.,** Isolierung und Testung einer kinetinartigen Substanz aus Kalluszellen von Laubmoossporophyten, *Z. Pflanzenphysiol.*, 54, 241, 1966.
39. **Beutelmann, P.,** Untersuchungen zur Biosynthese eines Cytokinins in Calluszellen von Laubmoossporophyten, *Planta,* 112, 181, 1973.
40. **Beutelmann, P. and Bauer, L.,** Purification and identification of a cytokinin from moss callus cells, *Planta,* 133, 215, 1977.
41. **Klein, B.,** Versuche zur Analyse der Protonemaentwicklung der Laubmoose. IV. Der Endogene Faktor H und seine Rolle bei der Morphogenese von *Funaria hygrometrica, Planta,* 73, 12, 1967.
42. **Bopp, M.,** Development of the protonema and bud formation in mosses, *J. Linn. Soc. (Bot.),* 58, 305, 1963.
43. **Gerhäuser, D. and Bopp, M.,** Cytokinin oxidases in mosses, in Abstr. XIV Int. Bot. Congr., Berlin, West Germany, July 24 to August 1, 1987, 102.
44. **Wang, T. L., Futers, T. S., McGeary, F., and Cove, D. J.,** Moss mutants and the analysis of cytokinin metabolism, in *The Biosynthesis and Metabolism of Plant Hormones,* Crozier, A. and Hillmann, J. R., Eds., Cambridge University Press, London, 1984, 135.
45. **Eberle, T., Wang, T. L., Cook, S., Wells, B., and Weiler, E. W.,** Immunoassay and ultrastructural localization of isopentenyladenine and related cytokinins using monoclonal antibodies, *Planta,* 172, 289, 1987.
46. **Skoog, F. and Armstrong, D. J.,** Cytokinins, *Annu. Rev. Plant Physiol.*, 21, 359, 1970.
47. **Hall, R. H.,** Cytokinins as a probe in developmental processes, *Annu. Rev. Plant Physiol.*, 24, 415, 1973.
48. **Zachau, H. G., Dutting, D., and Feldman, H.,** Nucleotidsequenzen zweier Serinspezifischer Transfer-Ribonucleinsauren, *Angew. Chem.*, 78, 392, 1966.
49. **Chen, C.-M. and Hall, R.,** The biosynthesis of N^6-(Δ^2-isopentenyl) adenosine in the tRNA of cultured tobacco pith tissue, *Phytochemistry,* 8, 1695, 1969.
50. **Bopp, M. and Erichsen, U.,** Metabolism of the cytokinins in mosses, in *Metabolism and Molecular Activities of Cytokinins,* Guern, J. and Peaud-Lenoel, C., Eds., Springer-Verlag, Berlin, 1981, 105.
51. **Erichsen, U., Knoop, B., and Bopp, M.,** Uptake, transport and metabolism of cytokinin in moss protonema, *Plant Cell Physiol.*, 19, 839, 1978.

52. **Bopp, M. and Gerhäuser, D.,** Localization of cytokinin responses in the moss *Funaria hygrometrica, Biol. Plant.,* 27, 265, 1985.
53. **Bopp, M., Gerhäuser, D., and Kessler, U.,** On the hormonal system of mosses, in *Plant Growth Substances,* Bopp, M., Ed., Springer-Verlag, Berlin, 1986, 263.
54. **Bhatla, S. C. and Chopra, R. N.,** Biological significance of cyclic adenosine 3′,5′-monophosphate in mosses, *Physiol. Plant.,* 57, 383, 1983.
55. **Handa, A. K. and Johri, M. M.,** Cyclic adenosine 3′:5′-monophosphate in moss protonema, *Plant Physiol.,* 59, 490, 1977.
56. **Langen, T. A.,** Protein kinases and protein kinase substrates, in *Advances in Cyclic Nucleotide Research,* Vol. 3, Greengard, P. and Robison, G. A., Eds., Raven Press, New York, 1973, 99.
57. **Robison, G. A., Butcher, R. W., and Sutherland, E. W.,** *Cyclic AMP,* Academic Press, New York, 1971.
58. **Kaul, R. and Sachar, R. C.,** On the presence of adenosine 3′,5′-cyclic monophosphate in moss *(Funaria hygrometrica), Biochem. Biophys. Res. Commun.,* 104, 126, 1982.
59. **Hintermann, R. and Parish, R. W.,** Determination of adenylate cyclase activity in a variety of organisms: evidence against the occurrence of the enzyme in higher plants, *Planta,* 146, 459, 1979.
60. **Wagner, R. C. and Bitensky, M. W.,** Adenylate cyclase, in *Electron Microscopy of Enzymes,* Vol. 2, Hayat, M. A., Ed., D Van Nostrand, New York, 1973, 110.
61. **Al-Azzawi, M. J. and Hall, J. L.,** Cytochemical localization of adenyl cyclase activity in maize root tips, *Plant Sci. Lett.,* 6, 285, 1976.
62. **Hilton, G. M. and Nesius, K. K.,** Localization of adenyl cyclase in meristems of young pea hypocotyls, *Physiol. Plant.,* 42, 49, 1978.
63. **Gardner, I. C., McNally, S. F., and Scott, A.,** Electron histochemical localisation of the adenyl cyclase activity in the root nodules of *Alnus glutinosa* L., *Plant Sci. Lett.,* 16, 387, 1979.
64. **Bhatla, S. C. and Chopra, R. N.,** Subcellular localisation of adenylate cyclase in the shoot apices of *Bryum argenteum* Hedw., *Ann. Bot. (London),* 54, 195, 1984.
65. **Smith, C. J., Brown, E. G., Newton, R. P., Al-Najafi, T., and Edwards, M. J.,** Adenylate cyclase activity in higher plants, *Biochem. Soc. Trans.,* 5, 1351, 1977.
66. **Brown, E. G., Edwards, M. J., Newton, R. P., and Smith, C. J.,** Plurality of cyclic nucleotide phosphodiesterase in *Spinacea oleracea*: subcellular distribution, partial purification and properties, *Phytochemistry,* 18, 1943, 1979.
67. **Sharma, S. and Johri, M. M.,** Uptake and degradation of cyclic AMP by chloronema cells, *Plant Physiol.,* 69, 1401, 1982.
68. **Fischer, U. and Amrhein, N.,** Cyclic nucleotide phosphodiesterase of *Chlamydomonas reinhardtii, Biochim. Biophys. Acta,* 341, 412, 1974.
69. **Amrhein, N.,** The current status of cyclic AMP in higher plants, *Annu. Rev. Plant Physiol.,* 28, 123, 1977.
70. **Lin, P.P.-C.,** Cyclic nucleotides in higher plants?, in *Advances in Cyclic Nucleotide Research,* Vol. 4, Greengard, P. and Robison, G. A., Eds., Raven Press, New York, 1974, 439.
71. **Banerji, S. and Johri, M. M.,** Cyclic nucleotide phosphodiesterase with unusual properties from moss protonema, in *Abstr. Proc. Cell Biology Conf.,* University of Delhi, Delhi, India, 1978, 52.
72. **Sharma, S.,** Characterization of Cyclic AMP Phosphodiesterases in the Moss *Funaria Hygrometrica,* Ph.D. thesis, Tata Institute of Fundamental Research, Bombay, India, 1981.
73. **Brewin, N. J. and Northcote, D. H.,** Partial purification of a cyclic AMP phosphodiesterase from soyabean callus. Isolation of a non-dialysable inhibitor, *Biochim. Biophys. Acta,* 320, 104, 1973.
74. **Jayaswal, R. K.,** Cell Density Dependent Changes in Metabolism of Chloronema Cell Cultures of *Funaria hygrometrica,* M.Sc. thesis, University of Bombay, Bombay, India, 1979.
75. **Aharoni, N. and Lieberman, M.,** Ethylene as a regulator of senescence in tobacco leaf discs, *Plant Physiol.,* 64, 801, 1979.
76. **Lieberman, M.,** Biosynthesis and action of ethylene, *Annu. Rev. Plant Physiol.,* 30, 533, 1979.
77. **Edwards, M. E. and Miller, J. H.,** Growth regulation by ethylene in fern gametophytes. II. Inhibition of cell division, *Am. J. Bot.,* 59, 450, 1972.
78. **Elmore, H. W. and Whittier, D. P.,** Ethylene production and ethylene-induced apogamous bud formation in nine gametophytic strains of *Pteridium aquilinum* (L.) Kuhn, *Ann. Bot. (London),* 39, 965, 1975.
79. **Cookson, C. and Osborne, D. J.,** The stimulation of cell extension by ethylene and auxin in aquatic plants, *Planta,* 144, 39, 1978.
80. **Rowher, F. and Bopp, M.,** Ethylene synthesis in moss protonema, *J. Plant Physiol.,* 117, 331, 1985.
81. **Jaffe, M. J.,** Evidence for the regulation of phytochrome-mediated processes in bean roots by the neurohormone, acetylcholine, *Plant Physiol.,* 46, 768, 1970.
82. **Hartmann, E.,** Über den Nachweis eines Neurohormones beim Laubmooscallus und seine Beeinflussung durch das Phytochrom, *Planta,* 101, 159, 1971.

83. **Hartmann, E. and Kilbinger, H.,** Gas-liquid-chromatographic determination of light-dependent acetylcholine concentrations in moss callus, *Biochem. J.*, 137, 249, 1974.
84. **Yunghans, H. and Jaffe, M. J.,** Rapid respiratory changes due to red light or acetylcholine during the early events of phytochrome-mediated photomorphogenesis, *Plant Physiol.*, 49, 1, 1972.
85. **Chauhan, E.,** Ultracytochemical localization of adenylate cyclase in the placental region of *Physcomitrium cyathicarpum, Cryptogamie Bryol. Lichenol.*, 8, 189, 1987.
86. **Fluck, R. A. and Jaffe, M. J.,** The acetylcholine system in plants, in *Commentaries in Plant Sciences*, Smith, H., Ed., Pergamon Press, Oxford, 1976, 119.
87. **Tanada, T.,** On the involvement of acetylcholine in phytochrome action, *Plant Physiol.*, 49, 2070, 1972.
88. **Bopp, M.,** Versuche zur Analyse von Wachstum und Differenzierung des Laubmoosprotonemas, *Planta*, 53, 178, 1959.
89. **Bopp, M. and Klein, B.,** Versuche zur Analyse der Protonemaentwicklung der Laubmoose. I. Endogene Wachstumsregulatoren in Protonema von *Funaria hygrometrica, Port. Acta Biol. Ser. A*, 7, 95, 1963.
90. **Kockel, H.,** Vorkommen und Wirkung des Faktors H bei verschiedener Laubmoosprotonemen, Diplomarbeit, Universität Hannover, Hannover, Federal Republic of Germany, 1967.
91. **von der Eltz, H.,** Nativer Morphoregulator (Faktor H) aus Moosprotonemen, Diplomarbeit, Universität Heidelberg, Heidelberg, Federal Republic of Germany, 1975.
92. **Rawat, M. S. and Chopra, R. N.,** Production of a morphoregulatory substance by secondary protonema of *Bryum klinggraeffii, Z. Pflanzenphysiol.*, 78, 372, 1976.
93. **Larpent-Gourgaud, M. and Aumâitre, M.-P.,** Intercellular exchanges and morphogenesis of protonema of *Ceratodon purpureus* Hedw., *Z. Pflanzenphysiol.*, 83, 467, 1977.
94. **Larpent-Gourgaud, M. and Aumâitre, M.-P.,** Action de l'orientation et de la qualité de la lumière sur le dévelopment des protonémas de Bryales, *Experientia*, 33, 1601, 1977.
95. **Sharma, S. and Johri, M. M.,** Partial purification and characterization of cyclic AMP phosphodiesterase from *Funaria hygrometrica, Arch. Biochem. Biophys.*, 217, 87, 1982.
96. **Cove, D. J.,** The role of cytokinin and auxin in protonemal development in *Physcomitrella patens* and *Physcomitrium sphaericum, J. Hattori Bot. Lab.*, 55, 79, 1984.
97. **Amrhein, N.,** Cyclic nucleotide phosphodiesterases in plants, *Z. Pflanzenphysiol.*, 72, 249, 1974.
98. **Bhatla, S. C.,** unpublished data.
99. **Kaminek,** personal communication with M. Bopp.
100. **Atzorn, A. and Bopp, M.,** unpublished data.

Chapter 6

CALCIUM AND CYTOKININ IN MOSSES

M. J. Saunders

TABLE OF CONTENTS

I. INTRODUCTION

Initiation of cell division in plants is under the control of a class of compounds called cytokinins, N^6-substituted adenines (see Reference 1 and references cited therein). I have recently postulated that these hormones may exert at least part of their effect on cells by modulating the intracellular calcium ion concentration ($[Ca^{2+}]_i$) in both a temporal and spatial manner within cells. Two calcium-dependent systems may be envisioned to be working in concert to effect physiological change, i.e., via calmodulin (CaM) activation and subsequent regulation of Ca/CaM-dependent protein kinases[2,3] and via production of the phosphatidylinositol (PI)-derived messengers inositol trisphosphate (IP_3) and diacylglycerol (DG, which activates protein kinase C).[4-10] Much of the evidence for the extension of these models to plant systems has come from research on cytokinin-induced cell division and bud formation in target caulonema cells of mosses[11-23] and research on cytokinin effects on calcium status of other cytokinin-responsive systems.[24]

II. CALCIUM AS A SECOND MESSENGER IN PLANTS

There is increasing evidence that in plant cells Ca^{2+} acts as a second messenger,[10,25-27] that a calcium/CaM-dependent system is operational,[28] that hormonal expression may be mediated by CaM,[13,29-31] and that calcium and CaM can regulate protein phosphorylation.[32-39] There have also been reports of cytokinin-dependent protein phosphorylation *in vivo*[40] in crown gall transformed cells[41] and of phosphorylation of cytokinin-binding proteins in wheat germ.[42]

There is less evidence for the presence of the PI cycle products IP_3 and DG in plants, although some initial evidence is beginning to accumulate.[9,43-47] The presence of a putative protein kinase C in plants has been described recently by Elliott and Skinner[24] and Tung et al.[48] However, specific phosphorylation of tyrosine residues in plants has not been seen, which would lead one to propose the existence of a protein kinase in plants akin to the Rous sarcoma virus *src* oncogene product that is a model system for stimulation of animal growth and development.[41,49]

III. CALCIUM AND MITOSIS

To understand how a quiescent cell can translate a rise in Ca^{2+} into a series of events culminating in mitosis, it is necessary to look at the regulation of Ca^{2+}.[50-54] Free Ca^{2+} levels within the cytosol of resting cells are maintained between 0.1 and 0.01 μM.[53,55] Weber[56] and Kretsinger[57] have postulated that this low level of free Ca^{2+} evolved very early as a necessary consequence of the high phosphate levels needed for energy metabolism via adenosine triphosphate (ATP); Ca^{2+} is extruded or sequestered to avoid calcium phosphate precipitation. Free Ca^{2+} levels are maintained at <0.1 μM via active sequestering by the mitochondria and a membrane reticulum and via active extrusion by plasma membrane Ca^{2+} ATPases. In addition, Ca^{2+} is bound to inner membrane surfaces and to intracellular proteins.[54] Because of the very low $[Ca^{2+}]_i$ the introduction of a relatively small number of ions can elicit a large increase in concentration. Cells have exploited this situation and use fluxes of Ca^{2+} to trigger events and/or transmit information. Besides existing at a low internal concentration, calcium is uniquely suited as a metabolic trigger since its weak polarizing power enables it to fit in irregular binding sites such as those in complex molecules like proteins.[51]

A rise in cytoplasmic free Ca^{2+} may have direct effects on membrane structure and certain Ca^{2+}-sensitive proteins, but the bulk of regulation of this ion occurs via the ubiquitous Ca^{2+}-dependent regulatory protein, CaM. Increasing intracellular Ca^{2+} increases binding of

Ca^{2+} to CaM; Ca^{2+} binding causes a conformational change in CaM to an active configuration and is followed by the binding of the Ca^{2+}-CaM complex to an enzyme, causing its activation.[58-62] CaM has been isolated from plants[63-65] and has been shown to regulate nicotinamide adenine dinucleotide (NAD) kinase in both plants[32] and fertilized sea urchin eggs[66] and to activate plant microsomal Ca^{2+} uptake.[67]

The role of CaM in the regulation of the cell cycle and mitosis remains to be established. CaM levels in cultured animal cells double in late G_1 or early S phase and do not appear to fluctuate during the rest of the cycle.[68] CaM has been localized in the spindle, and a role in microtubule depolymerization during chromosome motion to the poles has been postulated.[69-71] Injection of the Ca^{2+}-CaM complex into *Xenopus* oocytes can trigger meiosis in the absence of the maturation-inducing hormone.[72]

The effects of calcium on a cell may be dependent on the state of the cell and/or the duration of the signal. Triggering the common development event of mitosis may occur in some cells as the result of a Ca^{2+} transient; in others a sustained Ca^{2+} elevation is necessary. In some cells Ca^{2+} fluxes during mitosis appear to be regulated at the plasma membrane, and in others regulation may take place at endomembrane sites. Citing Ca^{2+} as a universal developmental trigger may appear to be a simplistic approach at first glance; however, the cell itself lends complexity to this single simple message by predetermined multiple sites for action.

IV. EVIDENCE FOR CALCIUM AND CYTOKININ RESPONSE COUPLING IN PLANTS

Current models for cellular response to external stimuli share common features with four main components: (1) there is a receptor-mediated interaction between the stimulus and (2) the activation of a secondary messenger which (3) subsequently has an amplifying or cascade effect within the cell, leading to (4) a physiological response.[73] If we look at cytokinin stimulation of cell division in plants as fitting this generalized model,[12,13,18] we can see that certain portions of the model have more evidence to sustain them than others. First, very little is known about cytokinin-binding proteins (receptors), although several laboratories are working on their characterization.[42,74-77] The relationship of these binding proteins to a physiological response is unclear.

There is more evidence concerning the second part of the model, the activation of a second messenger in response to cytokinin. We have shown in our laboratory that intracellular Ca^{2+} increases after cytokinin treatment.[18] Earlier research also demonstrated that Ca^{2+} is involved in cytokinin responses. The initial research in this area, by LeJohn and co-workers,[78-80] was on calcium transport in the water mold *Achlya*. *Achlya* is a filamentous coenocytic fungus that grows only in the presence of Ca^{2+}. Cytokinins stimulate Ca^{2+} release from a glycoprotein localized on the surface and enhance uptake of the ion into the cells.[78] This Ca^{2+} transport can be antagonized by Mg^{2+}.[79] The transport system consists of two components: a low-molecular-weight Ca^{2+}-binding glycopeptide present in the cell wall matrix and a Ca^{2+}-uptake component associated with the cell membrane. Cytokinin appears to act on both portions of the system; in osmotically shocked cells, without the glycopeptide, cytokinin still enhances Ca^{2+} uptake.[80]

Kinetin and Ca^{2+} cause a synergistic increase in ethylene production by mung bean hypocotyl segments. Tracer experiments show that kinetin greatly increases the uptake of $^{45}Ca^{2+}$ after 6 h of incubation.[81] A similar synergism between Ca^{2+} and cytokinin occurs in the delay of senescence in *Zea* leaf discs.[82] Apple leaves sprayed with cytokinin show an accumulation of $^{45}Ca^{2+}$.[83] In *Xantheum* cotyledons, dry weight gain is stimulated twofold by cytokinin; in cytokinin plus 1 to 10 m*M* $CaCl_2$ the increase in fresh weight is threefold.[84] Ralph et al.[40] found that Ca^{2+} will substitute for cytokinin in the leaf disc expansion assay for cytokinins using Chinese cabbage leaves and that cytokinins and Ca^{2+} affect membrane protein phosphorylation. Elliott[29,85] reported that the CaM inhibitor trifluoperazine (600 μ*M*) blocked cytokinin-induced betacyanin synthesis in *Amaranthus* cotyledons. Since CaM is activated by increases in [Ca^{2+}], she postulates that a rise in calcium is an intermediary in this process. Calcium at high concentrations (1 to 5 m*M*) inhibits betacyanin synthesis, as does the calcium ionophore A23187. However, pretreatment with ethyleneglycol-bis-(β-aminoethyl ether)-*N,N'*-tetraacetic acid (EGTA) to deplete the cell wall of calcium allows a stimulatory effect of both A23187 and increased Ca^{2+}.[24] A plasmalemma-enriched membrane preparation from soybean hypocotyls has been isolated that contains an ATP-dependent Ca^{2+} pump that is cytokinin modulated. The membranes isolated from meristematic regions have greater cytokinin sensitivity than those from mature regions.[86] These results of experiments on higher plants and fungi indicate that cytokinin may exert at least a portion of its activity by modulating the intracellular calcium concentration.

There are also data that substantiate the third part of the model, cellular amplification of the signal. Evidence for the mediation and amplification of the cytokinin-induced calcium increase by CaM has been provided by Elliott et al.[29,30] and Saunders and Hepler.[13] Activation of a calcium-dependent protein kinase after cytokinin treatment has not been described, but protein phosphorylation has been described by Ralph et al.[40] Two laboratories have presented initial evidence that IP_3 levels rise immediately after cytokinin treatment, implicating the PI cycle in the amplification response.[43,44]

The final part of the model, physiological response (i.e., cell division after cytokinin treatment), has been well documented[1,18] and will not be reviewed here.

It is clear that although there is evidence for the components of a receptor-mediated triggering system in plants, much of the evidence is fragmentary, and within any given system not all the components have been integrated (i.e., external stimulus, receptor binding, second messenger, amplification, and cellular response). I believe that the model system that has the most evidence at this time for the existence of all the components and the best potential for research on cellular signaling in plants is the cytokinin stimulation of asymmetrical cell division (bud formation) in the moss *Funaria hygrometrica*.

V. CALCIUM AND CYTOKININ IN *FUNARIA*

A. MORPHOLOGICAL RESPONSE TO CYTOKININ

The *Funaria* protonema grows as a filamentous mat of cells composed of two cell types: (1) chloronema cells with large chloroplasts and transverse cross walls; and (2) caulonema cells with small, spindle-shaped chloroplasts and oblique cross walls. The latter are the target cells for bud induction by exogenous cytokinin. The elongated target cells respond to cytokinin by a localized swelling at the distal end followed by nuclear migration to that region after 20 to 22 h of treatment. By 24 h of cytokinin treatment a small initial cell has been cut off from the original target cell. This initial cell divides in three planes, producing a mass of cells termed a bud. The ultrastructural details of these morphological changes are described by Conrad et al.[22]

B. EVIDENCE FOR ACTIVATION OF Ca^{2+} CHANNELS AND INCREASE IN INTRACELLULAR Ca^{2+} AFTER CYTOKININ TREATMENT

Previous research in my laboratory has produced evidence which substantiates the hypothesis that cytokinin-induced cell division in *Funaria* is calcium mediated. Increases in membrane-associated Ca^{2+} after cytokinin treatment were detected using chlorotetracycline (CTC) and correlated with the migrating nucleus, the site of the initial asymmetric division, and the rapidly dividing cells of the bud.[11,87] In addition to the use of CTC as a calcium indicator, localized distribution of Ca^{2+} in caulonema tip cells has been documented using proton microprobe and electron spectroscopic imaging techniques.[87]

Artificial induction of a rise in free calcium was accomplished using the Ca^{2+} ionophore A23187, which stimulates the initial division in the absence of cytokinin.[12] Prevention of a natural rise in [Ca^{2+}] after hormone treatment has been explored using Ca^{2+} antagonists in the presence of cytokinin. The Ca^{2+} transport inhibitors lanthanum, D 600, and verapamil and the intracellular Ca^{2+} antagonist TMB-8 block bud formation; extracellular Ca^{2+} is essential because blocking uptake with Ca^{2+}-transport inhibitors blocks both nuclear migration and subsequent division.[13] It is of interest that the buds formed in the presence of low concentrations of lanthanum and optimum cytokinin often reverted to filamentous growth.

Conrad and Hepler[23] extended the studies of the role of voltage-dependent Ca^{2+} channels during bud formation using dihydropyridines (DHP) that either stimulate or block movement of Ca^{2+} through potential-sensitive channels. DHP agonists stimulate bud initials in the absence of cytokinins, while DHP antagonists inhibit budding in the presence of optimal cytokinin. The authors suggest that one of the initial steps in the mode of action of cytokinin-induced budding is the increase of Ca^{2+} entry in target cells through voltage-dependent channels.

Ionic currents around cytokinin-treated cells have been explored using a nonintrusive vibrating probe.[19] These results indicate that (1) untreated caulonema cells have maximum inward current at the nuclear region; (2) addition of cytokinin induces an increase in inward current along the length of target cells that subsequently shifts to distal ends (presumptive bud site); and (3) after establishment of the growth zone at the presumptive bud site, inward current falls. This current has a Ca^{2+} component, since current falls to zero after treatment with Gd^{3+} (a competitive Ca^{2+} uptake inhibitor). Determination of the presumptive bud site is microfilament and not microtubule dependent. In the presence of microfilament inhibitors, buds form over the initial zone of high inward current, the nuclear region. Microtubule inhibitors do not affect determination of the bud site.

Free Ca^{2+} (determined from Indo 1 and Fura 2 fluorescence)[88] rises homogeneously within target cells immediately after cytokinin treatment, but after the establishment of a bulging growth zone free Ca^{2+} is spatially nonuniform (higher at the presumptive bud site) and decreases in cells of the developing bud.[14,18] There are problems with the use of these probes, however, and further research must be done to establish the amount of interference from fluorescence associated with the endomembranes that seem to concentrate the probes.

TMB-8 blocks bud formation by blocking release of Ca^{2+} from intracellular stores in *Funaria* (ascertained by monitoring CTC fluorescence). In related experiments on the effect of TMB-8 on cross wall formation it was demonstrated that TMB-8 kept phragmoplast microtubules polymerized and inhibited vesicle fusion within the plate and with the parental plasma membrane by keeping the intracellular [Ca^{2+}] low.[89]

CaM inhibitors do not block the determination of the presumptive bud site or the nuclear migration, but do block the transition of the nucleus from G_2 into M.[83] Nuclei will migrate into the region of the outgrowth at the distal end of target cells and remain there without dividing.[13]

Methylxanthines can block the bud-inducing effects of cytokinins. However, used alone they stimulate cell division in caulonema cells in a dose-dependent manner and also stimulate greening, creating chloronema cells.[90] Subsequent experiments have shown that Ca^{2+} uptake is essential for this action and that dibutyryl cyclic AMP can mimic the response. A possible mechanism of action is an increase in cyclic AMP levels as methylxanthines block phosphodiesterase activity.[91]

The relationship between cytokinin and the calcium status of the tissue for developmental response has been described by other researchers working on mosses. Bopp,[16] Cove and Ashton[17], and Spiess et al.[15] found that the number of cytokinin-induced buds reflected the calcium concentration present in the medium. Markmann-Mulisch and Bopp[21] recently reviewed the evidence for the role of Ca^{2+} as a cytokinin effector in mosses. In this review they differentiate between the initial stages of bud formation and the subsequent morphological changes after the initial outgrowth. They conclude that Ca^{2+} does play a role in the initial steps, but they prefer to term that portion of the developmental process "side branch formation". However, since the cytokinin concentration used in the cited experiments is sufficient to cause all subsequent outgrowth of target cells to develop into buds, this is probably an unnecessary discriminatory point. Brandes and Kende[92] showed many years ago that if the cytokinin is removed from developing buds of *Funaria* they will revert to filamentous growth. There is a delicate balance between bud formation and filamentous growth controlled by the cytokinin and Ca^{2+} concentration that can undergo developmental switching if either the cytokinin or the Ca^{2+} concentration falls below the threshold level.

C. EVIDENCE FOR THE PHOSPHATIDYLINOSITOL CYCLE

Preliminary investigations into the role of IP_3 in cytokinin-induced cell division have indicated that an increase in IP_3 can be detected 30 s after cytokinin treatment.[43] Protonemata were prelabeled with ^{32}P for 2 h and then incubated in 1 m*M* benzyladenine (BA) or in nutrient medium alone as a control. Tissue was then homogenized and extracted in a chloroform-methanol mixture following the procedure of Hajra et al.[93] The chloroform fraction was dried under argon, then redissolved in 1 ml chloroform and stored at $-20°C$. IP_3 was isolated from the methanol fraction by anion exchange chromatography on Bio-Rad® AG1-X8 Econo-Columns in the formate form, following the procedure of Berridge et al.[94] After evaporating most of the methanol under a stream of argon, the fraction was placed on a 1 ml column and free inositol was eluted with 10 ml distilled water. The phosphate esters were then eluted sequentially using 5 m*M* disodium tetraborate/60 m*M* sodium formate (for glycerophosphoinositol), 0.1 *M* formic acid/0.2 *M* ammonium formate (for inositol 1-phosphate), 0.1 *M* formic acid/0.4 *M* ammonium formate (for inositol 1,4-bisphosphate), and 0.1 *M* formic acid/0.1 *M* ammonium formate (for inositol 1,4,5-trisphosphate). IP_3 fractions were then assayed for ^{32}P content in a scintillation counter. The earliest a rise in IP_3 can be detected is 30 s after BA treatment.

Conrad[20] examined the role of protein kinase C by analyzing the effects of the tumor-promoting phorbol ester 12-*O*-tetradecanoylphorbol 13-acetate (TPA), in the presence and absence of the Ca^{2+} ionophore A23187, on bud formation in *Funaria*. She also examined the effects of 5-hydroxytryptamine (5-HT) and neomycin, agents that stimulate or inhibit phosphatidylinositol 1,4-bisphosphate (PIP_2) hydrolysis, respectively, in an attempt to raise or lower the production of IP_3 and DG artificially. When TPA or A23187 is used alone there is a slight stimulation of budding. However, when used together there is a dramatic increase in the number and development of buds, indicating that the receptor-mediated transducing effects for bud initiation can be bypassed by drugs activating the Ca^{2+}/CaM and protein kinase C branches of the Ca^{2+} messenger system. In the presence of 5-HT the number of target caulonema cells increased, but the buds that were formed did not develop normally and often reverted to filamentous growth. Neomycin only stimulated budding in the presence of cytokinin and had no effect when the hormone was absent. Conrad concludes that long-term cytokinin effects are capable of being mediated through DG alone and, therefore, are not dependent on IP_3-induced Ca^{2+} release.

D. CYTOKININ RECEPTORS

Affinity-purified polyclonal antibodies to BA riboside linked to bovine serum albumin have been produced that show cross-reactivity to cytokinin. *Funaria* cells are prepared for immunocytochemical processing using techniques similar to those developed by Lloyd's group for immunocytochemical staining of *Physcomitrella patens* cells.[95] *Funaria* cells are fixed and attached to glass cover slips, and the cell walls are partially digested in Driselase. They are then incubated in BA or BA riboside, incubated in primary antibody and fluores-

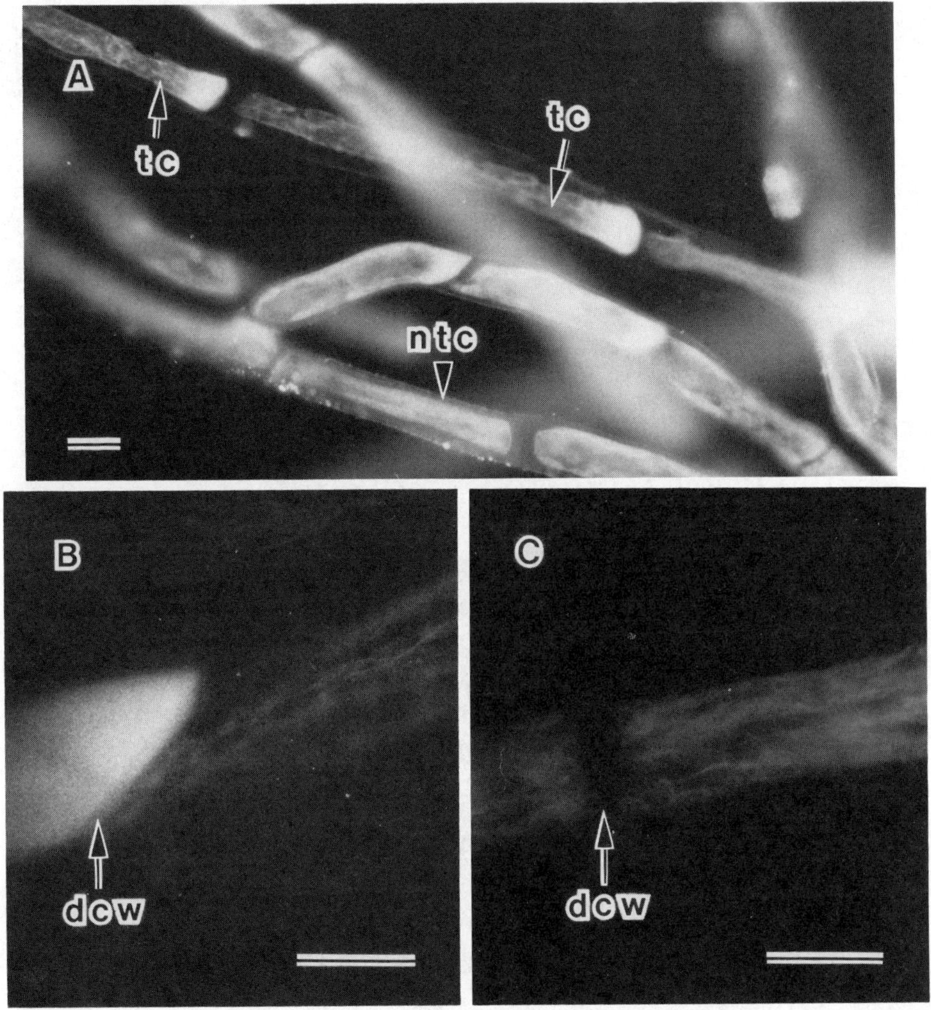

FIGURE 1. Fluorescence micrographs of immunocytological localization of benzyladenine riboside binding in *Funaria*. (A) Note the bright fluorescence at distal ends of target caulonema cells (tc) as contrasted with nontarget chloronema cells (ntc). (B) Higher magnification of zone of bright fluorescence in a target caulonema cell near the distal cross wall (dcw). (C) Higher magnification of fluorescence near the distal cross wall (dcw) in a chloronema cell. (Bars = 50 μm.)

cently tagged second antibody. Cytokinin binding or uptake is localized to the distal end of target cells. Very little binding is seen in nontarget cells (Figure 1). Similar results are obtained using commercially available monoclonal antibodies to dihydrozeatin riboside using a zeatin preincubation before the immunocytochemistry. If the cells are labeled immunocytochemically for microtubules using a commercially available antibody to B tubulin, an initial depolymerization of microtubules is seen immediately after cytokinin treatment (Figure 2). We hypothesize that this may be an indirect indication of a change in the cytoplasmic free calcium status and that microtubule depolymerization is involved in both nuclear migration and the extensive cytoplasmic rearrangements seen in target caulonema cells after hormone treatment. A similar result has also been described after cytokinin treatment in *Physcomitrella patens*.[96]

E. PROTEIN PHOSPHORYLATION PATTERNS

Protein patterns and protein phosphorylation in chloronema and caulonema cells have

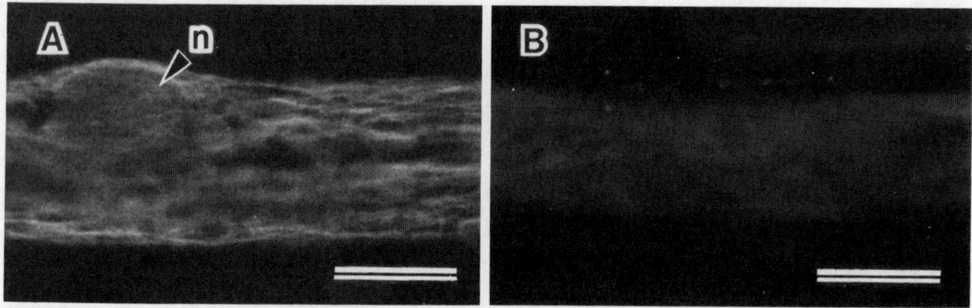

FIGURE 2. Fluorescence micrographs of *Funaria* caulonema cells of immunocytological localization of tubulin. (A) Note microtubules in cell and nonstained nucleus (n) before cytokinin treatment; (B) note loss of fluorescence labeling of microtubules in cell 10 min after treatment with 1 μ*M* BA. (Bar = 50 μm.)

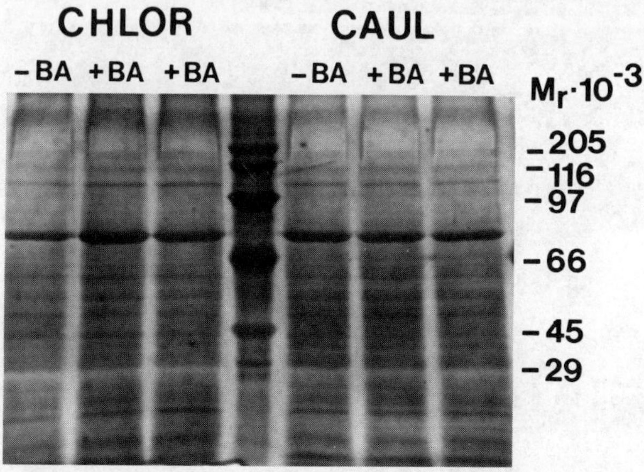

FIGURE 3. SDS PAGE gel (12%) of *Funaria* chloronema and caulonema cells with and without a 1-h incubation in benzyladenine (BA). No changes in protein synthesis are detectable.

been analyzed before and after cytokinin treatment (Figures 3 and 4). Protonemata were separated into chloronemata and caulonemata, preincubated in ^{32}P (50 μCi/ml phosphate-free Laetsch's medium) for 4 h ± 20 μ*M* BA for 1 h. Homogenates were prepared from the labeled cells using a Dounce homogenizer and 440 μl of 10 m*M* phosphate buffer (pH 6.2, containing 5 mm EGTA, 100 m*M* sodium fluoride, and 50 μg/ml leupeptin) on ice. Cell debris was removed from the homogenate by centrifugation. Lipid was removed by Freon extraction and centrifugation. Protein was precipitated with 20% trichloroacetic acid, washed with acetone, and lyophilized. Sodium dodecyl sulfate polyacrylamide gel electrophoresis (SDS PAGE) was performed according to Laemmli[97] in a Bio-Rad® mini gel apparatus using 12% acrylamide gels. Gels were silver stained, autoradiographed, and analyzed using an LKB® laser densitometer. No visible changes in the synthesis of proteins could be observed on these one-dimensional gels. However, a 50-kDa and a 47-kDa protein were phosphorylated only in caulonema cells treated with BA.

This observed rise in intracellular calcium may trigger cell division in *Funaria* by affecting the phosphorylation/dephosphorylation of proteins. Two calcium-dependent systems may be envisioned to be working in concert to effect these changes: i.e., via CaM activation and subsequent regulation of Ca/CaM-dependent protein kinases and via production of the phosphatidylinositol-derived messengers inositol trisphosphate and diacylglycerol.

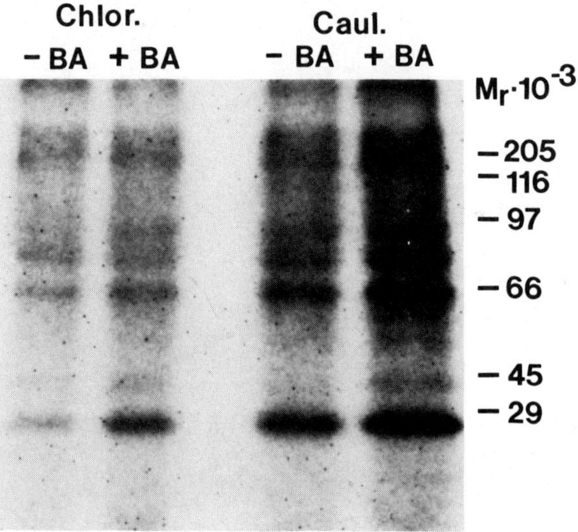

FIGURE 4. Autoradiograph of SDS PAGE gel of ^{32}P-labeled proteins in chloronema and caulonema cells of *Funaria* with and without a 1-h incubation in 1 μM benzyladenine (BA). Note the phosphorylation of two proteins in the BA-treated caulonema cells with apparent molecular weights of 50 and 43 kDa.

VI. MODEL FOR CYTOKININ-INDUCED CELL DIVISION IN MOSSES

Incorporating the results from experiments on *Funaria* and other mosses, the following model is proposed for cytokinin-induced cytomorphogenesis (Figure 5).

1. In the absence of cytokinin, moss protonemata grow as filamentous mats consisting of three cell types — elongating and actively dividing tip cells, green nondividing cells (chloronemata), and cytokinin-target cells (caulonemata). A cyclic AMP-dependent pathway is responsible for chloronemata formation. Because of the tip-growing, filamentous nature of the system a molecular polarity must exist within the caulonema cells. Localization of cytokinin receptors on the plasma membrane may be determined by this gradient and be clustered at the distal end. Open Ca^{2+} channels show maximal concentration over the nucleus, which lies midway along the long axis, next to the side wall anchored by a web of microtubules and arrested in G_2 of the cell cycle. The caulonema cell before cytokinin treatment is partially polarized, but in a quiescent state.
2. Addition of cytokinin (CK) presumably results in its binding to hormone receptors (R). This affects Ca^{2+} channel proteins and opens closed channels along the length of the cell. The membrane permeability to Ca^{2+} increases, and the ion moves passively and homogeneously into the cell down its steep concentration gradient.
3. Calcium influx becomes localized at distal ends of target cells as channels accumulate there by a microfilament-based process. Ca^{2+} influx would thus be localized to the distal end, whereas the efflux pumps may be either randomized or concentrated at the proximal end of the cell. Ca^{2+} permeability may also affect other ion transport and the membrane potential. Intracellular $[Ca^{2+}]$ would now increase within the cell in a gradient that is highest at the distal end. Because of active sequestering by mitochondria and intracellular membranes and active extrusion at the plasma membrane, the ability

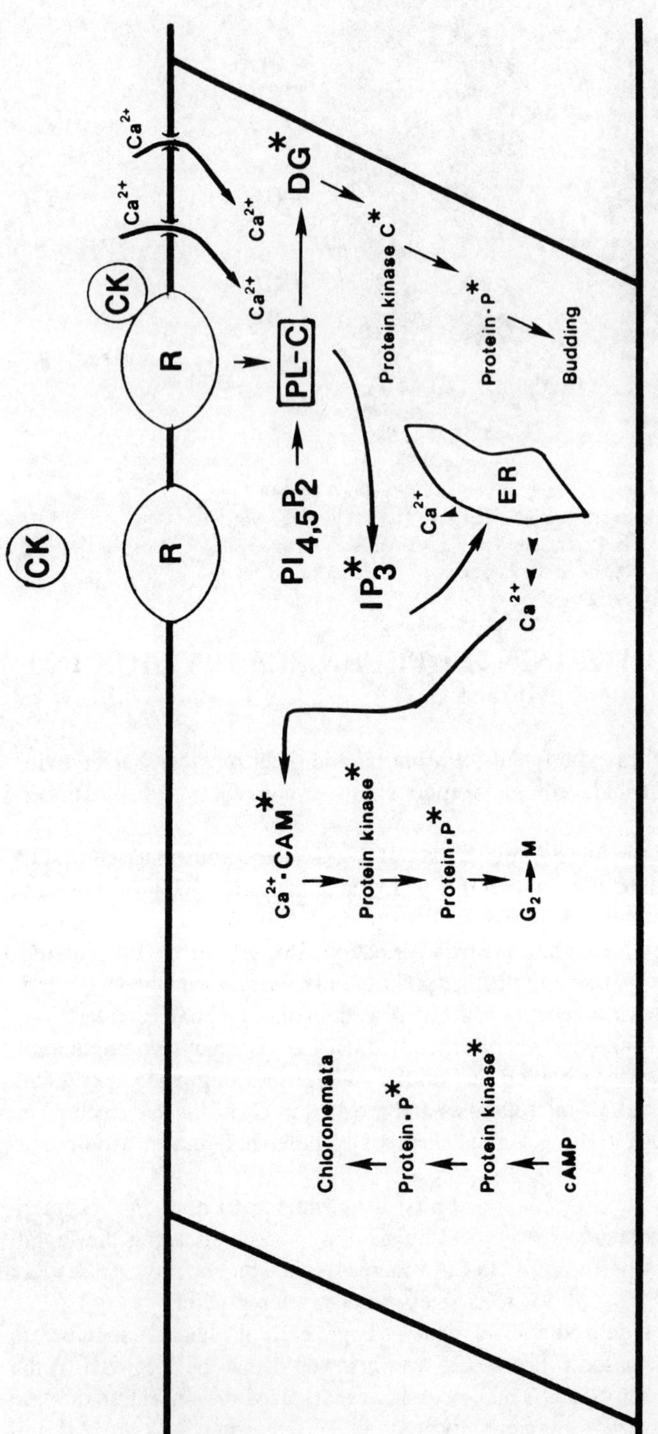

FIGURE 5. Model for cytokinin stimulation of bud formation in target caulonema cells of mosses. Abbreviations: CK — cytokinin; R — hormone receptor; PL-C — phospholipase C; $PI_{4,5}P_2$ — phosphatidylinositol 4,5-bisphosphate; IP_3 — inositol trisphosphate; DG — diacylglycerol; ER — endoplasmic reticulum; CAM — calmodulin.

of Ca^{2+} to diffuse through the cytoplasm is limited, and one could imagine that, initially, high $[Ca^{2+}]_i$ would be localized near the surface at the distal end of the target cell. Supporting evidence comes from the fact that the first detectable morphological changes are localized wall expansion and an increase in vesicles next to the plasma membrane in that region of the cell.[98] Local high $[Ca^{2+}]_i$ at the plasma membrane would facilitate vesicle fusion to the plasma membrane at the new growth zone, the presumptive bud site.

4. Accompanying Ca^{2+} influx across the plasma membrane is a second mechanism for Ca^{2+} release. The binding of cytokinin to a plasma membrane receptor activates a phospholipase C (PL-C) that stimulates the hydrolysis of phosphatidylinositol 4,5-bisphosphate (PI 4,5-P2), releasing IP_3 and DG. IP_3 is then responsible for calcium release from the endoplasmic reticulum-like membrane compartment (ER).
5. The zone of high $[Ca^{2+}]$ is first created within the outgrowth marking the presumptive bud site and later extends further down the long axis of the cell. At this point Ca^{2+} channels are no longer activated or concentrated at the distal end, and the source of Ca^{2+} is only from internal stores within concentrated ER at the distal end triggered by IP_3 binding. Calcium concentrations, therefore, vary both temporally and spatially within the cytoplasm of the target cell. There are multiple receptors for calcium within the cell, and a wide variety of molecular changes occur in response to the regionally elevated $[Ca^{2+}]$. These include (a) activation of CaM, which in turn can activate a variety of enzymes by protein phosphorylation; (b) activation of protein kinase C by DG and subsequent protein phosphorylation, leading to increased or decreased activity of certain metabolic pathways; (c) increased microtubule depolymerization, releasing the nucleus from its anchorage on the side wall; and (d) stimulation of an actomyosin microfilament system that directs membrane flow and/or organelle movement. It is attractive to imagine that several of these processes work together to cause the profound cytoplasmic rearrangements that occur following cytokinin induction.
6. When the nucleus reaches the presumptive bud site, mitosis is triggered. CaM is essential for the G_2-to-M transition, possibly by a phosphorylation of nuclear lamin protein mechanism.[99] The mitotic apparatus creates its own microdomain, and ionic fluxes are regulated within it. After mitosis the newly formed cell plate fuses with caulonema walls at a position which may be determined by the concentration of Ca^{2+} channels, producing a small bud initial cell.
7. Since the initial division is asymmetrical, the partitioning of membrane components between the caulonema cell and the bud initial cell is unequal. The plasma membrane of the initial cell may be enriched in cytokinin receptors; thus, continued stimulation by cytokinin affects only the initial cell — the caulonema cell reverts to its nondividing state. The small volume and large concentration of organelles and endomembranes of the initial cell would lead to highly regulated sequestration and release of Ca^{2+}. The relationship between the division rate and the growth rate of the cells may provide geometrical constraints that determine the pattern of division and give rise to the organized mass of cells termed a bud.

REFERENCES

1. **Letham, D. S.,** Cytokinins, in *Phytohormones and Related Compounds — A Comprehensive Treatise,* Vol. 1, Letham, D. S., Goodwin, P. B., Higgins, T. J. V., Eds., Elsevier/North-Holland, Amsterdam, 1978, 205.
2. **Cheung, W. Y.,** Calmodulin plays a pivotal role in cellular regulation, *Science,* 207, 19, 1980.

3. **Manalan, A. S. and Klee, C. B.,** Calmodulin, in *Advances in Cyclic Nucleotide and Protein Phosphorylation Research,* Greengard, P. and Robison, G. A., Eds., Raven Press, New York, 1984, 227.
4. **Berridge, M. J. and Irvine, R. F.,** Inositol triphosphate, a novel second messenger in cellular signal transduction, *Nature (London),* 312, 315, 1984.
5. **Nishizuka, Y.,** Turnover of inositol phospholipids and signal transduction, *Science,* 225, 1365, 1984.
6. **Rasmussen, H., Kojima, I., Kumifeo, K., Zawalich, W., and Apfeldorf, W.,** Calcium as intracellular messenger: sensitivity modulation, C-kinase pathway and sustained cellular response, in *Advances in Cyclic Nucleotide and Protein Phosphorylation Research,* Greengard, P. and Robison, G. A., Eds., Raven Press, New York, 1984, 159.
7. **Takai, Y., Kikkawa, U., Kaibuchi, K., and Nishizuka, Y.,** Membrane phospholipid metabolism and signal transduction for protein phosphorylation, in *Advances in Cyclic Nucleotide and Protein Phosphorylation Research,* Greengard, P. and Robison, G. A., Eds., Raven Press, New York, 1984, 119.
8. **Aub, D. L. and Putney, J. W., Jr.,** Metabolism of inositol phosphates in parotid cells: implication for the pathway of the phosphoinositide effect and for the possible messenger role of inositol trisphosphate, *Life Sci.,* 34, 1347, 1984.
9. **Poovaiah, B. W., Reddy, A. S. N., and McFadden, J. J.,** Calcium messenger system: role of protein phosphorylation and inositol bisphospholipids, *Physiol. Plant.,* 69, 569, 1987.
10. **Poovaiah, B. W. and Reddy, A. S. N.,** Calcium messenger system in plants, *Crit. Rev. Plant Sci.,* 6, 47, 1987.
11. **Saunders, M. J. and Hepler, P. K.,** Localization of membrane-associated calcium following cytokinin treatment in *Funaria* using chlorotetracycline, *Planta,* 152, 272, 1981.
12. **Saunders, M. J. and Hepler, P. K.,** Ca^{2+} ionophore A23187 stimulates cytokinin-like mitosis in *Funaria, Science,* 217, 943, 1982.
13. **Saunders, M. J. and Hepler, P. K.,** Calcium antagonists and calmodulin inhibitors block cytokinin-induced bud formation in *Funaria, Dev. Biol.,* 99, 41, 1983.
14. **Saunders, M. J.,** Cytokinin activates and redistributes plasma membrane channels creating a zone of high free calcium that predicts the site of cell division, *J. Cell Biol.,* 101, 4a, 1985.
15. **Spiess, L.D., Lippincott, B., and Lippincott, J. A.,** Influence of octopine, calcium and compounds that effect calcium transport on zeatin-induced bud formation by *Pylaisiella selwynii, Am. J. Bot.,* 71, 1416, 1984.
16. **Bopp, M.,** Developmental physiology of bryophytes, in *New Manual of Bryology,* Vol. 1, Schuster, R. M., Ed., Hattori Botanical Laboratory, Nichinan, Miyazaki, Japan, 1983, 276.
17. **Cove, D. J. and Ashton, N. W.,** The hormonal regulation of gametophytic development in bryophytes, in *The Experimental Biology of Bryophytes,* Dyer, A. F. and Duckett, J. G., Eds., Academic Press, London, 1984, 177.
18. **Saunders, M. J.,** Evidence for calcium involvement in cytokinin-induced cell division, in *Molecular and Cellular Aspects of Calcium in Plant Development,* Trewavas, A. J., Ed., Plenum Press, New York, 1986, 188.
19. **Saunders, M. J.,** Cytokinin activation and redistribution of plasma membrane ion channels in *Funaria:* a vibrating-microelectrode and cytoskeletal inhibitor study, *Planta,* 167, 402, 1986.
20. **Conrad, P. A.,** An Ultrastructural and Physiological Analysis of Cytokinin-induced Bud Formation in the Moss *Funaria hygrometrica,* Ph.D. thesis, University of Massachusetts, Amherst, 1987.
21. **Markmann-Mulisch, U. and Bopp, M.,** The hormonal regulation of protonema development in mosses. IV. The role of Ca^{2+} as cytokinin effector, *J. Plant Physiol.,* 129, 155, 1987.
22. **Conrad, P. A., Steucek, G. L., and Hepler, P. K.,** Bud formation in *Funaria:* organelle redistribution following cytokinin treatment, *Protoplasma,* 131, 211, 1986.
23. **Conrad, P. A. and Hepler, P. K.,** The effect of 1,4-dihydropyridines on the initiation and development of gametophore buds in the moss *Funaria, Plant Physiol.,* 86, 684, 1988.
24. **Elliott, D. C. and Skinner, J. D.,** Calcium-dependent, phospholipid-activated protein kinase in plants, *Phytochemistry,* 25, 39, 1986.
25. **Marme, D.,** The role of Ca^{++} in signal transduction of higher plants, in *Plant Growth Substances,* Wareing, P. F., Ed., Academic Press, New York, 1982, 419.
26. **Roux, S. J. and Slocum, R. D.,** Role of calcium in mediating cellular functions important for growth and development in higher plants, in *Calcium and Cell Function,* Cheung, W. Y., Ed., Academic Press, New York, 1982, 409.
27. **Hepler, P. K. and Wayne, R. O.,** Calcium and plant development, *Annu. Rev. Plant Physiol.,* 36, 397, 1985.
28. **Dieter, P.,** Calmodulin and calmodulin-mediated processes in plants, *Plant Cell Environ.,* 7, 371, 1984.
29. **Elliott, D. C.,** Inhibition of cytokinin-regulated responses by calmodulin binding compounds, *Plant Physiol.,* 72, 215, 1983.
30. **Elliott, D. C., Batchelor, S. M., Cassar, R. A., and Marinos, N. G.,** Calmodulin binding drugs affect responses to cytokinin, auxin and gibberellic acid, *Plant Physiol.,* 72, 219, 1983.

31. **Raghothama, K. G., Mizrahi, Y., and Poovaiah, B. W.,** Effects of calmodulin inhibitors on auxin induced elongation, *Plant Physiol.*, Suppl. 72, 144, 1983.
32. **Anderson, J. M., Carbonneau, H., Jones, H. P., McCann, R. O., and Cormier, M. J.,** Characterization of the plant nicotinamide adenine dinucleotide kinase activator protein and its identification as calmodulin, *Biochemistry,* 19, 3113, 1980.
33. **Marme, D. and Dieter, P.,** Role of Ca^{2+} and calmodulin in plants, in *Calcium and Cell Function,* Vol. 4, Cheung, W. Y., Ed., Academic Press, New York, 1983, 263.
34. **Hetherington, A. and Trewavas, A.,** Calcium-dependent protein kinase in pea shoot membranes, *FEBS Lett.*, 145, 67, 1982.
35. **Salimath, B. P. and Marme, D.,** Protein phosphorylation and its regulation by calcium and calmodulin in membrane fraction from zucchini hypocotyls, *Planta,* 158, 560, 1983.
36. **Graziana, A., Ranjeva, R., and Boudet, A. M.,** Provoked changes in cellular calcium controlled protein phosphorylation and activity of quinate: NAD^+ oxidoreductase in carrot cells, *FEBS Lett.*, 156, 325, 1983.
37. **Blowers, D. P., Hetherington, A., and Trewavas, A.,** Isolation of plasma-membrane-bound calcium/calmodulin regulated protein kinase from pea using western blotting, *Planta,* 166, 208, 1985.
38. **Ranjeva, R., Refeno, G., Boudet, A. M., and Marme, D.,** Activation of plant quinate: NAD^+ 3-oxidoreductase by Ca^{2+} and calmodulin, *Proc. Natl. Acad. Sci. U.S.A.*, 80, 5222, 1983.
39. **Veluthambi, K. and Poovaiah, B. W.,** Calcium and calmodulin-regulated phosphorylation of soluble and membrane proteins from corn coleoptiles, *Plant Physiol.*, 76, 359, 1986.
40. **Ralph, R. K., Buillivant, S., and Wojcik, S. J.,** Effects of kinetin on phosphorylation of leaf membrane proteins, *Biochim. Biophys. Acta,* 421, 319, 1976.
41. **Elliott, D. C. and Geytenbeek, M.,** Identification of products of protein phosphorylation in T37-transformed cells and comparison with normal cells, *Biochim. Biophys. Acta,* 845, 317, 1985.
42. **Polya, G. M. and Davies, J. R.,** Resolution and properties of a protein kinase catalyzing the phosphorylation of a wheat germ cytokinin-binding protein, *Plant Physiol.*, 71, 482, 1983.
43. **Talbot, J. and Saunders, M. J.,** Fluctuation of IP_3 concentration during cytokinin-induced bud formation in *Funaria, Plant Physol.*, 80, 113a, 1986.
44. **Conrad, P. A. and Hepler, P. K.,** The PI cycle and cytokinin-induced bud formation in *Funaria, Plant Physiol.*, 80, 60a, 1986.
45. **Morse, M. J., Crane, R. C., and Satter, R. L.,** Phosphatidylinositol turnover in *Samanea pulvini:* a mechanism of phototransduction, *Plant Physiol.*, 80, 92a, 1986.
46. **Morre, D. J., Gripshover, B., and Morre, J. T.,** Phosphatidylinositol turnover in isolated soybean membrane stimulated by the synthetic growth hormone, 2,4-dichlorophenosyl acetic acid, *J. Biol. Chem.*, 259, 15364, 1984.
47. **Sandelius, A. S. and Morre, D. J.,** Calcium-calmodulin requirements of phosphatidylinositol turnover stimulated by auxin, in *Molecular and Cellular Aspects of Calcium in Plant Development,* Trewavas, A. J., Ed., Plenum Press, New York, 1986, 351.
48. **Tung, H. Y., Wayne, R., Roux, S., and Thompson, G.,** The Ca^{2+}-phospholipid dependent protein kinase (protein kinase C) is present in the phototactic green alga, *Dunaliella salina, Planta,* submitted, 1990.
49. **Sefton, B. M. and Hunter, T.,** Tyrosine protein kinases, in *Advances in Cyclic Nucleotide and Protein Phosphorylation Research,* Greengard, P. and Robison, G. A., Eds., Raven Press, New York, 1984, 192.
50. **Baker, P. F.,** The regulation of intracellular calcium, in *S.E.B. Symp. XXX, Calcium in Biological Systems,* Duncan, C. J., Ed., Cambridge University Press, London, 1976, 219.
51. **Williams, J. P.,** Calcium chemistry and its relation to biological function, in *S.E.B. Symp. XXX, Calcium in Biological Systems,* Duncan, C. J., Ed., Cambridge University Press, London, 1976, 1.
52. **Carafoli, E. and Crompton, M.,** The regulation of intracellular calcium, *Curr. Top. Membr. Transp.*, 10, 151, 1978.
53. **Kretsinger, R. H.,** The information role of calcium in the cytosol, *Adv. Cyclic Nucleotide Res.*, 11, 1, 1979.
54. **Kretsinger, R. H.,** Mechanisms of selective signalling by calcium, *Neurosci. Res. Program Bull.*, 19, 217, 1981.
55. **Williamson, R. E. and Ashley, C. C.,** Free Ca^{2+} and cytoplasmic streaming in the alga *Chara, Nature (London),* 296, 647, 1982.
56. **Weber, A.,** Synoposis of the presentations, in *S.E.B. Symp. XXX, Calcium in Biological Systems,* Duncan, C. J., Ed., Cambridge University Press, London, 1976, 445.
57. **Kretsinger, R. H.,** Why does calcium play an informational role unique in biological systems?, in *9th Jerusalem Symp. Quantum Chemistry and Biochemistry,* Pullman, B. and Goldblum, N., Eds., D. Reidel, Boston, 1976, 257.
58. **Klee, C. B., Crouch, T. H., and Richman, P. G.,** Calmodulin, *Annu. Rev. Biochem.*, 49, 489, 1980.
59. **Weiss, B. and Levin, R. M., Eds.,** Calmodulin and cell functions, *Ann. N.Y. Acad. Sci.*, 356, 1, 1980.
60. **Means, A. R. and Dedman, J. R.,** Calmodulin — an intracellular calcium receptor, *Nature (London),* 285, 73, 1980.

61. **Siegel, F. L., Carafoli, E., Kretsinger, R. H., MacLennan, D. H., and Wasserman, R. H., Eds.,** *Calcium Binding Proteins: Structure and Function,* Elsevier, Amsterdam, 1980.
62. **Means, A. R., Tash, J. S., and Chafouleas, J. G.,** Physiological implications of the presence, distribution and regulation of calmodulin in eucaryotic cells, *Physiol. Rev.,* 62, 1, 1982.
63. **Charbonneau, H., Jarret, H. W., McCann, R. O., and Cormier, M. J.,** Calmodulin in plants and fungi, in *Calcium Binding Proteins: Structure and Function,* Siegel, F. L., Carafoli, E., Kretsinger, R. H., MacLennan, D. H., and Wasserman, R. H., Eds., Elsevier, Amsterdam, 1980, 155.
64. **Grand, R. J. A., Nairn, A. C., and Perry, S. V.,** The preparation of calmodulins from barley (*Hordeum* sp.) and basidiomycete fungi, *Biochem. J.,* 185, 755, 1980.
65. **Watterson, D. M., Iverson, D. B., and Van Eldik, L. J.,** Spinach calmodulin: isolation, characterization and comparison with vertebrate calmodulins, *Biochemistry,* 19, 5762, 1980.
66. **Epel, D., Patton, C., Wallace, R. W., and Cheung, W. Y.,** Calmodulin activates NAD kinases of sea urchin eggs: an early event of fertilization, *Cell,* 23, 543, 1981.
67. **Dieter, P. and Marme, D.,** Calmodulin activation of plant microsomal Ca^{2+} uptake, *Proc. Natl. Acad. Sci. U.S.A.,* 77, 7311, 1980.
68. **Chafouleas, J. G., Bolton, W. E., Hidaka, H., Boyd, A. E., and Means, A. R.,** Calmodulin and the cell cycle: involvement in regulation of cell cycle progression, *Cell,* 28, 41, 1982.
69. **Welsh, M. J., Dedman, J. R., Brinkley, B. R., and Means, A. R.,** Calcium-dependent regulator protein: localization in the mitotic apparatus of eucaryotic cells, *Proc. Natl. Acad. Sci. U.S.A.,* 75, 1867, 1978.
70. **Welsh, M. J., Dedman, J. R., Brinkley, B. R., and Means, A. R.,** Tubulin and calmodulin. Effects of microtubule and midrofilament inhibitors on localization in the mitotic apparatus, *J. Cell Biol.,* 81, 624, 1979.
71. **Lambert, A. and Vantard, M.,** Calcium and calmodulin as regulators of chromosome movement during mitosis in higher plants, in *Molecular and Cellular Aspects of Calcium in Plant Development,* Trewavas, A. J., Ed., Plenum Press, New York, 1986, 104.
72. **Wasserman, W. J. and Smith, L. D.,** Calmodulin triggers the resumption of meiosis in amphibian oocytes, *J. Cell Biol.,* 89, 389, 1981.
73. **Rasmussen, H. and Barrett, P. Q.,** Calcium messenger system: an integrated view, *Physiol. Rev.,* 64, 938, 1984.
74. **Fox, J. F. and Erion, J. L.,** A cytokinin-binding protein from higher plant ribosomes, *Biochem. Biophys. Res. Commun.,* 64, 694, 1975.
75. **Gardner, G., Sussman, M. R., and Kende, H.,** In vitro cytokinin binding to a particulate cell fraction from protenemata of *Funaria hygrometrica, Planta,* 143, 67, 1978.
76. **Brinegar, A. C., Stevens, A., and Fox, J. E.,** Biosynthesis and degradation of a wheat germ cytokinin-binding protein during embryogenesis and germination, *Plant Physiol.,* 79, 706, 1985.
77. **Miassod, R.,** Uptake of [^{14}C]-8-azido-N^6-benzyladenine, a radioactive photosensitive cytokinin, by the cells of the moss *Funaria hygrometrica,* in *Metabolism and Molecular Activities of Cytokinins,* Guern, J. and Peaud-Lenoel, C., Eds., Springer-Verlag, Berlin, 1981, 162.
78. **LeJohn, H. B. and Stevenson, L. E.,** Cytokinins and magnesium ions may control the flow of metabolites and calcium ions through fungal cell membranes, *Biochem. Biophys. Res. Commun.,* 54, 1061, 1973.
79. **LeJohn, H. B., Cameron, L. E., Stevenson, R. M., and Meuser, R. U.,** Influence of cytokinins and sulfhydryl group-reacting agents on calcium transport in fungi, *J. Biol. Chem.,* 294, 4016, 1974.
80. **LeJohn, H. B. and Cameron, L. E.,** Cytokinins regulate calcium binding to a glycoprotein from fungal cells, *Biochem. Biophys. Res. Commun.,* 54, 1053, 1973.
81. **Lau, O.-L. and Yang, S. F.,** Interaction of kinetin and calcium in relation to their effect on stimulation of ethylene production, *Plant Physiol.,* 55, 738, 1975.
82. **Poovaiah, B. W. and Leopold, A. C.,** Deferral of leaf senescence with calcium, *Plant Physiol.,* 52, 236, 1973.
83. **Shear, C. B. and Faust, M.,** Calcium transport in apple trees, *Plant Physiol.,* 45, 670, 1970.
84. **Leopold, A. C., Poovaiah, B.W., de la Fuente, R. C., and Williams, R. J.,** Regulation of growth with inorganic solutes, in *Plant Growth Substances,* Hirokawa, Tokyo, 1973, 780.
85. **Elliott, D. C.,** Calmodulin inhibitor prevents plant hormone response, *Biochem. Int.,* 1, 290, 1980.
86. **Kubowicz, B. D., Vanderhoef, L. N., and Hanson, J. B.,** ATP-dependent calcium transport in plasmalemma preparations from soybean hypocotyls, *Plant Physiol.,* 69, 187, 1982.
87. **Wacker, I., Reiss, H., Schnepf, E., Traxel, K., Bauer, R., and Polar, K.,** Distribution of calcium- and phosphorus-rich globules induced by glutaraldehyde/tannic acid fixation in the caulonema tip cell of the moss, *Funaria hygrometrica:* light microscopy, transmission electron microscopy (TEM), proton microprobe (PIXE), and electron spectroscopic imaging (ESI), *Eur. J. Cell Biol.,* 40, 94, 1986.
88. **Grynkiewicz, C., Polnie, M., and Tsien, R. Y.,** A new generation of Ca^{2+} indicators with greatly improved fluorescence properties, *J. Biol. Chem.,* 260, 3440, 1985.
89. **Saunders, M. J. and Jones, K. J.,** Distortion of cell plate formation by the purported intracellular calcium antagonist TMB-8, *Protoplasma,* 144, 92, 1988.

90. **Saunders, M. J. and Boullion, K. J.,** Calcium regulation of cytokinesis in *Funaria, Plant Physiol.,* 80, 60a, 1986.
91. **Butcher, R. W. and Sutherland, E. W.,** Adenosine 3′5′-phosphate in biological materials. I. Purification and properties of 3′5′-phosphodiesterase and use of this enzyme to characterize 3′5′-phosphate in human urine, *J. Biol. Chem.,* 237, 1244, 1962.
92. **Brandes, H. and Kende, H.,** Studies on cytokinin-controlled bud formation in moss protonemata, *Plant Physiol.,* 43, 827, 1968.
93. **Hajra, A. K., Seguin, E. B., and Agranoff, B. F.,** Rapid labelling of mitochondrial lipids by labelled orthophosphate and adenosine triphosphate, *J. Biol. Chem.,* 7, 1609, 1968.
94. **Berridge, M. J., Dawson, R. M. C., Downes, C. P., Heslop, J. P., and Irvine, R. F.,** Changes in the levels of inositol phosphates after agonist-dependent hydrolysis of membrane phosphoinositides, *Biochem. J.,* 212, 473, 1983.
95. **Doonan, J. H., Cove, D. J., and Lloyd, C. W.,** Immunofluorescence microscopy of microtubules in intact cell lineages of the moss *Physcomitrella patens.* I. Normal and CIPC treated cells, *J. Cell Sci.,* 75, 131, 1985.
96. **Doonan, J. H., Cove, D. J., Corke, F. M. K., and Lloyd, C.,** Pre-prophase band of microtubules, absent from tip-growing moss filaments, arises in leafy shoots during transition to intercalary growth, *Cell Motility Cytoskeleton,* 7, 138, 1987.
97. **Laemmli, U. K.,** Cleavage of structural proteins during the assembly of the head of bacteriophage T4, *Nature (London),* 227, 680, 1970.
98. **Conrad, P. A. and Hepler, P. K.,** Ultrastructural changes associated with cytokinin-induced bud formation in the moss *Funaria, J. Cell Biol.,* 99, 244a, 1984.
99. **Ottaviano, Y. and Gerace, L.,** Phosphorylation of the nuclear lamina during interphase and mitosis, *J. Biol. Chem.,* 260, 624, 1985.

Chapter 7

PHYSIOLOGY OF MOSS-BACTERIAL ASSOCIATIONS

Luretta D. Spiess, Barbara B. Lippincott, and James A. Lippincott

TABLE OF CONTENTS

I. INTRODUCTION

The beneficial effects that result from interactions between plants and microorganisms whether in a casual, mutualistic, or symbiotic association are of considerable ecological and agricultural importance. These interactions are receiving increased attention because of the potential gains in plant productivity and agriculture that can depend on these associations.[1] Similar interactions have been described between various lower plants and microorganisms, and these are more amenable to laboratory investigation due to their smaller size and shorter growth cycles. Such interactions provide model systems for studies of the ways in which two organisms influence each other and provide information of interest to the ecology and evolution of these organisms. In this review we will consider briefly the kinds of associations found between microorganisms and plants in general and proceed to a more detailed consideration of bacteria and mosses, where we have described a series of moss-bacterium interactions which illustrate some of the ways and possible mechanisms by which microorganisms influence moss growth and development.

II. EXAMPLES OF PLANT-MICROORGANISM INTERACTIONS

A. FLOWERING PLANT ASSOCIATIONS

Vascular plants, which originated in the late Silurian, may have been facilitated in their development by mycorrhizal associates which increased their nitrogen and phosphorus uptake efficiency. As a result new habitats were exploited and the development of complex tissues resulted.[2] Stubblefield et al.[3] have found fossil fungi in a Triassic root sample which are very similar to modern vesicular-arbuscular endomycorrhizae.

Many microbe-plant associations have been described that have beneficial or harmful effects on plant growth (for reviews, see References 4 and 5). The symbiotic and mycorrhizal associations with vascular plants and the common pathogens of plants represent the more highly developed extreme of these relationships. One of the more complex of these interactions is that involving an *Agrobacterium* sp. and host plants where bacterial plasmid genetic material is transferred and incorporated directly into the host genome. The plant cells proliferate and produce unusual amino acids and elevated levels of auxins and cytokinins as a direct result of these newly acquired plasmid genes. The abnormal amino acids or opines[117] in turn promote bacterial growth and genetic recombination.[6,7] The natural genetic engineering abilities of this group of bacteria have provided the best vector system to date for introducing new genetic information into plants.[8-10] The investigations of Spiess et al.[79-82] to be described in Section III raise the possibility that this group of bacteria may also transfer plasmid DNA to mosses.

Plant growth hormones produced by bacteria have been suggested as a cause of the effects of a phenomenon termed bacterization, the increase in yield obtained by the addition of bacteria to seeds of crop plants, and for the natural promotive effects of rhizoplane bacteria found in association with plant roots.[11] As examples, the rhizosphere species of *Pseudomonas, Azotobacter,* and *Bacillus* commonly found in association with roots produce plant hormones.[11,12] The cytokinins isopentanyladenine and zeatin are synthesized by *Agrobacterium tumefaciens,* the rhizobia synthesize zeatin and ribosyl zeatin, and *Corynebacterium fasciens,* the witches broom pathogen, produces ribosyl zeatin.[13] Similarly, agrobacteria produce auxin,[13] as do the rhizobia[14] and *Pseudomonas syringae* subsp. *savastanoi,* the causal agent of olive gall.[15] The crown gall tumors induced by agrobacteria produce increased amounts of auxin and cytokinin. The *Agrobacterium* genes responsible for this increased synthesis have been described recently, as have a similar set of phytohormone biosynthetic genes in *P. syringae* subsp. *savastanoi.*[16]

Azospirillum brasilense, a nitrogen-fixing bacterium found in the rhizosphere of various grass species, was shown to produce indoleacetic acid (IAA), three cytokinins, and a small amount of gibberellin when grown in liquid culture.[17] Pearl millet grown in liquid culture with these hormones formed roots similar to those caused by inoculation with *A. brasilense*.[17] Inoculation of sugar beet seeds with bacteria that produce a large concentration of IAA in culture caused a decrease in root elongation but an increased shoot-root ratio. Sugar beets responded favorably to other strains which produced less IAA.[18] Some reservation must be used in interpreting the effect on plant growth of any hormone produced in culture by a bacterium, since an approach for directly examining the concentrations and effects of such hormones in the rhizosphere has not been perfected.[18]

Competition between different rhizosphere bacteria can also influence plant growth response, either directly by the production of growth-promoting products or indirectly by the inhibition of harmful organisms.[11] Roots may be colonized by a variety of bacteria that hinder root growth. Toxigenic but nonparasitic species of *Pseudomonas, Enterobacter, Klebsiella, Citrobacter, Flavobacterium, Achromobacter,* and *Arthrobacter* groups have been described.[19] Reversal of the effects of these bacteria on sugar beets by plant-growth-promoting rhizobacteria was demonstrated in greenhouse experiments.[5] Other organisms may protect against these deleterious effects by forming antibiotics or siderophores (high-affinity iron transport compounds), by physically displacing harmful bacteria, or by producing growth-promoting substances. About 8 to 15% of bacteria in the rhizosphere have been found to be harmful, while only 2 to 5% promote growth. Most of these beneficial bacteria produce broad-spectrum antibiotics *in vitro*.[20] However, there seems to be little correlation between the ability of these microorganisms to exhibit antibiosis in agar plate tests and their capacity to control pests in the field.[20] The contribution of these substances to the beneficial effects of these bacteria still has to be clearly demonstrated.

The black root rot of tobacco caused by *Thielaviopsis basicola* can be suppressed by certain strains of *Pseudomonas*,[21] and *Agrobacterium radiobacter* strain K84 can greatly reduce crown gall formation caused by *Agrobacterium tumefaciens*.[22,23] The effectiveness of strain 84 has been shown to depend on its ability to produce a nucleotide bacteriocin,[24,25] and other agrobacteria have been found to produce similar bacteriocins.[23,26] Other bacteria have shown promise as biological control agents, but it has been difficult to show consistent positive results in field studies.[27]

Some beneficial bacteria produce antibiotics or phytotoxic compounds, many of which belong to the class of nitrogen-containing heterocyclics such as phenazines and pyrrolnitrin-type antibiotics. They also produce unusual amino acids and peptides of unknown function in the physiology and ecology of these organisms.[5]

The possibility that the growth-promoting effects of some bacteria depend on production of iron-chelating siderophores has been investigated. The *Pseudomonas fluorescens-putida* group is often used as an inoculant on crop plants because it increases yields by antagonizing deleterious fungi and bacteria. The production of siderophores which reduce the available iron for growth of the harmful organisms appears necessary for this response.[28,29] The increase in plant growth by root-colonizing fluorescent pseudomonads in the rhizosphere is also thought to be due to siderophore production under iron-limiting conditions.[30] Evidence that supports this hypothesis includes the observation that addition of iron abolishes the effect, siderophore-negative mutants do not promote growth, and addition of siderophores promotes growth. The promotion of growth and suppression of inhibition also may be due to competition for carbon compounds or the production of hormones, antibiotics, or bacteriocins.[29]

Siderophores are produced by most bacteria and fungi and have been found in soil, the concentration and type depending on the particular microorganisms. There are at least three ways in which siderophores may affect plant life: the chelates solubilize and transport ferric ion, which is very important in plant nutrition; they may facilitate plant disease; and they may discourage growth or metabolism of competing microorganisms.[31]

Erwinia mutants that are resistant to bacteriocins have reduced virulence, although they secrete normal amounts of pectinase and cellulase. Mutants deficient in iron assimilation also have reduced virulence and, under iron-limiting conditions, three proteins are found to be missing in the outer membrane. These proteins are thought to constitute the siderophore receptor.[32]

The symbiotic nitrogen fixation associated with the interactions of *Rhizobium* and *Frankia* with host plants needs only to be mentioned here and clearly results from complex endophytic relationships.[33-35] In what may be a kind of associative (nonendophytic) symbiosis, *Azospirillum lipoferum,*[36] *Azospirillum brasilense,*[37] and *Azotobacter paspali*[38] are nitrogen-fixing bacteria found in association with the roots of certain tropical grasses.[39] In addition to fixing nitrogen, various rhizosphere species synthesize alginates, poly-β-hydroxybutyrate, pigments, and plant hormones.[40]

An apparent mutualistic relationship exists between the alga *Beggitoa* and rice plants. Sulfides released by *Desulfovibrio* spp. that are toxic to rice plants are decomposed by the alga, and peroxides produced by the alga, which might be hazardous to their growth, are in turn decomposed by the rice root catalase.[41]

In the mutualistic symbiosis of *Acremonium* endophytes that infect perennial ryegrass and tall fescue, benefits to the fungus include nutrition (soluble nitrogen and carbon compounds), protection, and improved dissemination via the grass seed. The benefits to the plant host are suggested by the observation that endophyte-free cultivars do not survive as well and, under controlled conditions, infected plants exhibit a higher rate of photosynthesis, increased fresh weight and production of more tillers during regrowth, use water more efficiently, and exhibit a striking resistance to herbivores (insects and grazing animals). Siegel et al.[42] conclude from these observations that infected plants are at an advantage during periods of stress.

Interactions directly resulting from surface contact also may initiate elaborate plant responses, e.g., the fungal and bacterial elicitation of the hypersensitive response and of phytoalexin production.[43,44] These responses result from an induced synthesis of many enzymes. As demonstrated in parsley, the process involves changes in transcription of as many as 20 or more genes in the induction of phytoalexin production.[43] While the elicitors produced by microorganisms appear to be primarily complex cell wall carbohydrates, a variety of agents have been found to elicit similar changes, e.g., heavy metals, icosapentanoic acid, and arachidonic acid.[45] These suggest that microorganisms also may influence these generalized responses in either a positive or negative way by altering the levels of these compounds or metals available to the plant.

B. LOWER PLANT ASSOCIATIONS

Lichens successfully occupy several unique ecological niches as a result of the symbiotic relationship between a fungus and a unicellular alga. The algal cells produce photosynthates for the fungus, and the fungus absorbs moisture and minerals and serves to protect that alga from high light intensity and desiccation.[46] Apparently only as a result of this interaction, lichens can produce substances that function as antibiotics, as antiherbivores, and as inhibitors of seed or moss spore germination, thereby reducing natural competition.[47]

Bacteria and possibly algae and other fungi play a significant role in the growth and fruiting process of mushrooms. *Agaricus brunnescens,* which fails to fruit on sterilized substrates, does so when *Pseudomonas putida* is present.[48] The growth and development of both the fungal mycelia and the bacteria are stimulated by extracts from the other. Not all strains of *P. putida* cause fruiting bodies to form in *A. brunnescens,* nor do all mushrooms respond similarly to this bacterium. Other microorganisms implicated in the phenomenon of fungal fruiting are *Bacillus megaterium, Arthrobacter terregens,* and *Rhizobium meliloti,* as well as yeasts and microalgae.[49,50] The removal of inhibitors, the production of growth factors, or the alteration of gene transcription through surface interactions may play a still to be documented role in these responses.[51]

High nitrogen secretion rates have been noted for the fern *Azolla* and several cycads as the result of an apparent symbiotic relationship with cyanobacteria.[52] Algal cells living in the mucilaginous cavities of the gametophytic thalli of at least four genera of liverworts transfer substantial amounts of ammonium compounds to the liverwort, while the algal cells depend upon the liverwort for a carbon source.[53] *Nostoc* colonies occur in overlapping thalli and in rolled thallus margins of the liverwort *Porella*. Although the algal colonies can be moved to some extent, they are difficult to separate completely from the liverwort. Evidence for symbiotic nitrogen fixation associated with this relationship was found for 58 of the 68 samples tested.[54] The liverworts *Anthoceros, Phaeoceros,* and *Blasia* have slime cavities containing *Nostoc,* the cells surrounding the cavities showing attentuated growth and the production of branched filaments from hyaline cells which penetrate the algal colonies.[55,56] In *Blasia,* these bryophyte filaments develop labyrinthine wall ingrowths characteristic of transfer cells, possibly facilitating exchange of metabolites.[56]

Several strains of bacteria isolated from the liverwort *Scapania nemorosa* stimulated its growth and development.[57] Pink-pigmented facultative methylotrophs which produce vitamin B_{12} were isolated from several other bryophytes,[58] as well as from flowering plants and occasionally from soil and water.[59] Vitamin B_{12} added to axenic cultures of gametophytes of *Jungermannia leiantha* and *Gymnocolea inflata* increased growth and branching, suggesting that bacteria may be an important source of the vitamin and may affect normal developmental patterns of these mosses.[59] Vitamin B_{12}-dependent enzymes occur in several families of flowering plants[60] and also may function in lower plants. Vitamin B_{12} is widely distributed in organisms that fix atmospheric nitrogen, including cyanobacteria, nodules of leguminous and nonleguminous symbionts, *Rhizobium* spp., *Azotobacter vinelandii,* and *Clostridium pasteurianum*. The small amounts of vitamin B_{12} found in higher plant tissue appear to be of microbial origin.[61]

Blue-green algae epiphytic on the leaves of mosses are of importance in nitrogen fixation in the Arctic.[62-64] The algae *Anabaena variabilis* and *Nostoc muscorum,* growing in association with the moss *Funaria hygrometrica,* cause rapid growth of the moss and an increase in the number of gametophores, while *Funaria* in turn has a growth-promoting effect on the cyanobacter. A survey 6 years after the volcanic island Surtsey formed off the coast of Iceland showed that wherever *Funaria* was found *Anabaena* colonies were also growing, suggesting some positive interaction between the two plants which could be important to their colonization of the island.[65]

Water-filled hyaline cells of *Sphagnum lindbergii* contain numerous *Nostoc* filaments.[66] In the Swedish Lapland, blue-green algae which grow on *Sphagnum* are the main source of nitrogen for the moss.[67,68] *Drepanocladus* found in pools in subarctic Sweden present yet another association of moss and blue-green algae.[67] The high nitrogen fixation rates associated with *Funaria*-colonized areas in Tasmania have been attributed to the epiphytic microorganisms on the moss.[69] Nothing is known about the physiological relationships between this moss and its epibionts.

A green alga, ''Chlorochytrium'', occurs as an endophyte in *Bryum capillare*.[70] Algae, fungi, and bacteria are commonly found as epiphytes on mosses, but only suggestive structural evidence indicates these relationships with the moss to be of physiological importance.[71] On Signy Island, off the coast of Antarctica, 11 species of algae (yellow green and green) are associated with the moss *Polytrichum alpestre*.[72] A study of the leaf ultrastructure of *Polytrichum commune* revealed the presence of a *Chlorococcum* sp. (green alga), *Trichoderma viride* (fungus), and an unidentified bacterium.[71]

Various fungal-moss associations have also been documented.[71,73-76] Chopra and associates[77] showed that inoculation of moss cultures of *Barbula gregaria* and *Timmiella anomala* with *Rhodotorula* (yeast) or *Aerobacter* increased protonemal growth and bud formation. Observations by others showing that fungi can increase protonemal filament growth and bud formation in a variety of mosses are summarized by Chopra et al.[77]

Indirect evidence of moss-bacterium interactions may be inferred from the fact that antibiotics active against at least one bacterium were found in extracts of 55% of the mosses tested.[78] The genera which seem most active in antibiotic production are *Atrichum, Cratoneuron, Dicranum, Polytrichum,* and *Sphagnum.* Extracts of *Funaria hygrometrica* had no effect against any test organism, perhaps because this moss exploits habitats opened by fires and, hence, has little competition. The compound(s) responsible for antibiotic activity have not been identified.[78]

III. MOSS-ASSOCIATED BACTERIA AND MOSS DEVELOPMENT

Detailed studies by Spiess et al.[79-82] show that mosses in nature harbor tightly adhering bacteria (MAB, or moss-associated bacteria), many of which can affect growth and gametophore development of the moss in its protonemal stage. Habitat as well as species of moss and possibly major changes in locale may influence the kinds of bacteria found adhering to a moss in nature.

Pylaisiella selwynii and *Heterophyllium haldaneanum* were collected from the same decaying oak log at an Illinois site, *Atrichum undulatum* from the ground ca. 2 m distant, and similar mosses were collected at a Wisconsin site. *Pylaisiella* without sporangia was also collected from a Minnesota site.[82] Results from 283 strains of MAB isolated from these three species of mosses from several locales and tested on their host or a different moss showed that at least 40% of MAB isolated from each source could promote moss growth or development. Only a few bacterial isolates were inhibitory. *Pylaisiella* and *Heterophyllium,* although of different taxonomic families,[83] respond positively to ca. 50 to 80% of the MAB regardless of the moss source of the isolates. There was a high degree of correlation between the ability of a bacterial strain to promote development in *Pylaisiella* and in *Heterophyllium.* Similarity of growth habit and/or locus of collection rather than taxonomic relatedness seems to account for this correlation.

Comparison of the response of *Pylaisiella selwynii* from Illinois and that of *Pylaisiella* obtained from Wisconsin to MAB from *P. selwynii* or *Atrichum* shows that *Pylaisiella* from both sources responds similarly to these MAB. From 70 to 76% of these strains had the same effects on both *Pylaisiella* samples. However, *Pylaisiella* MAB from Minnesota were different in their effects on the Wisconsin *P. selwynii* from the MAB obtained from the Wisconsin moss. Difference in time of collection, the locale, or both may account for this difference. Mosses at different stages in their development and at different times of the year may conceivably harbor different bacteria.

Atrichum undulatum, a soil-growing species of moss which belongs to a different family from that of either *Pylaisiella* or *Heterophyllium,* was found to respond to fewer of the MAB isolates. More MAB strains inhibited growth and development of this moss, and this was true of *Atrichum* collected in both Illinois and Wisconsin. The response of *Atrichum* to the bacteria isolated from *Atrichum* also differed from that of either *Pylaisiella* or *Heterophyllium.* Similarly, *Pylaisiella* and *Heterophyllium* differentiate between their bacterial flora and that of *Atrichum.*[82]

Although MAB were isolated from aerial gametophores of these mosses, the small size of the mosses places them near the soil surface and they develop from gametophore buds formed at or below this surface. Their epiflora, therefore, may be more nearly equivalent to rhizosphere flora. The effects of these MAB on protonemal development then may be considered to arise from rhizosphere bacteria which interact with protonemal filaments in this surface interface. Bacteria beneficial for higher plant growth appear to be associated

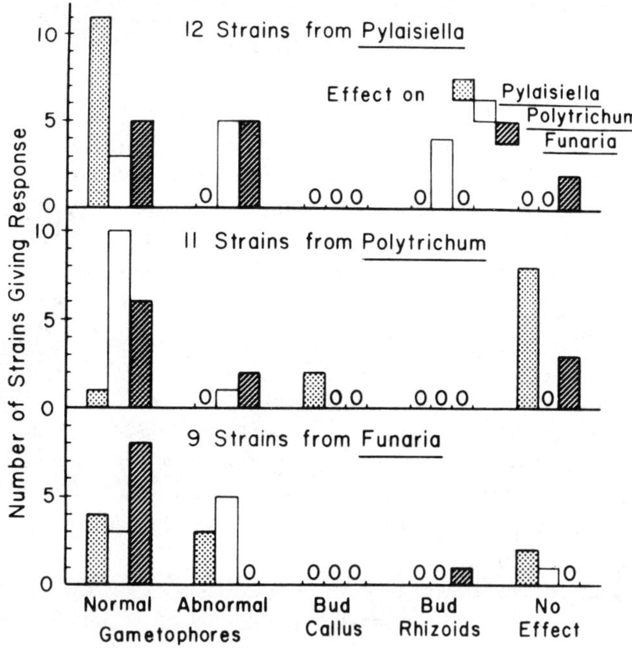

FIGURE 1. Developmental response, in liquid medium, of the aseptically cultured protonemal stage of three mosses to several strains of bacteria isolated from each of the three mosses in nature. The bacterial isolates include 12 strains obtained from *Pylaisiella selwynii,* 11 strains from *Polytrichum commune,* and 9 strains from *Funaria americana.* When none of the tested isolates produced a particular response this is indicated by a zero (0). The bacterial isolates from each moss were more effective on the host moss than on the other two mosses, suggesting there is a selective association between moss and bacteria in nature.

primarily with plant roots and also show specificity in their effectiveness on different species.[5] Specificity in the moss-MAB relationship was clearly shown in these studies, with more MAB generally proving beneficial to their host moss species and few of those adhering MAB being harmful to the mosses.[82]

Similar specificity in moss response to MAB was found in earlier studies[79-81] based on the laboratory response of *Pylaisiella selwynii, Funaria hygrometrica,* and *Polytrichum commune* to MAB from the same genera growing in different sites and habitats. Media considered to be selective for agrobacteria were used to isolate the MAB strains in these tests. When tested on their host of origin, 11 of 12 *Pylaisiella selwynii* MAB, 10 of 11 *Polytrichum commune* MAB, and 8 of 9 *F. hyrometrica* MAB increased gametophore formation (Figure 1).[81] Less than half of the MAB isolates showed similar results when tested on a nonhomologous moss. Of these 32 strains none was found to be without some developmental effect on its host moss, and only about 25% of the tests of these MAB isolates on nonhomologous mosses showed them to be without effect. Three isolates were found that promoted normal gametophore formation on all three mosses. In addition to promotion of growth and branching, some strains promoted abnormal gametophores, buds that later developed into rhizoids, and a few strains induced caulonemata on *Pylaisiella selwynii,* which typically does not exhibit this stage in culture.[79-81]

IV. EFFECTS OF AGROBACTERIA AND RHIZOBIA ON MOSS DEVELOPMENT

Early experiments showed that agrobacteria and rhizobia promoted growth and gametophore formation in *Pylaisiella selwynii*.[84,85] Thus, it was of interest to determine if these genera were typical of the MAB isolated on *Agrobacterium*-selective media.[79] All of these MAB were found to be Gram-negative short motile rods of approximately the size range of agrobacteria.[86] With one exception, the MAB grew on a minimal glucose-salts medium[87] and, hence, were prototrophic. In a number of additional tests most isolates were indistinguishable from agrobacteria and/or pseudomonads. However, none of the MAB strains active in moss development caused tumors when inoculated on bean leaves[88] or produced ketolactose, a unique characteristic of a major group of agrobacteria.[89] Several MAB isolates enhanced proliferation around the cambial area of sterile carrot slices, and a few initiated *de novo* root formation on the carrot discs,[79] although none of these responses was comparable to those obtained with virulent strains of *A. tumefacians* or *A. rhizogenes*. Five of the seven MAB strains tested reduced the number of *A. tumefaciens*-induced tumors formed on bean leaves, indicating that many of the MAB can compete for the same plant sites essential for *Agrobacterium* tumor initiation.

In a survey of the effects of 35 strains of *Agrobacterium* on *Pylaisiella selwynii*, 7 strains of avirulent *A. radiobacter* and 4 nontumorigenic strains of *A. tumefaciens* were found to have relatively little effect on moss development. All 24 virulent strains tested induced developmental effects ranging from normal gametophores to abnormal gametophores and/or callus masses (Figure 2). *A. rubi* (two strains) induced massive production of callus and abnormal gametophores.[85] The 5 strains of *A. rhizogenes* tested and 3 of the 17 virulent strains of *A. tumefaciens* induced rhizoid-like filaments and only occasional gametophores. Scanning and transmission electron microscope studies of agrobacteria and mosses show details of these moss responses to these bacteria (Figure 3).[90,91]

Representatives of five species of *Rhizobium* induced bud formation and normal gametophore development on *Pylaisiella selwynii*,[85] while *Corynebacterium fasciens*, a cytokinin-producing pathogen which causes witches broom in higher plants,[92] induced masses of brown callus similar to that produced by a high concentration of cytokinin.[85] A strain of *Pseudomonas syringae* induced buds and abnormal gametophores on *Pylaisiella*, and one strain of *P. syringae* subsp. *savastanoi* induced a few normal gametophores while a second had no effect.[115]

While some mosses when cultured with agrobacteria show growth and developmental changes comparable to those obtained with *Pylaisiella selwynii*, many others do not. *Heterophyllium haldaneanum* and *Entodon sedutrix* produced buds and gametophores in response to agrobacteria. *Atrichum undulatum, Polytrichum commune, Funaria hygrometrica, Thuidium delicatum,* and *Climacium americanum,* however, showed no overt response to the presence of *Agrobacterium tumefaciens*.[93]

V. PHYSIOLOGY OF MOSS RESPONSE TO BACTERIA

These moss-bacteria results pose the following questions: what is (are) the mechanism(s) by which bacteria alter moss growth and development? Conversely, what is the basis for the differences in response of different mosses such as *Pylaisiella, Polytrichum,* or *Funaria* to the same bacterium?

Many of the differences in responses of the different mosses may derive from differences in growth habits and nutrient requirements, reflecting their adaptation to different ecological niches. Table 1 illustrates several differences between the three mosses that have been used in some of these studies. The moss response to these bacteria can be assumed to result from

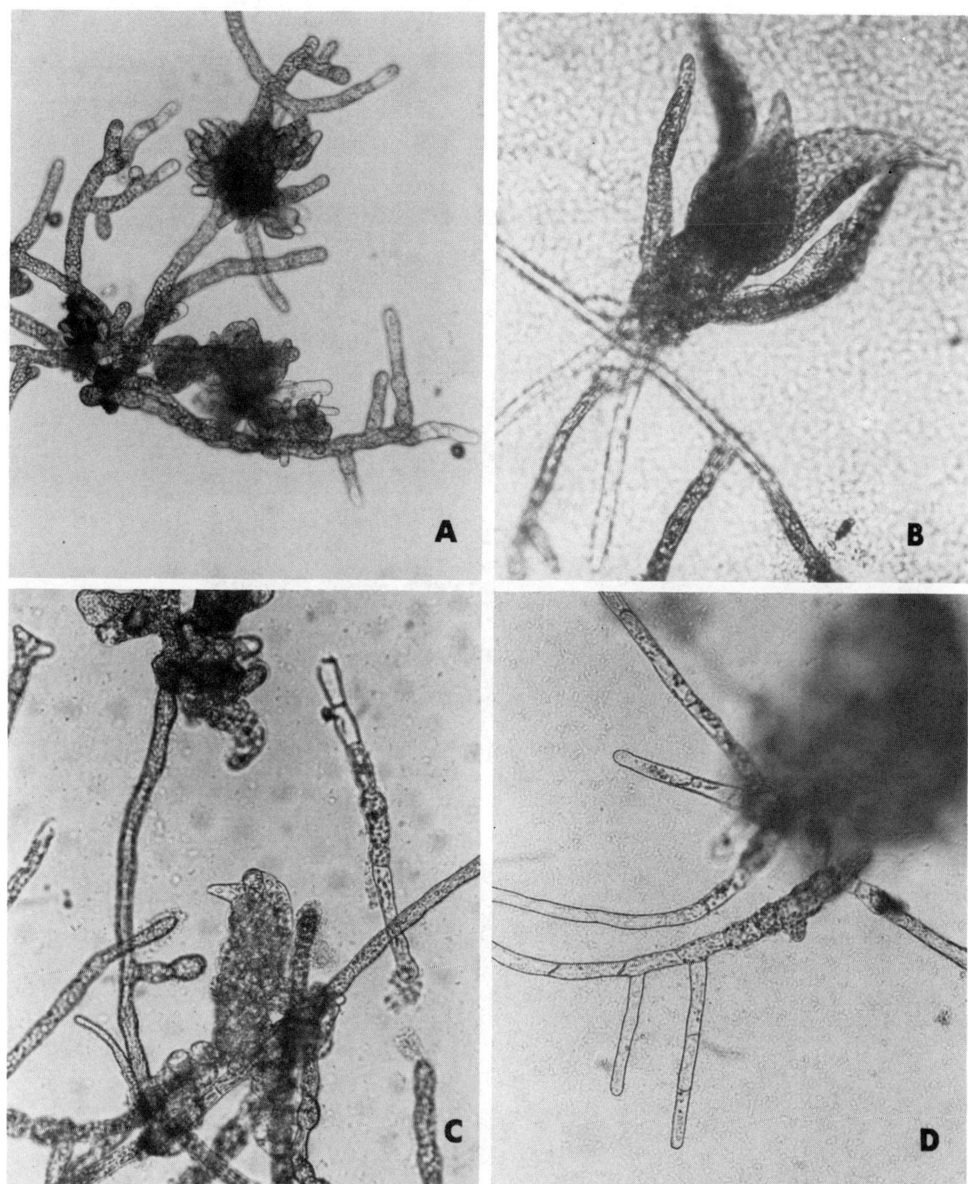

FIGURE 2. Light photomicrographs of representative responses of *Pylaisiella selwynii* protonema when cultured with phytohormones or bacteria. (A) Callus buds obtained with kinetin plus indoleacetic acid (each 10 μM); (B) normal gametophore induced by a bacterium isolated from *Pylaisiella*; (C) abnormal gametophores induced by *Agrobacterium rubi* strain 13334; (D) rhizoid development from aborted buds induced by *Agrobacterium rhizogenes* strain TR107. (Magnification × 60.)

one or more substances produced by the latter either normally or as a result of a moss product and to which different mosses may show different sensitivity. Mosses that do not show a developmental response when cultured with agrobacteria (*Atrichum, Polytrichum, Funaria, Thuidium,* and *Climacium*) are unresponsive to the cytokinin ribosyl zeatin, whereas those mosses that respond (*Pylaisiella, Heterophyllium,* and *Entodon*) also show increased bud callus formation in response to this cytokinin nucleoside. All of these mosses produce bud callus or gametophores in response to added zeatin,[93,94] suggesting that differences in either cytokinin uptake sites or sites of action exist in these mosses.

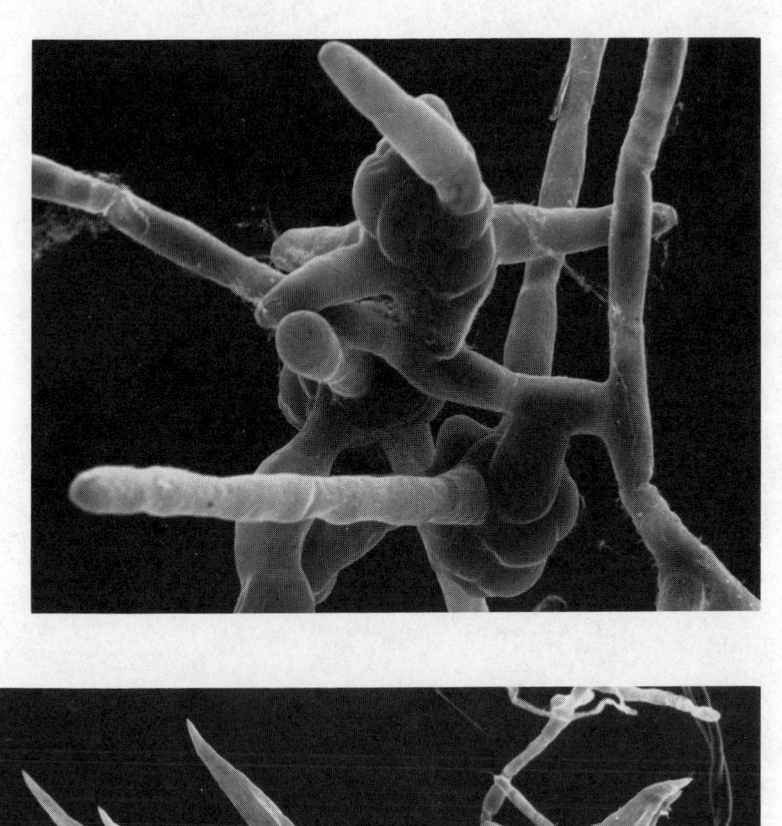

B

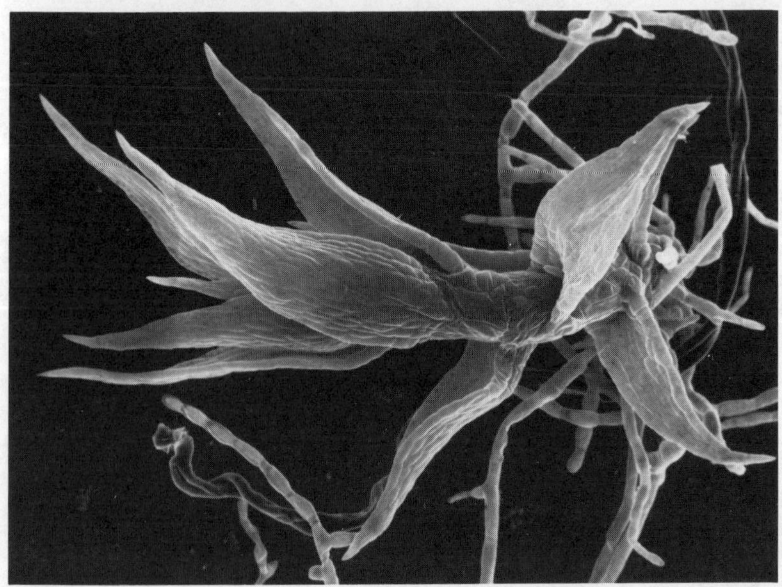

A

FIGURE 3. Scanning electron photomicrograph of (A) normal gametophore and (B) abnormal gametophores of *Pylaisiella selwynii* induced by agrobacteria. (Magnification: normal × 144; abnormal × 510.)

TABLE 1
Some Physiological and Developmental Differences between Three of the Mosses Tested for Response to Bacteria

	Pylaisiella	*Polytrichum*	*Funaria*
Growth medium	Nebel & Naylor	Bold	Bold
Optimal pH	5.0	6.5	6.5
Habitat	Tree, oak logs	Soil	Lime-rich soil
Bud development	On filaments	Center of plant	On filaments
Caulonemal stage	No	Yes	Yes
Calcium requirement	High	Unknown	Low
Appearance of gametophores (days after germination)	28[a]	17—19	17—19

[a] Only a few plants form gametophores in culture.

Other substances produced by bacteria were found to affect *Pylaisiella.* Bud initiation was enhanced 3- to 20-fold by vitamin B_{12}, and the time of appearance of the buds was advanced 6 to 12 days compared to control plants. Differences in the ratio of added IAA and cytokinin also influence the number and structure of callus buds produced by *Pylaisiella selwynii.*[95] Common amino acids including alanine, lysine, tryptophan, arginine, serine, histidine, and methionine increase the growth of *Pylaisiella* protonema.[96] In addition, the unusual amino acid octopine, which is produced by certain *Agrobacterium*-induced tumor tissues,[13] greatly increases protonemal growth, and when combined with a cytokinin it increases the number of callus-forming buds on *Pylaisiella.*[96] At appropriate concentrations of octopine plus cytokinin, normal gametophores develop. Lysopine, which is produced by certain crown gall tissues, also increases the number of buds, many of which become normal gametophores.[97]

The combination of octopine and cytokinin is no more effective than cytokinin alone on *Polytrichum* or *Funaria,*[80,81] but at reduced calcium concentrations *Funaria* also responds to octopine.[98] *Polytrichum* was not tested on low-calcium medium. Moss responses to lysopine and octopine with appropriate concentrations of cytokinin are comparable to those induced by various strains of agrobacteria, suggesting that variations in similar products synthesized by bacteria in combination with moss could account for the many differences in moss response induced by different bacterial strains.

These bacteria-moss effects depend on physical contact between the two organisms in all but a few instances, and this feature may contribute to the specificity of these interactions. Separation of bacteria from moss by bacteria-retaining filters in a parabiotic chamber eliminates the ability of the bacteria to induce gametophores.[99] Heat-killed *Agrobacterium tumefaciens,* bacterial lipopolysaccharide (LPS), or plant cell wall preparations to which the bacterium adheres when added with viable agrobacteria all reduce gametophore production in a concentration-dependent manner. These substances are inactive if added several hours after the viable bacteria, consistent with a mode of inhibition that results from preventing adherence of the viable bacteria to the moss.[91,99,100] Site binding is also essential for disease induction by agrobacteria in higher plants.[101]

A release of stable active substances into the medium after bacterial attachment and interaction with the moss does not appear to be involved, since filter-sterilized medium from such cultures does not promote growth and development of a second moss inoculum[115] and bacteria plus moss on one side of a parabiotic chamber does not affect the moss growing in the opposite chamber.[99] Rhizoid induction on *Pylaisiella* by *Agrobacterium rhizogenes,* in contrast, takes place when these bacteria are separated from the moss in parabiotic chambers.[85] Similarly, when tested in parabiotic chambers, one of four MAB strains that promoted gametophores was effective in the absence of direct physical contact.[79]

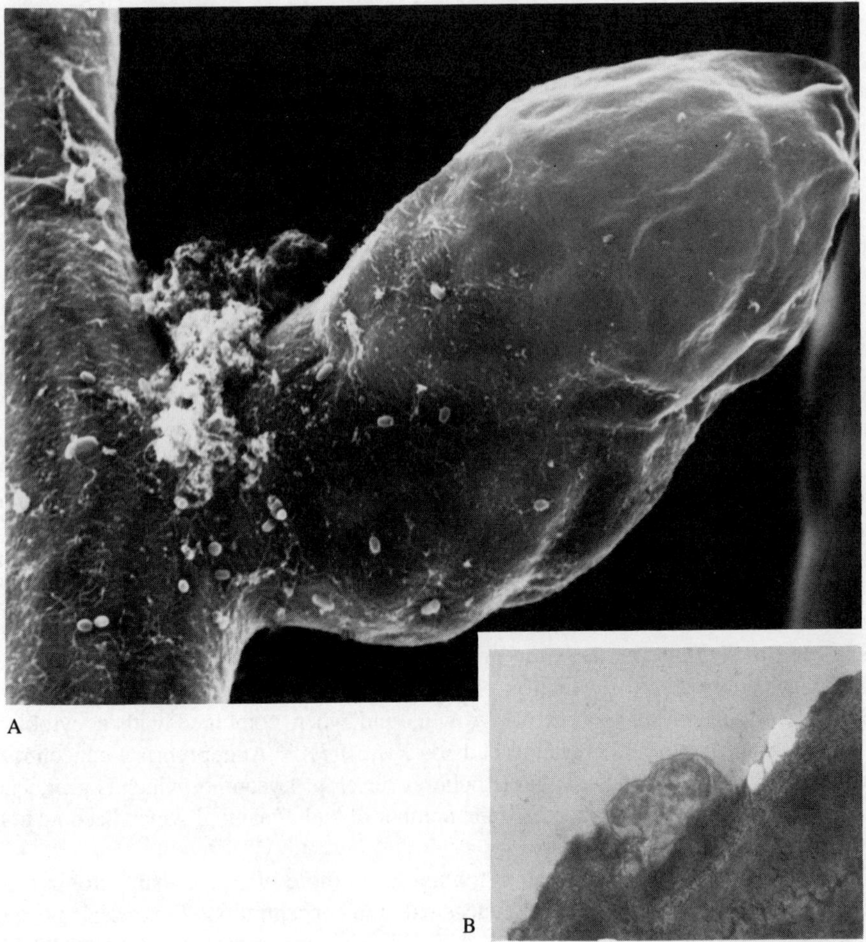

FIGURE 4. (A) Scanning electron photomicrograph showing agrobacteria adhering to the base of a developing *Pylaisiella selwynii* gametophore bud (magnification × 2,340) and (B) a transmission electron photomicrograph showing the tight adherence of a bacterium to the protonemal cell wall (magnification × 11,500).

Scanning electron microscopy demonstrated variations in the external structure of cell walls of *Pylaisiella* and *Funaria* which might contribute to differences in response to agrobacteria. *Funaria* has a smooth cell wall in all stages of development, whereas the outer wall of the *Pylaisiella* protonema is rough and differs in this respect from that of the early germ tube and later gametophores of this moss.[90] Transmission electron micrographs of *Pylaisiella* with *Agrobacterium tumefaciens* show the bacterium embedded in the cell wall of the protonema (Figure 4).[91] Inoue et al.[102,103] found significant differences in the cell wall composition of six species of moss as well as differences in the staining responses of juvenile and mature moss cell walls. Such differences could account for differences in moss response to particular strains of bacteria.

Pectin (and especially its polygalacturonic acid component) inhibits bacterial promotion of gametophore formation and is proposed to be the major cell wall constituent to which these bacteria adhere.[91] Cellulose is inactive in this test, consistent with this proposal.[91] Cell walls of *Polytrichum,* a moss that ordinarily does not respond to agrobacteria, are much less inhibitory in competition tests than those from *Pylaisiella* protonema, but on treatment with pectinesterase become more inhibitory, suggesting that methylated pectin does not support

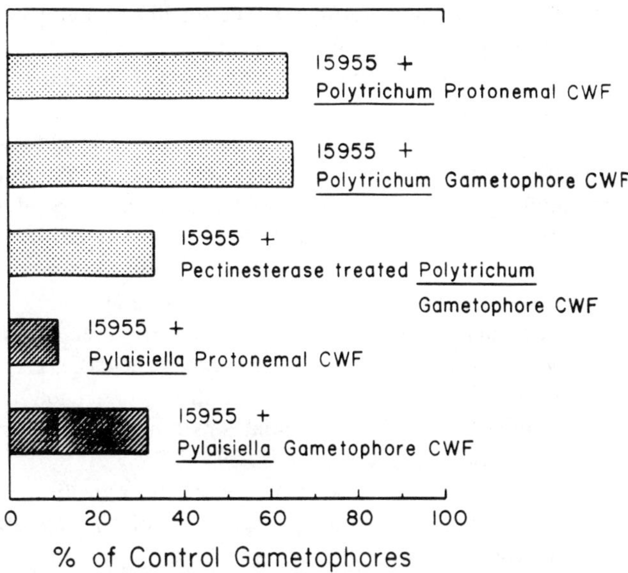

FIGURE 5. The effect of isolated moss cell wall preparations (CWF) on the induction of *Pylaisiella selwynii* gametophores by *Agrobacterium tumefaciens* strain ATCC 15955. Equal amounts of CWF were added to the protonemal cultures just before addition of bacteria and appeared to inhibit bacterial gametophore induction by binding bacteria, thus preventing them from adhering to the moss. *Pylaisiella* is responsive to agrobacteria, whereas *Polytrichum commune* responds only after treatment with pectinesterase. These responses correlate with the effect of CWF from these two mosses and after pectinesterase treatment as illustrated here.

bacterial adherence (Figure 5). After treating *Polytrichum commune* with pectinesterase, gametophore production is promoted by *Agrobacterium*.[91] Similar differences in pectin methylation have been found to correlate with susceptibility to *Agrobacterium* transformation in higher plants.[104]

The effects of octopine on moss development may be partially explained by its ability to chelate divalent cations, since certain other chelating agents such as A23187 and ethyleneglycol-bis-(β-aminoethyl ether)-*N*,*N*′-tetraacetic acid (EGTA) have similar effects in increasing zeatin-induced buds.[98,105] Since the chelator ethylenediaminetetraacetic acid (EDTA) does not promote buds on *Pylaisiella* and has a much lower affinity for calcium than EGTA, the possible involvement of calcium was suggested. Saunders and Hepler[106-108] reported an increase in membrane-associated calcium at presumptive bud sites on *Funaria* after cytokinin treatment, and they noted that calcium antagonists and calmodulin inhibitors block cytokinin-induced bud formation.

Reduction in calcium concentration of the *Pylaisiella* culture medium by a factor of 1000 greatly reduces the number of zeatin-induced callus buds, but similar reductions in magnesium levels have no effect on this response.[98] Agents known to inhibit calcium transport across membranes or calmodulin responses also inhibit this response. These include cobalt, lanthanum, zinc and cadmium chlorides, verapamil, perhexilene maleate, chlorpromazine, calmidazolium, W7, and phenothiazine.[98,105,115] In all cases these inhibitions could be overcome at least partially by octopine or EGTA. These results appear to complement those obtained earlier showing that cyclic adenosine-3′,5′-monophosphate (cAMP) and cyclic guanosine-3′,5′-monophosphate (cGMP) also antagonize cytokinin-induced bud callus formation in this moss.[109] They are also consistent with a mechanism of external induction of gametophore formation that depends on transmembrane signals involving calcium and cyclic nucleotides as second messengers. Johri and colleagues[110,111] have demonstrated that cAMP is present in *Funaria* and that it may be involved in the regulation of caulonemal vs. chloronemal filament formation.

Despite the many data demonstrating a role of calcium in the induction of moss gametophores, this mechanism cannot account in any direct way for the effects of octopine and several other chelating agents on this process. Titration experiments show that octopine is a good chelator for heavy metal ions, but has little or no binding affinity for calcium.[116] Also, the metal ion chelators iminodiacetic acid (IDA) and citric acid promote gametophore formation on *Pylaisiella,* but have little or no calcium chelating affinity,[112,115,116] suggesting that chelation of ions other than calcium may be involved. However, other good chelators, nitrilotriacetic acid (NTA), 1,2-diaminocyclohexanetetraacetic acid (CDTA), and 8-hydroxyquinoline, inhibit like EDTA. Iron and copper ions are strongly chelated by all these agents as well as octopine and EGTA, and these essential ions have been shown to affect moss development[113] and the induction of flowering in *Lemna*.[114]

The possible role of divalent cations other than calcium on zeatin-induced bud callus formation by *Pylaisiella* has been investigated.[115] Omission of all trace elements (Bo, Fe, Zn, Mn, Cu, Se, Mo, I) from the *Pylaisiella* culture medium results in a 34% reduction in the number of callus buds induced by cytokinin, and both octopine and EGTA remain promotive in the absence of these trace ions. If only Fe^{3+} is deleted a 24% reduction occurs, if only Cu^{2+} is omitted a 98% reduction is obtained, but if both are omitted there is only a 20% reduction. Thus, iron and copper appear to be important trace element cations for zeatin-induced gametophore formation, and at least part of the effect of copper is to counter potential inhibitory effects of iron. These differences due to changes in trace elements do not appear to result from protonemal growth differences, as only a 30% reduction in size occurred if only Cu, only Fe, or both were omitted.[115] Protonemal growth also decreases with a decrease in calcium concentration in the medium,[115] although not to the extent that it would appear to account for the decreased production of gametophores.[98]

The effects of cyclic nucleotides, calcium levels, and inhibitors and promoters of calcium uptake on *Pylaisiella* gametophore formation focus attention on the possibility that these effects, as well as those induced by the agrobacteria and MAB, depend on effects mediated through one or more of the transmembrane signaling processes. Compounds that affect inositol phosphate metabolism also affect the formation of these buds, some inhibiting and others promoting their formation, further implicating this type of mechanism.[115] Thus, the interaction of bacterial products with moss membrane receptors may account directly for the response to adhering bacteria, in much the same way as fungal or bacterial carbohydrate elicitors promote phytoalexin synthesis and associated changes in higher plants.[44,45] The ultimate result of the appropriate membrane signals would be changes in gene expression essential for gametophore formation and development.

VI. SUMMARY AND PROSPECTUS

The activities of the *Agrobacterium*-induced crown gall products octopine and lysopine, as well as EGTA, IDA, and A23187, on moss development seem to result from effects that circumvent or complement transmembrane signaling processes. There is no evidence to suggest that EGTA can function as an ion porter like A23187 and, since several other compounds which have metal ion affinities similar to that of EGTA are inactive or even inhibitory, neither calcium binding nor chelation appears sufficient to account for this activity. A possible explanation is that the unique structure of compounds such as EGTA and octopine allows them to interact directly with a particular membrane receptor site and alter its activity, whereas the structure of the inactive chelating agents is sufficiently different that they fail to bind these same sites. Chelation may still be necessary for activity, but insufficient in the absence of appropriate site binding to have any developmental effect.

Bacteria adhering to moss in nature may affect protonemal growth and gametophore initiation and, thus, may be an important ecological factor in the growth and development of mosses. The association is clearly not fortuitous given the specificity involved. Most of the bacteria studied must attach to the moss cell wall to produce these changes, although a few are active at a distance. Many products such as hormones, amino acids, cyclic nucleotides, and siderophores that are produced by microorganisms can initiate moss developmental changes and are suggestive of ways in which bacteria may bring about these changes. Much may still be learned relevant to bacteria-plant interactions and the physiology and biochemistry of plant growth in general from the continued study of these moss-bacteria interactions.

REFERENCES

1. **Crawford, M.,** OSTP ponders plant research initiatives, *Science,* 231, 212, 1986.
2. **Stubblefield, S. P., Taylor, T. N., and Trappe, J. M.,** Fossil mycorrhizae: a case for symbiosis, *Science,* 237, 59, 1987.
3. **Stubblefield, S. P., Taylor, T. N., and Trappe, J. M.,** Vesicular-arbuscular mycorrhizae from the Triassic of Antarctica, *Am. J. Bot.,* 74, 1904, 1987.
4. **Schroth, M. N. and Hancock, J. G.,** Selected topics in biological control, *Annu. Rev. Microbiol.,* 35, 453, 1981.
5. **Schroth, M. N. and Hancock, J. G.,** Disease-suppressive soil and root-colonizing bacteria, *Science,* 216, 1376, 1982.
6. **Gelvin, S. B.,** Plant tumorigenesis, in *Plant-Microbe Interactions,* Vol. 1, Kosuge, T. and Nester, E. W., Eds., Macmillan, New York, 1984, 343.
7. **Kerr, A.,** The impact of molecular genetics on plant pathology, *Annu. Rev. Phytopathol.,* 25, 87, 1987.
8. **Barton, K. A. and Brill, W. J.,** Prospects in plant genetic engineering, *Science,* 219, 671, 1983.
9. **Horsch, R. B., Fry, J. E., Hoffmann, N. L., Eichholtz, D., Rogers, S. G., and Fraley, R. T.,** A simple and general method for transferring genes into plants, *Science,* 227, 1229, 1985.
10. **Klee, H., Horsch, R., and Rogers, S.,** *Agrobacterium*-mediated plant transformation and its further applications to plant biology, *Annu. Rev. Plant Physiol.,* 38, 467, 1987.
11. **Brown, M. E.,** Seed and root bacterization, *Annu. Rev. Phytopathol.,* 12, 181, 1974.
12. **Greene, E. M.,** Cytokinin production by microorganisms, *Bot. Rev.,* 46, 25, 1980.
13. **Nester, E. W. and Kosuge, T.,** Plasmids specifying plant hyperplasias, *Annu. Rev. Microbiol.,* 35, 531, 1981.
14. **Triplett, E. W., Heitholt, J. J., Evensen, K. B., and Blevins, D. G.,** Increase in internode length of *Phaseolus lunatus* L. caused by inoculation with a nitrate reductase deficient strain of *Rhizobium* sp., *Plant Physiol.,* 67, 1, 1981.
15. **Smidt, M. L. and Kosuge, T.,** The role of indole-3-acetic acid accumulation by alpha methyl tryptophan-resistant mutants of *Pseudomonas savastanoi* in gall formation on oleanders, *Physiol. Plant Pathol.,* 13, 203, 1978.
16. **Morris, R. O.,** Genes specifying auxin and cytokinin biosynthesis in phytopathogens, *Annu. Rev. Plant Physiol.,* 37, 509, 1986.
17. **Tien, T. M., Gaskins, H. H., and Hubbell, D. H.,** Plant growth substances produced by *Azospirillum brasilense* and their effect on growth of pearl millet *(Pennisetum americanum* L.*), Appl. Environ. Microbiol.,* 37, 1016, 1979.
18. **Loper, J. E and Schroth, M. N.,** Influence of bacterial sources of indole-3-acetic acid on root elongation of sugar beet, *Phytopathology,* 76, 386, 1986.
19. **Suslow, T. V. and Schroth, M. N.,** Role of deleterious rhizobacteria as minor pathogens in reducing crop growth, *Phytopathology,* 72, 111, 1982.
20. **Baker, K. F.,** Evolving concepts of biological control of plant pathogens, *Annu. Rev. Phytopathol.,* 25, 67, 1987.
21. **Stutz, E. W., Defago, G., and Kern, H.,** Naturally occurring fluorescent pseudomonads involved in suppression of black root rot of tobacco, *Phytopathology,* 76, 181, 1986.
22. **Moore, L. W. and Warren, G.,** *Agrobacterium radiobacter* strain 84 and biological control of crown gall, *Annu. Rev. Phytopathol.,* 17, 163, 1979.

23. **Kerr, A. and Tate, M. E.,** Agrocins and the biological control of crown gall, *Microbiol. Sci.*, 1, 1, 1984.
24. **Roberts, W. P., Tate, M. E., and Kerr, A.,** Agrocin 84 is a 6-*N*-phosphoramidate of an adenosine nucleotide analogue, *Nature (London)*, 265, 379, 1977.
25. **Tate, M. E., Murphy, P. J., Roberts, W. P., and Kerr, A.,** Adenine N^6-substituent of agrocin 84 determines its bacteriocin-like specificity, *Nature (London)*, 280, 697, 1979.
26. **Hendson, M., Askjaer, L., Thomson, J. A., and Van Montagu, M.,** Broadhost-range agrocin of *Agrobacterium tumefaciens, Appl. Environ. Microbiol.*, 45, 1526, 1983.
27. **Schroth, M. N., Loper, J. E., and Hildebrand, D. C.,** Bacteria as biocontrol agents of plant disease, in *Current Perspectives in Microbial Ecology*, Klug, M. J. and Reddy, C. A., Eds., American Society for Microbiology, Washington D.C., 1984, 362.
28. **Kloepper, J. W., Leong, J., Teintze, M., and Schroth, M. N.,** Enhanced plant growth by siderophores produced by plant growth-promoting rhizobacteria, *Nature (London)*, 286, 885, 1980.
29. **Leong, J.,** Siderophores: their biochemistry and possible role in the biocontrol of plant pathogens, *Annu. Rev. Phytopathol.*, 24, 187, 1986.
30. **Schippers, B., Bakker, A. W., and Bakker, P. A. H. M.,** Interactions of deleterious and beneficial rhizosphere microorganisms and the effect of cropping practices, *Annu. Rev. Phytopathol.*, 25, 339, 1987.
31. **Neilands, J. B. and Leong, S. A.,** Siderophores in relation to plant growth and disease, *Annu. Rev. Plant Physiol.*, 37, 187, 1986.
32. **Kotoujansky, A.,** Molecular genetics of pathogenesis by soft-rot Erwinias, *Annu. Rev. Phytopathol.*, 25, 405, 1987.
33. **Berry, A. M.,** The actinorhizal infection process: review of recent research, in *Current Perspectives in Microbial Ecology*, Klug, M. J. and Reddy, C. A., Eds., American Society for Microbiology, Washington D.C., 1984, 222.
34. **Tjepkema, J. D., Schwintzer, C. R., and Benson, D. R.,** Physiology of actinorhizal nodules, *Annu. Rev. Plant Physiol.*, 37, 209, 1986.
35. **Djordjevic, M. A., Gabriel, D. W., and Rolfe, B. G.,** *Rhizobium* — the refined parasite of legumes, *Annu. Rev. Phytopathol.*, 25, 145, 1987.
36. **Smith, R. L., Bouton, J. H., Schank, S. C., Quesenberry, R. H., Typer, M. E., Gaskins, M. H., and Littel, R. C.,** Nitrogen fixation in grasses inoculated with *Spirillum lipoferum, Science*, 193, 1003, 1976.
37. **Kapulnik, Y., Okon, Y., Kigel, J., Nur, I., and Henis, Y.,** Effects of temperature, nitrogen fertilization, and plant age on nitrogen fixation by *Setaria italica* inoculated with *Azospirillum brasilense* (strain cd), *Plant Physiol.*, 68, 340, 1981.
38. **Dobereiner, J., Day, J. M., and Dart, P. J.,** Associative symbioses in tropical grasses: characterization of microorganisms and dinitrogen-fixing sites, in *Proc. 1st Int. Symp. Nitrogen Fixation*, Vol. 2, Newton, W. E. and Nyman, C. J., Eds., Washington State University Press, Pullman, 1972, 518.
39. **Benson, D. R.,** Consumption of atmospheric nitrogen, in *Bacteria in Nature*, Vol. 1, Leadbetter, E. R. and Poindexter, J. S., Eds., Plenum Press, New York, 1985, 155.
40. **Kennedy, C. and Toukdarian, A.,** Genetics of azotobacters: applications to nitrogen fixation and related aspects of metabolism, *Annu. Rev. Microbiol.*, 41, 227, 1987.
41. **Pitts, G. A., Allan, I., and Hallis, J. F.,** Beggiatoa: occurrence in the rice rhizosphere, *Science*, 178, 990, 1972.
42. **Siegel, M. R., Latch, G. C. M., and Johnson, M. C.,** Fungal endophytes of grasses, *Annu. Rev. Phytopathol.*, 25, 293, 1987.
43. **Darvill, A. G. and Albersheim, P.,** Phytoalexins and their elicitors — a defense against microbial infection in plants, *Annu. Rev. Plant Physiol.*, 35, 243, 1984.
44. **Ebel, J.,** Phytoalexin synthesis: the biochemical analysis of the induction process, *Annu. Rev. Phytopathol.*, 24, 235, 1986.
45. **Halverson, L. J. and Stacey, G.,** Signal exchange in plant-microbe interactions, *Microbiol. Rev.*, 50, 193, 1986.
46. **Ahmadjian, V.,** *The Lichen Symbiosis*, Ginn/Blaisdell, Boston, 1967.
47. **Lawrey, J. D.,** Biological role of lichen substances, *Bryologist*, 89, 111, 1986.
48. **Hayes, W. A., Randle, P. E., and Last, F. T.,** The nature of the microbial stimulus affecting sporophore formation in *Agaricus bisporus* (Lange) Sing., *Ann. Appl. Biol.*, 64, 177, 1969.
49. **Park, J. Y. and Agnihotri, V. P.,** Bacterial metabolites trigger sporophore formation in *Agaricus bisporus, Nature (London)*, 222, 984, 1969.
50. **Curto, S. and Favelli, F.,** Stimulative effect of certain micro-organisms (bacteria, yeasts, microalgae) upon fruit-body formation of *Agaricus bisporus* (Lange) Sing., *Mushroom Sci.*, 8, 67, 1972.
51. **Stamets, P. and Chilton, J. S.,** *The Mushroom Cultivator, A Practical Guide to Growing Mushrooms at Home*, Agarikon Press, Olympia, WA, 1983.
52. **Sprent, J. I.,** *The Biology of Nitrogen-Fixing Organisms*, McGraw-Hill, London, 1979.

53. **Rodgers, G. A. and Stewart, W. D. P.,** The cyanophyte-hepatic symbiosis. I. Morphology and physiology, *New Phytol.*, 78, 441, 1977.
54. **Dalton, D. A. and Chatfield, J. M.,** A new nitrogen-fixing cyanophyte-hepatic association: *Nostoc* and *Porella, Am. J. Bot.*, 72, 781, 1985.
55. **Ridgeway, J. E.,** The biotic relationship of *Anthoceros* and *Phaeoceros* to certain Cyanophyta, *Ann. Mo. Bot. Gard.*, 54, 95, 1967.
56. **Duckett, J. G., Prasad, A. K. S. K., Davies, D. A., and Walker, S.,** A cytological analysis of the *Nostoc*-bryophyte relationship, *New Phytol.*, 79, 349, 1977.
57. **Basile, D. V., Slade, L. L., and Corpe, W. A.,** An association between a bacterium and a liverwort, *Scapania nemorosa, Bull. Torrey Bot. Club,* 96, 711, 1969.
58. **Corpe, W. A. and Basile, D. V.,** Methanol-utilizing bacteria associated with green plants, *Dev. Ind. Microbiol.*, 23, 483, 1982.
59. **Basile, D. V., Basile, M. R., Li, Q.-Y., and Corpe, W. A.,** Vitamin B_{12}-stimulated growth and development of *Jungermannia leiantha* Grolle and *Gymnocolea inflata* (Huds.) Dum. (Hepaticae), *Bryologist,* 88, 77, 1985.
60. **Poston, J. M.,** Coenzyme B_{12}-dependent enzymes in potatoes: leucine 2,3-aminomutase and methylmalonyl CoA mutase, *Phytochemistry,* 14, 401, 1978.
61. **Evans, H. J. and Kliewer, M.,** Vitamin B_{12} compounds in relation to the requirements of cobalt for higher plants and nitrogen-fixing organisms, *Ann. N.Y. Acad. Sci.*, 112, 735, 1963.
62. **Alexander, V. and Shell, D. M.,** Seasonal and spatial variation of nitrogen fixation in the Barrow, Alaska tundra, *Arct. Alp. Res.*, 5, 77, 1973.
63. **Granhall, V. and Selander, H.,** Nitrogen fixation in a subarctic mire, *Oikos,* 24, 8, 1973.
64. **Jordan, D. C., McNicol, P. J., and Marshall, M. R.,** Biological nitrogen fixation in the terrestrial environment of a high Arctic ecosystem (Truelove Lowland, Devon Island, N.W.T.), *Can. J. Microbiol.*, 24, 643, 1978.
65. **Rodgers, G. A. and Henriksson, E.,** Associations between the blue-green algae *Anabaena variabilis* and *Nostoc muscorum* and the moss *Funaria hygrometrica* with reference to the colonization of Surtsey, *Acta Bot. Isl.*, 4, 10, 1976.
66. **Granhall, V. and Hofsten, A. V.,** Nitrogenase activity in relation to intercellular organisms in *Sphagnum* moss, *Physiol. Plant.*, 38, 88, 1976.
67. **Brasiler, K., Granhall, V., and Stenstron, T. A.,** Nitrogen fixation in wet minerotrophic moss communities of a subarctic mire, *Oikos,* 31, 236, 1978.
68. **Brasiler, K.,** Fixation and uptake of nitrogen in *Sphagnum* blue-green algal associations, *Oikos,* 34, 239, 1980.
69. **Scheirer, D. C. and Brasell, H. M.,** Epifluorescence microscopy for the study of nitrogen fixing blue-green algae associated with *Funaria hydrometrica* (Bryophyta), *Am. J. Bot.*, 71, 461, 1984.
70. **Reese, W. D.,** "Chlorochytrium," a green alga endophytic in musci, *Bryologist,* 84, 75, 1981.
71. **Scheirer, D. C. and Dolan, H. A.,** Bryophyte leaf epiflora: an SEM and TEM study of *Polytrichum commune* Hedw., *Am. J. Bot.*, 70, 712, 1983.
72. **Broady, P. A.,** The Signy Island terrestrial reference sites. VII. The ecology of the alga of site I, a moss turf, *Br. Antarct. Surv. Bull.*, 45, 47, 1977.
73. **Parke, H. and Linderman, R. G.,** Association of vesicular-arbuscular mycorrhizal fungi with the moss *Funaria hygrometrica, Can. J. Bot.*, 58, 1898, 1980.
74. **Rabitin, S. C.,** The occurrence of the vesicular-arbuscular mycorrhizal fungus *Glomus tenuis* with moss, *Mycologia,* 72, 191, 1980.
75. **Redhead, S. A.,** Parasitism of bryophytes by agarics, *Can. J. Bot.*, 59, 63, 1981.
76. **Grasso, S. M. and Scheirer, D. C.,** Scanning electron microscopic observations of a moss-fungus association, *Bryologist,* 84, 348, 1981.
77. **Chopra, R. N., Kumra, P. K., and Rekhi, A.,** Influence of *Rhodotorula* and *Aerobacter* on protonemal growth and bud initiation in two mosses, *Curr. Sci.*, 47, 735, 1978.
78. **Banerjee, R. D. and Sen, S. P.,** Antibiotic activity of bryophytes, *Bryologist,* 82, 141, 1979.
79. **Spiess, L. D., Lippincott, B. B., and Lippincott, J. A.,** Bacteria isolated from moss and their effect on moss development, *Bot. Gaz. (Chicago),* 142, 512, 1981.
80. **Spiess, L. D., Lippincott, B. B., and Lippincott, J. A.,** Bacterial-moss interaction in the regulation of protonemal growth and development, *J. Hattori Bot. Lab.*, 53, 215, 1982.
81. **Spiess, L. D., Lippincott, B. B., and Lippincott, J. A.,** Moss growth and development is facilitated by natural bacterial flora, in *Being Alive on Land,* Margaris, N. S., Arianoustou-Farragitaki, M., and Oechel, W. C., Eds., Junk, The Hague, 1984, 271.
82. **Spiess, L. D., Lippincott, B. B., and Lippincott, J. A.,** Specificity of moss response to moss associated bacteria: some influences of moss species, habitat and locale, *Bot. Gaz. (Chicago),* 147, 418, 1986.
83. **Crum, H. A. and Anderson, L. E.,** *Mosses of Eastern North America,* Columbia University Press, New York, 1981.

84. **Spiess, L. D., Lippincott, B. B., and Lippincott, J. A.,** Development and gametophore initiation in the moss *Pylaisiella selwynii* as influenced by *Agrobacterium tumefaciens, Am. J. Bot.*, 58, 726, 1971.
85. **Spiess, L. D., Lippincott, B. B., and Lippincott, J. A.,** Comparative response of *Pylaisiella selwynii* to *Agrobacterium* and *Rhizobium* species, *Bot. Gaz. (Chicago)*, 138, 35, 1977.
86. **Allen, O. N. and Holding, A. J.,** *Agrobacterium*, in *Bergey's Manual of Determinative Bacteriology*, 8th ed., Buchanan, R. E. and Gibbons, N. E., Eds., Williams and Wilkins, Baltimore, 1974, 264.
87. **Monod, J.,** Sur une mutation spontanée affectant le pouvoir de synthèse de la methioniné chez une bacterie coliforme, *Ann. Inst. Pasteur (Paris)*, 72, 879, 1946.
88. **Lippincott, J. A. and Heberlein, G. T.,** The quantitative determination of the infectivity of *Agrobacterium tumefaciens, Am. J. Bot.*, 52, 856, 1965.
89. **Lippincott, J. A., Lippincott, B. B., and Starr, M. P.,** The genus Agrobacterium, in *The Prokaryotes*, Starr, M. P., Stolp, H., Truper, H. G., Balows, A., and Schlegel, H. G., Eds., Springer-Verlag, Heidelberg, 1981, 842.
90. **Spiess, L. D., Turner, J. C., Malhberg, P. G., Lippincott, B. B., and Lippincott, J. A.,** Adherence of agrobacteria to moss protonema and gametophores viewed by scanning electron microscopy, *Am. J. Bot.*, 64, 1200, 1977.
91. **Spiess, L. D., Lippincott, B. B., and Lippincott, J. A.,** Role of the moss cell wall in gametophore formation induced by *Agrobacterium tumefaciens, Bot. Gaz. (Chicago)*, 145, 302, 1984.
92. **Klambt, D., Thies, G., and Skoog, F.,** Isolation of cytokinins from *Corynebacterium fasciens, Proc. Natl. Acad. Sci. U.S.A.*, 56, 52, 1966.
93. **Spiess, L. D.,** Developmental effects of zeatin, ribosyl-zeatin and *Agrobacterium tumefaciens* B6 on certain mosses, *Plant Physiol.*, 58, 107, 1976.
94. **Spiess, L. D.,** Comparative activity of isomers of zeatin and ribosyl-zeatin on *Funaria hygrometrica, Plant Physiol.*, 55, 583, 1975.
95. **Spiess, L. D., Lippincott, B. B., and Lippincott, J. A.,** Effect of hormones and vitamin B_{12} on gametophore development in the moss *Pylaisiella selwynii, Am. J. Bot.*, 60, 708, 1973.
96. **Spiess, L. D., Lippincott, B. B., and Lippincott, J. A.,** Promotion of *Pylaisiella selwynii* growth and gametophore formation by octopine and cytokinin, *Physiol. Plant*, 51, 99, 1981.
97. **Spiess, L. D., Lippincott, B. B., and Lippincott, J. A.,** Influence of certain plant growth regulators and crown gall related substances on bud formation and gametophore development of the moss *Pylaisiella selwynii, Am. J. Bot.*, 59, 233, 1972.
98. **Spiess, L. D., Lippincott, B. B., and Lippincott, J. A.,** Influence of octopine, calcium and compounds that affect calcium transport on zeatin-induced bud formation by *Pylaisiella selwynii, Am. J. Bot.*, 71, 1416, 1984.
99. **Spiess, L. D., Lippincott, B. B., and Lippincott, J. A.,** The requirement of physical contact for moss gametophore induction by *Agrobacterium tumefaciens, Am. J. Bot.*, 63, 324, 1976.
100. **Whatley, M. H. and Spiess, L. D.,** Role of bacterial lipopolysaccharide in attachment of *Agrobacterium* to moss, *Plant Physiol.*, 60, 765, 1977.
101. **Lippincott, B. B. and Lippincott, J. A.,** Bacterial attachment to a specific wound site as an essential stage in tumor initiation by *Agrobacterium tumefaciens, J. Bacteriol.*, 97, 620, 1969.
102. **Inoué, S., Ishida, A., and Kodama, M.,** Polysaccharide composition of the bryophyte cell wall, *Proc. Bryol. Soc. Jpn.*, 2, 169, 1980.
103. **Inoué, S., Ishida, A., and Kodama, M.,** Cellulose and uronic acid contents of cell wall isolated from gametophytes of some mosses and liverworts, *J. Hattori, Bot. Lab.*, 49, 141, 1981.
104. **Rao, S. S., Lippincott, B. B., and Lippincott, J. A.,** *Agrobacterium* adherence involves the pectic portion of the host cell wall and is sensitive to the degree of pectin methylation, *Physiol. Plant.*, 56, 374, 1982.
105. **Spiess, L. D., Lippincott, B. B., and Lippincott, J. A.,** Facilitation of moss growth and development by bacteria *J. Hattori Bot. Lab.*, 55, 67, 1984.
106. **Saunders, M. J. and Hepler, P. K.,** Localization of membrane associated calcium following cytokinin treatment in *Funaria* using chlorotetracycline, *Planta*, 152, 272, 1981.
107. **Saunders, M. J. and Hepler, P. K.,** Calcium ionophore A23187 stimulates cytokinin-like mitosis in *Funaria, Science*, 217, 943, 1982.
108. **Saunders, M. J. and Hepler, P. K.,** Calcium antagonists and calmodulin inhibitors block cytokinin-induced bud formation in *Funaria, Dev. Biol.*, 99, 41, 1983.
109. **Spiess, L. D.,** Antagonism of cytokinin induced callus in *Pylaisiella selwynii* by nucleosides and cyclic nucleotides, *Bryologist*, 82, 47, 1979.
110. **Handa, A. K. and Johri, M. M.,** Involvement of cyclic adenosine-3′,5′-monophosphate in chloronema differentiation in protonema cultures of *Funaria hygrometrica, Planta*, 144, 317, 1979.
111. **Sharma, S. and Johri, M. M.,** Uptake and degradation of cyclic AMP by chloronema cells, *Plant Physiol.*, 69, 1401, 1982.

112. **Sillen, L. G. and Martell, A. E.,** *Stability Constants of Metal Ion Complexes,* Suppl. 1, Special Publ. 25, Chemical Society of London, London, England, 1971.
113. **Rahbar, K. and Chopra, R. N.,** Effect of chelating agents on bud induction and accumulation of iron and copper by the moss *Bartramidula bartramioides, Physiol. Plant,* 59, 148, 1983.
114. **Khurana, J. P. and Maheshwari, S. C.,** Floral induction in short-day *Lemna paucicostata* 6746 by 8-hydroxyquinoline, under long days, *Plant Cell Physiol.,* 25, 77, 1984.
115. **Spiess, L. D., Lippincott, B. B., and Lippincott, J. A.,** unpublished data, 1987.
116. **Lippincott, J. A. and Lippincott, B. B.,** unpublished data, 1987.
117. **Gelvin, S. B.,** Plant tumorigenesis, in *Plant-Microbe Interactions,* Vol. 1, Kosuge, T. and Nester, E., Eds., Macmillan, New York, 1984.

Chapter 8

SEXUAL REPRODUCTION IN BRYOPHYTES IN RELATION TO PHYSICAL FACTORS OF THE ENVIRONMENT

R. E. Longton

TABLE OF CONTENTS

I. INTRODUCTION

Bryophytes are unique among land plants in that the gametophyte generation is photosynthetically dominant, with the sporophyte permanently attached to and partially parasitic upon the gametophyte. It is therefore convenient to consider the process of sexual reproduction as beginning with the development of gametangia upon the gametophyte and proceeding through gametangial maturation, fertilization, sporophyte development, and sporogenesis to the release of spores. Thereafter, spore germination and, commonly, protonema growth and bud induction must be accomplished before a new generation of gametophytes is established. All these phases are affected by physical factors of the environment which interact with inherent characteristics of the biotypes concerned to regulate the seasonal timing of the various events and to determine the success of the reproductive process.

There are two essential approaches to elucidating mechanisms of environmental control over biological processes. The first involves field studies, in the present case investigating matters such as seasonal patterns of gametangial and sporophytic development in relation to climatic and microclimatic conditions. Such work is valuable in generating hypotheses concerning the factors controlling various stages in the reproductive process. Since individual environmental factors seldom vary in isolation, the hypotheses require testing by experiment. The second approach, therefore, involves laboratory studies in which the influence on reproductive development of individual factors, or combinations of factors, is investigated by experiments under controlled conditions. Ideally, this work should lead to inquiry into the mode of action of the environmental control thus demonstrated, whether by direct effects on rates of cell division and cell enlargement or indirectly by influencing the supply of auxins and other growth regulators.

In work on the environmental control of reproductive development in bryophytes there has been little coordination so far between field and laboratory studies, which have largely been carried out by different investigators looking at different species. The value of a more integrated approach is demonstrated below in relation to a few species such as *Plagiomnium* (= *Mnium*) *undulatum* (Sections II.B.3.b, III.A, and III.B.2.a). Similarly, while much has been learned about the influence of chemical growth regulators on bryophyte reproduction, as reviewed in Chapter 9 of the present volume, little is known yet about environmental control over the concentration and distribution of these compounds within the plant.

II. THE FIELD APPROACH

A. REPRODUCTIVE PHENOLOGY

1. Phenological Patterns in Temperate, Perennial Mosses

There have been two main phases in studies of seasonal patterns of reproductive development. In early investigations[1-8] a number of authors reported dates of fertilization and spore release in a wide range of European and North American mosses. While lacking in detail these studies, as reviewed by Zander,[9] demonstrated the occurrence of seasonal periodicity in many species, with major differences between species and minor differences between the dates for a given taxon in climatically different regions. Thus, Arnell[1] recognized two patterns among species of *Polytrichum,* one with capsule dehiscence occurring during the summer about 1 year after fertilization and the other with more rapid development leading to capsule dehiscence during the winter or early spring, only 7 to 10 months after fertilization. Fertilization in these and other mosses was shown commonly to occur 1 to 2 months earlier in Germany than in Sweden, a result attributed to development in Sweden being delayed by prolonged winter snow cover.

During the second phase, in the past 30 years, more intensive studies of several species have been undertaken by scoring sequential samples from individual populations, or single

TABLE 1
Stages of Gametangial and Sporophytic Development in Mosses[10]

Stage	Index value	Event marking beginning of stage
Gametangia		
Juvenile (J)	1	Gametangia become visible at ×20; pale green
Immature (I)	2	Gametangia reach half length of dehisced gametangia
Mature (M)	3	Apices of gametangia rupture; archegonia become receptive for fertilization, and liberation of antherozoids from antheridia begins
Dehisced (D)	4	Development of brown coloration begins in gametangia with ruptured apices
Aborted (A)		Development of brown or hyaline coloration begins in gametangia with unruptured apices in J or I stages
Sporophytes		
Swollen venter (SV)	1	Venter of archegonium begins to swell
Early calyptra in perichaetium (ECP)	2	Calyptra becomes recognizable as a distinct structure
Late calyptra in perichaetium (LCP)	3	Calyptra becomes half exserted from bracts
Early calyptra intact (ECI)	4	Calyptra becomes fully exserted from bracts
Late calyptra intact (LCI)	5	Swelling of capsule begins
Early operculum intact (EOI)	6	Annulus becomes red or brown
Late operculum intact (LOI)	7	Capsule becomes brown
Operculum fallen (OF)	8	Operculum falls
Empty and fresh (EF)	9	>75% of spores are shed
Aborted (A)		Apex of sporophyte withers prior to spore formation, usually in ECP, LCP, or ECI stages

samples collected from many populations within a region at different times of the year, according to schemes such as that outlined in Table 1.[10,11] By averaging the index values (Table 1), numerical indices summarizing the developmental state of antheridia, archegonia, and sporophytes within each sample (maturity indices) can be calculated.[12,13] The results can then be expressed either graphically (Figures 1 and 2)[14] or in tabular form.

Table 2 summarizes data for a number of species selected from those subjected to recent critical study (usually over 2 or more years), in order to reflect the wide range of phenological patterns represented among northern temperate mosses. Thus, gametangia are first observed from spring (e.g., *Entodon cladorrhizans*) through summer (*Forsstroemia trichomitria*) to autumn (*Pleurozium schreberi*) and winter (*Brachythecium rutabulum*). Male and female gametangia may appear at the same time (*Forsstroemia trichomitria*), but more commonly juvenile antheridia are recorded several months before archegonia (*Polytrichum alpestre*), even in monoecious taxa (*Forsstroemia producta*), suggesting that different factors may stimulate induction of the two sexes. The time of gametangial appearance shown in Table 2 was generally determined by dissection and examination at magnifications of ×20 to ×25[10] or by squashing shoot apices under a cover glass before observation at magnifications up to ×70.[20] It is possible that gametangia are initiated some time before they become recognizable by these methods, but Newton's[21] careful histological study of *Mnium hornum,* in which she recognized antheridia and archegonia at the one-cell stage, suggested that in this species most gametangia are initiated in early spring as suggested by dissecting microscope studies (Table 2).

Fertilization may begin within 1 to 2 months of the first appearance of archegonia (*Entodon cladorrhizans*), or it may be long delayed as in *Forsstroemia producta*, reflecting slower gametangial maturation. Gametangia initiated during the autumn usually overwinter in the juvenile or immature stages. The principal period of fertilization thus ranges from spring to autumn (Table 2, and in most species considered in broader surveys such as those

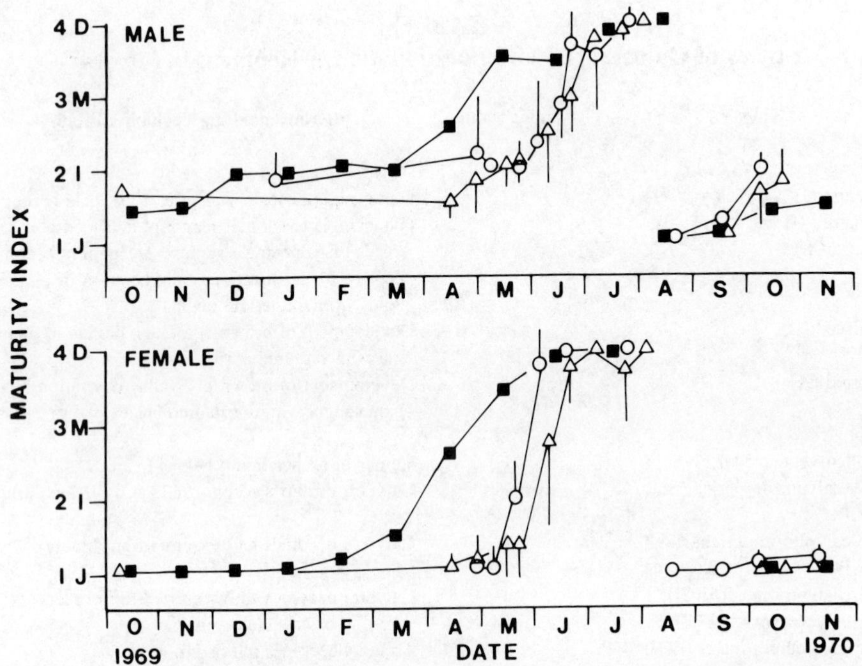

FIGURE 1. Gametangial phenology of *Pleurozium schreberi* at two sites (○ and △) in the boreal forest region of southeastern Manitoba, Canada, showing maturity indices compared with monthly indices for Great Britain (■). Vertical lines indicate 1 SD. (From Longton, R. E., *Monogr. Syst. Bot. Mo. Bot. Gard.*, 11, 51, 1985. With permission.)

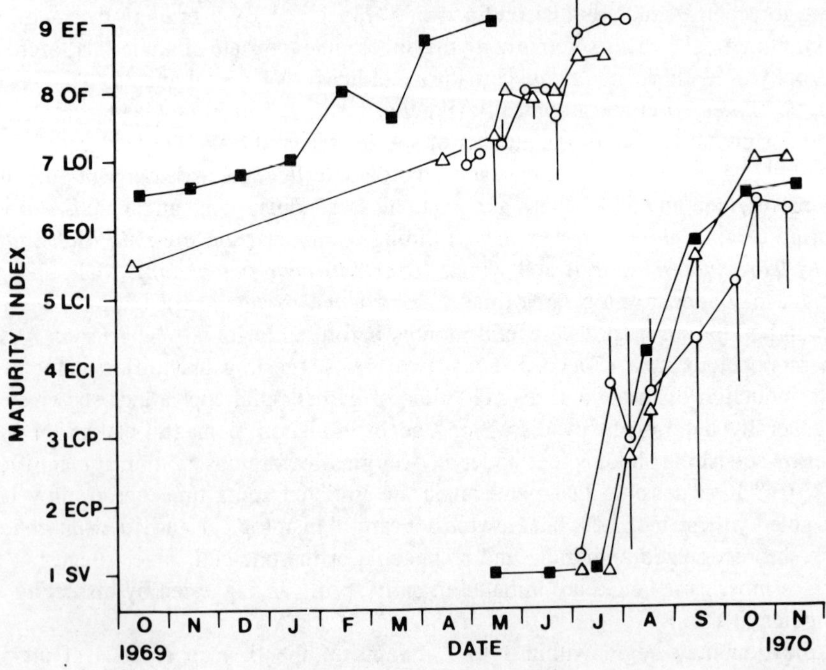

FIGURE 2. Sporophytic phenology of *Pleurozium schreberi* at two sites (○ and △) in the boreal forest region of southeastern Manitoba, Canada, showing maturity indices compared with monthly indices for Great Britain (■). Vertical lines indicate 1 SD. (From Longton, R. E., *Monogr. Syst. Bot. Mo. Bot. Gard.*, 11, 51, 1985. With permission.)

TABLE 2
Reproductive Phenology of Selected Perennial Mosses in Temperate Regions

Species	Gametangial appearance		Principal period of fertilization	Sporophyte resting phase	Spore liberation	Approximate time from fertilization to spore liberation (months)	Location	Ref.
	Male	Female						
Brachythecium rutabulum	Dec	Feb	Jul—Aug	Oct—Dec (LO1)	Dec—Feb	6	Great Britain	10
Bryum argenteum	Oct—Nov	Oct—Mar	Apr—Jun	None	Dec—May	6—9	Great Britain	15
Entodon cladorrhizans	Mar	May	Jun	Nov—Jan (LO1)	Jan—Mar	6—9	Pennsylvania	13
Atrichum undulatum	Jan—Feb	Mar—Apr	May—Jun	Oct—Dec (LO1)	Dec—May	7	Great Britain	15
Pleurozium schreberi	Aug	Oct	Apr—May	May—Jul (SV): Oct—Jan (LO1)	Jan—Apr	9—11	Great Britain	16
Mnium hornum	Feb—Mar	Mar	Apr—Jul	Oct—Feb (EC1)	Apr—May	10—12	Great Britain	10
Polytrichum alpestre	Sep—Oct	Mar—Apr	Jun—Jul	Aug—Feb (ECP/LCP)	Jun—Jul	12	Great Britain	15
Pohlia nutans	(Sep—)Feb	(Sep—)Mar	Apr—May	Oct—Mar (EC1)	Jun—Jul	14	Great Britain	17
Pylasia polyantha	—	—	Aug(?)	Mar—Jun (EC1)	Dec—Feb	14—18	Great Britain	10
Forsstroemia trichomitria	Jun	Jun	Sep—Nov	Oct—Jun (SV)	Feb—Mar	15—17	Virginia	18,19
Forsstroemia producta	Dec	Mar	Aug—Sep	Aug—Jun (SV)	Feb—Mar	17—18	Virginia	18

of Arnell[1] and van der Wijk[22]). It is possible that this reflects an adaptation to overcome problems of sperm motility under cold conditions, but this point has yet to be investigated experimentally, and winter fertilization occurs in some species, e.g., *Ulota bruchii.*[22]

The period required for sporophyte development commonly ranges from 6 months (*Brachythecium rutabulum*) to 18 months with two annual cycles developing concurrently (*Forsstroemia producta*). Van der Wijk[22] suggested that sporophyte development in *Ptilium crista-castrensis* takes almost 3 years. Sporophyte development in most species shows a resting phase when little morphological change can be detected for several months. This may involve embryos still wholly or partly enclosed within the calyptra (*Forsstroemia* spp., *Polytrichum alpestre*), sporophytes with setal elongation complete but before capsule development (*Mnium hornum*), or mature sporophytes containing viable spores (*Brachythecium rutabulum*). Such resting phases normally embrace all or part of a winter season (Table 2), but may occur in spring (*Pylasia polyantha*) or extend for up to 9 months (*Forsstroemia trichomitria*). Finally, spore liberation also may occur at any time from spring (*Mnium hornum*) through summer (*Pohlia nutans*) and autumn (*Oxystegus tenuirostris*)[9] to winter (*Brachythecium rutabulum*).

Many temperate perennial mosses thus show rather clearly defined seasonal patterns of gametangial and sporophytic development. It should be noted, however, that the cycles may be modified by climatic variation from year to year, with development subject to delay by extreme temperatures or drought. Thus, in Britain dry summer conditions may delay fertilization until autumn in *Atrichum undulatum* (see Table 2), while in *Bryum argenteum* early stages of sporophyte development were retarded during a dry summer in southern England, but not under wetter conditions in northern Wales.[15,23]

2. Phenological Patterns in Specific Environments

It is clear from Table 2 that an almost infinite variety of phenological patterns is evident among temperate mosses. The environmental factors controlling the different phases of development are therefore likely to vary between species. However, common features sometimes may be detected among groups of species occupying particular habitats or showing similar patterns of geographical distribution. In these cases, it is likely that the reproductive cycles are adapted to particular environments as a result of natural selection influencing environmental regulation of reproductive processes. Thus, spore release occurs in winter or early spring in *Atrichum undulatum, Brachythecium rutabulum, Forsstroemia* spp., and other species considered in Table 2 that typically occur in deciduous woodland. Noting a similar pattern in *Atrichum angustatum* in Illinois, Zehr[24] speculated that the absence of leaves in the canopy might favor effective spore dispersal. Conversely, summer spore release occurs in many species of more open habitats, e.g., *Pohlia nutans, Polytrichum alpestre* (Table 2), and *Tortula muralis.*[15]

Another example occurs among aquatic mosses, in which fertilization can be inhibited if the plants are fully submerged, particularly in lakes or other large volumes of water or in fast-flowing streams. Glime[25,26] has noted that gametangial maturation and fertilization in *Fontinalis* spp. in the eastern U.S. typically occur during autumn, when water levels are at their lowest and parts of the moss communities are exposed but remain moist. This interpretation is consistent with reports that *Fontinalis antipyretica* fails to produce sporophytes when submerged, but does so in abundance when exposed.[27] In some populations of *Fontinalis* spp. in fast-flowing streams the male plants appear to be uncommon but abundantly fertile, and Glime thought that sperm may be dispersed considerable distances by spray to emergent female plants.

Glime and Vitt[26] suggest that spores of *Fontinalis* are released by capsule abrasion during periods of high water levels in spring. In contrast, a comprehensive survey of the Pottiaceae by Zander[9] suggested that spore release in hygrophilic species tends to occur in summer or autumn, when water levels are low and the capsules are most likely to be exposed, whereas in more xeric taxa the spores are released during the season of maximum rainfall, when conditions might favor germination and establishment. The ecological significance of seasonality in spore release clearly remains a matter of speculation.

Fertilization in epiphytes and epiliths growing in areas with relatively dry climates could be restricted by lack of water, and here it has been noted that gametangial maturation may be spread over several months within a population, possibly increasing the probability of adequate moisture being available at some time during the potential fertilization period. Monoecious, epiphytic species of *Forsstroemia* studied by Stark[18] were partially protogynous in that some of the archegonia matured before the first antheridia on many shoots, even though in *F. producta* the antheridia were initiated before the archegonia (Table 2): this arrangement may increase the incidence of cross-fertilization. In some epiliths the period over which gametangia mature within a population is so extended that gametangial development appears to lack clearly defined seasonality. Examples include *Orthotrichum anomalum* in Arizona[28] and *Grimmia pulvinata* and *Tortula muralis* in Britain.[15]

Reproductive phenology in tropical West African mosses appears to be adapted to the seasonal alternation of wet and dry weather. In all species studied by Egunyomi[29] and Odu[30] juvenile gametangia were recorded from March to May, at the beginning of the rainy season, and fertilization occurred from May to July, in some cases continuing until September. Mature sporophytes were normally present by the end of the rains in October and, in contrast to Zander's[9] observations on xeric members of the Pottiaceae, spores were liberated during relatively dry weather between November and March, when conditions were likely to be favorable for dispersal. The complete reproductive cycle from gametangial initiation to spore release was completed within 12 months, with no significant resting phase, in contrast to the longer cycles characteristic of many temperate mosses (Table 2). However, the precise timing of reproductive development may again vary between habitats differing in water availability.[29]

3. Phenological Variation in Widely Distributed Mosses

Reproductive phenology in several widely ranging species has been compared in temperate, boreal forest, and polar localities. The results have generally indicated essentially similar seasonal cycles at the contrasting sites, but with minor variation attributable to climatic differences, as indicated for *Pleurozium schreberi* in Figures 1 and 2. Fruiting populations are rare in much of Great Britain, but where they occur antheridia and archegonia typically appear in August and October, respectively. The gametangia overwinter in the juvenile or immature stages, with rapid development in early spring leading to fertilization during April and May. Young sporophytes in the swollen venter stage are recorded from May onward, but further development is delayed until August. Rapid maturation then results in fully developed capsules in the operculum intact stage being present from October. The sporophytes subsequently undergo a second resting phase before the opercula fall and spore release occurs from January to April, some 9 to 11 months after fertilization (Table 2).[16]

It is clear from Figures 1 and 2 that the reproductive cycle of *Pleurozium schreberi* in coniferous woodland of central-southern Canada is basically similar, but with subtle differences. These may be related to the severe winter conditions at the Canadian sites, where (in contrast to Great Britain) there is prolonged frost and snow cover commonly lasting from October to April. Juvenile archegonia appeared at the same time as the antheridia during warm weather in August at one of the Canadian sites, but at a second site the archegonia were first recorded in October, as in Britain. However, the spring phase of rapid gametangial development was delayed until after snowmelt, resulting in fertilization occurring 2 to 3 months later than in Britain. Therefore, sporophytes in the swollen venter stage were not recorded until July. Development then proceeded immediately without a resting phase, however, and the capsules reached the operculum intact stage during the autumn as in the British populations. Capsule dehiscence was also delayed until after snow clearance, so most spores were released in May and June.

Pleurozium schreberi reaches its greatest abundance in boreal forest regions, whereas *Polytrichum alpestre* is a bipolar moss widespread in mild-, cool-, and cold-polar regions,[31] as well as throughout the boreal forest zone and in some temperate localities. Reproductive phenology of *Polytrichum alpestre* in arctic and antarctic regions[12,20,32] is basically similar to that at British sites (Table 2). In Britain fertilization occurs later in spring in *Polytrichum alpestre* than in *Pleurozium schreberi,* and in the former there is no significant delay in fertilization in areas with severe winter conditions. It has been shown that gametangial maturation following the spring thaw in late May at Churchill in the Canadian mild-arctic occurs more rapidly than at a boreal forest site where the snow clears 4 to 6 weeks earlier. June/July is the principal period of fertilization both at these Canadian sites and in Wales (Table 2), where there is no prolonged winter snow cover. Following fertilization and early sporophyte development the sporophytes overwinter in the calyptra in the perichaetium stage, and this resting phase is extended at the polar sites. In Wales sporophyte development is commonly resumed in March and the spores are shed in June and July, whereas at the mild-arctic site setal elongation is delayed until June and spore release until July and August. Rather curiously, sporophytes at the boreal forest sites show a second resting phase in the operculum intact stage so that, although setal elongation begins following the thaw in April and May, the spores normally are not released until September and October.

The basic similarity in reproductive phenology in different populations of *Polytrichum alpestre* is remarkable given the major climatic differences between the sites. Thus, mean daily temperatures recorded during midsummer in the surface layers of the moss turfs range from 18 to 20°C in the boreal forest populations to 10 to 12°C at Churchill and only 8 to 10°C at a cool-antarctic site, where summer conditions are similar to those experienced by *Polytrichum alpestre* in Wales during the winter.[12,20] The rapid gametangial development in *Polytrichum alpestre* following the spring thaw at Churchill could be an adaptation to the short summer growing season. Since this phase of development occurs at Churchill under cooler, intuitively less favorable microclimatic conditions than the slower development at the boreal forest sites, it seems likely that genetic differentiation is involved. This has yet to be confirmed experimentally in *Polytrichum alpestre,* but Clarke and Greene[33] have demonstrated genetically determined variation in rates of reproductive development between polar and temperate populations of *Pohlia nutans* by collateral cultivation under uniform conditions (Section III.C).

4. Reproductive Phenology in r-Selected Species

Most of the species discussed so far adopt the perennial stayer or perennial shuttle life strategies as defined by During.[34] Such taxa occur in habitats normally available for prolonged occupancy, and they show many of the characteristics of evolutionarily conservative, K-selected species,[35] sensu MacArthur and Wilson.[36] Species showing the colonist, fugitive, and annual shuttle strategies occupy transient habitats; these strategies are more opportunistic, with the species concerned tending toward r-selection. Seasonality of reproductive development in such species is commonly less clearly defined than in the perennials.

Atrichum undulatum and *Bryum argenteum* can be regarded as colonists, and it has already been noted that the timing of their gametangial and sporophytic development, respectively, can be modified significantly by variation in climate (Section II.A.1). The timing of gametangial development in *B. argenteum* is also more variable than in long-lived perennials such as *Polytrichum alpestre,*[15] but the later stages of sporophytic development in both these species show distinct seasonality. A more extreme situation is seen in annual shuttle species such as *Pottia truncata*. Van der Wijk[22] reported that in the Netherlands fertilization can occur any time from June to November and that, as sporophyte maturation takes only 2 months, sporophytes at all stages of development can be found together during the autumn and early winter.

Van der Wijk[22] also noted that fertilization in the fugitive *Funaria hygrometrica* can occur at any time and that, although sporophyte maturation takes longer than in *Pottia,* all stages of sporophyte development are again commonly intermixed within a population. Weitz and Heyn[37] reported that this pattern in *F. hygrometrica* is typical only of localities having summer rain with the sporophytes maturing synchronously elsewhere (e.g., in the Mediterranean region), and they demonstrated that the difference in behavior results from genetic differentiation within the species. Although sporophyte development in *F. hygrometrica* is slower than in *Pottia truncata,* mature spores may be present within 8 to 10 months of the gametophytes becoming established both in Britain and at Mediterranean sites.[35]

5. Reproductive Phenology in Hepatics

There have been few detailed studies of reproductive phenology in hepatics, but it is clear that, as in mosses, different seasonal patterns of development may occur among northern temperate species. Zehr[24] reported that gametangia of *Nowellia curvifolia* in Illinois are formed over an extended period during summer, when fertilization occurs. Capsule, seta, and foot become differentiated in the sporophyte during the autumn, with capsule enlargement occurring in spring prior to setal elongation and spore release occurring in May and June. In contrast, Illinois populations of *Lophocolea heterophylla* produce gametangia throughout much of the year: as a result, sporophytes in various stages of development can be found at any time, and setal elongation and spore release occur whenever adequate moisture is available.[24] This pattern is in marked contrast to the behavior of *L. heterophylla* in Britain, where capsules overwinter enclosed within the calyptra and perianth and setal elongation and capsule dehiscence are restricted to a short period in spring.

Reproduction in the Marchantiales is complicated by the production of gametangia in specialized receptacles. Taylor and Hollensen[38] report that initiation of male receptacles in Michigan populations of *Conocephalum conicum* occurs in August, differentiation of antheridia is completed during October, and, in contrast to the leafy hepatics discussed above, the immature antheridia overwinter. Development of both the antheridiophore and sex organs resumes in spring, and the antherozoids are released during June. Initiation of archegonia also occurs in August, with maturation completed after a period of winter dormancy, before fertilization in June. The archegoniophore remains immersed within the thallus until shortly before fertilization. It develops significantly during the summer so that its cap protrudes above the thallus by October. Sporophyte development also proceeds during this period. Capsule, seta, and foot become differentiated in July; meiosis occurs in September and the spores mature during October, prior to a further winter resting phase. Finally, the archegoniophore stalk elongates in March and April and capsule dehiscence occurs in April — some 10 months after fertilization and 21 months after gametangial initiation.

The reproductive phenology of *Conocephalum conicum* appears to be broadly similar in Britain,[39] but several species of *Riccia* operate with an annual shuttle strategy. They complete their life cycle from spore germination through vegetative growth and gametangial and sporophytic development to spore maturation within a few months during winter in temperate regions or following seasonal rains in deserts.[35]

B. FAILURE IN SEXUAL REPRODUCTION

1. Frequency of Sporophyte Production in Mosses

Not all bryophytes produce spores regularly, and in some cases sporophytes are unknown throughout the range of a species. Such taxa must reproduce entirely by asexual means. A survey of the current British Moss Flora[40] indicated that sporophytes are considered to be frequent to common in ca. 40% of species, occasional in ca. 15%, and rare in ca. 25%. Some 20% of species are not known to fruit within the British Isles, but less than 5% of British species fail to produce sporophytes anywhere within their range.[41,42] Comparable surveys have suggested that only 10 to 20% of the moss species fail to produce sporophytes in New Zealand, another temperate region, and under tropical conditions in Gautemala.[42] However, the proportion of species without sporophytes increases to 60 to 70% in the cool-polar regions, to more than 75% in the cold-antarctic, and to ca. 90% in the frigid-antarctic.[31,43-46] Sporophytes of one frigid-antarctic moss developed in laboratory culture, although they were lacking in the field populations.[47]

It is not uncommon for the frequency of sporophyte production to vary within the range of a given species, often declining toward the periphery of the area of distribution. Thus, sporophytes of *Pleurozium schreberi* are widespread in the boreal forest zone, but become rare in temperate regions to the south and in the mild-arctic to the north.[14,48,49] Several bipolar species of *Polytrichum* fruit freely in temperate to cool-polar regions, but rarely if at all under more severe environmental conditions in the cold-antarctic.[32] *Pleurochaete squarrosa* does not produce sporophytes in Britain, at the northern extremity of its range, but does so freely in southern Europe.[42] However, sporophytes appear to be rare throughout the range of other species such as *Rhytidium rugosum*.

2. Sporophyte Production, Reproductive System, and Evolution

Gemmel[50] noted in 1950 that the majority of British mosses regarded as rare in fruit or with sporophytes unknown were dioecious; sporophyte production was frequent in other dioecious species and in the majority of monoecious taxa. This relationship has since been confirmed for British mosses based on more modern data.[42,51] Similarly, literature surveys suggest that most New Zealand and Guatemalan mosses that do not produce sporophytes are dioecious,[42] while field and herbarium studies have demonstrated a similar relationship between sexuality and sporophyte production in a range of temperate and polar habitats.[44,46,52-54] Thus, the proportions of populations with sporophytes in two contrasting forest types in northern Michigan were found to be ca. 75% and 85% for monoecious mosses compared with ca. 20% and 12.5% for dioecious species.[55] Rarity of sporophytes is similarly associated with dioecism in hepatics, particularly in areas affected by Pleistocene glaciation.[42]

Reproductive systems also impinge on life history strategies. Thus, dioecism, often associated with at least local rarity or absence of sporophytes, is prevalent in K-selected species with long-lived, perennial gametophytes occupying stable habitats (e.g., *Pleurozium schreberi, Rhytidium rugosum*). In contrast, there are strong trends toward monoecism, with retention of high fertility, among colonist, fugitive, and shuttle species which form transient populations under more rapidly changing conditions (e.g., *Funaria* and *Riccia* spp.).[35,42]

These considerations appear to influence the distribution and evolutionary flexibility of the taxa concerned,[53] but the relationships are by no means clearly defined. Among British mosses, those species not known to produce sporophytes anywhere within their range typically have narrow or disjunct, possibly relict distribution patterns and strikingly few recognized infraspecific variants worldwide.[42] Similarly, in the hepatic genus *Plagiochila*, species in section *Bidentes* rely on asexual reproduction and have restricted distribution patterns, whereas a high incidence of sporophyte production is combined with wide distribution and substantial morphological variation in section *Asplenioides*.[56] It has also been shown that fruiting dioecious moss species are significantly more widely distributed within the British Isles and are represented in Britain by a higher mean number of infraspecific taxa than are monoecious species.[42,50,51]

These findings are usually attributed to the bypassing of meiosis in nonfruiting species and the potential for inbreeding in monoecious forms. However, Lefebvre[57] has noted that monoecious species of *Plagiothecium* in Belgium are generally more variable and widely distributed than dioecious species, a fact that he related to abundant spore production in the monoecious species and an assumption that some outbreeding occurred. Similarly, the most rapidly evolving groups in the arctic moss flora appear to be highly fertile, monoecious colonists and shuttle species.[31]

3. Factors Contributing toward Rarity of Sporophytes

a. Spatial Separation of Sexes

Gemmel[50] suggested that rarity or absence of sporophytes in dioecious species theoretically could arise by restriction of fertilization either through the presence of large numbers of incompatible lines or through spatial separation of male and female plants. More recent studies have provided evidence of intraspecific incompatibility only between widely separated populations of the hepatic *Riella americana*,[58] but have established failure of male and female gametangia to develop in close proximity as a major cause of the rarity of sporophytes that characterizes many dioecious taxa.

This is not surprising, as fertilization ranges are generally short. Sporophytes seldom occur more than 10 cm from the nearest male plants in terrestrial mosses lacking discoid perigonia.[48,59-61] The effectiveness of discoid perigonia as splash cups[62] is apparently variable, the maximum fertilization distance recorded in the field ranging from ca. 2.5 cm in *Breutelia chrysocoma* to 200 cm and, exceptionally, 380 cm in *Dawsonia superba*.[12,63-66] However, in Rohrer's[55] survey dioecious species with antheridia in splash cups showed significantly enhanced sporophyte production compared with other dioecious mosses. Fertilization ranges up to several meters have also been recorded in some hydric species (Section II.A.2) and among epiphytes in humid regions, especially where the male plants occur higher on the trees than the females.[42] In both these cases the sperm are likely to be carried to the female plants by water.

Examples of taxa where sporophyte production is limited by the spatial separation of male and female plants include *Arnellia fennica* on Ellesmere Island,[67] *Stephensoniella brevipedunculata* in India,[68] and *Bryoxiphium norvegicum* in both the U.S. and Japan.[52,69] Thus, Hague and Welch[69] examined 42 American collections of *B. norvegicum* and found sporophytes in only one: of the remainder 14 contained only antheridia, 8 only archegonia, and 19 were sterile. No archegonia have been found in populations of *B. norvegicum* in Iceland and Greenland.[70] There are other, more striking cases where the sexes are separated at a continental level. Examples include the moss *Tortula pagorum,* known only as females in North America and males in Europe, with sporophytes recently recorded in Australia,[71] and the hepatic *Plagiochila corniculata,* in which antheridia occur in Europe and archegonia in the Appalachian region. Sporophytes are commonly produced in the closely related, possibly conspecific taxon *P. bidens* of the neotropics.[42]

b. Rarity of Antheridia

Rarity of gametangia of one or both sexes is a feature of many bryophyte populations. In extreme cases either antheridia (e.g., *Tortula rhizophylla*) or archegonia (e.g., *Barbula mamillosa*) are unknown throughout the range of a species, while in some mosses (e.g., *Campylopus schwarzii*) no gametangia of either sex have been recorded.[42] It seems likely that these species are dioecious, since failure to produce gametangia is more frequent in dioecious than in monoecious forms.[54] In species that are clearly dioecious the sex ratios are often unbalanced, and rarity of antheridia, or less commonly of archegonia, is a major factor contributing to local or general rarity of sporophytes. It is seldom clear, however, whether the sexual imbalance arises through rarity of male plants or failure in antheridial formation, due to the presence of sterile plants bearing no gametangia. Sex ratios may vary within the range of a species, and the contrasting patterns that have been detected suggest different forms of environmental control, as the following examples demonstrate.

Syrrhopodon texanus is a dioecious moss endemic to eastern North Amercia, where it occurs as two largely allopatric populations. The first occupies the Atlantic and Gulf coastal plain, while the second occurs in the interior highlands and the Cumberland and southern Appalachian Mountains. Reese[72] has shown that archegonial formation is widespread throughout these areas, whereas antheridia, and therefore sporophytes, are rare and restricted to the coastal plain. The pattern here, as in *Tortula pagorum, Plagiochila corniculata,* arctic populations of *Bryoxiphium norvegicum* (Section II.B.3.a), and several other species,[42] could well be the result of biotype depletion by unfavorable conditions associated with Pleistocene glaciation, with the inland population of *S. texanus* having developed from one or a small number of surviving female clones.

A different situation prevails in *Polytrichum alpestre*. This species fruits freely throughout much of its bipolar range, except near its southern limit in the cold-antarctic, a region where mean temperatures in the surface layer of the *P. alpestre* turfs reach only 0 to 3°C in summer (compare with data in Section II.A.3). Here, *P. alpestre* is abundant and highly productive vegetatively,[31] but most populations examined were sterile or developed only archegonia. Antheridia occurred only at occasional, generally well-insolated north-facing sites throughout the region, and there is evidence that they developed on the perennial gametophytes during some years, but not in others, with their initiation stimulated by relatively warm microclimatic conditions. Sporophytes were less widespread than antheridia, and their occurrence was thus limited by other factors, including antheridial abortion (Table 1) and failure in fertilization or sporophyte development.[32]

In this case the rarity of antheridia is undoubtedly due in part to failure in their induction or early development on the male plants, but since the sex of sterile plants could not be determined the ratio of male to female plants is unclear. However, the wide, if scattered, distribution of male plants in an area where most mosses are regarded as postglacial immigrants[31] suggests that the rarity of antheridia in the cold-antarctic reflects the influence of contemporary environmental conditions rather than historical effects of the Pleistocene.

A comparable situation exists in the hepatic *Lunularia cruciata,* a species often regarded as native to the Mediterranean region, but widely distributed and possibly introduced elsewhere.[73] Gametangia of both sexes are produced freely in areas of Mediterranean climate in Europe and California, but sporophytes are rarely seen due to inadequate summer moisture to support fertilization and/or sporophyte development.[73,74] Sporophytes are also rare in temperate regions, but here it is because environmental conditions are unfavorable for gametangial formation. In Britain most populations remain sterile, but archegonia (and less commonly antheridia) are occasionally recorded, particularly in the south and west (Section III.B.1.a).[73,75]

Pleurozium schreberi (Section II.A.3) commonly develops antheridia, archegonia, and sporophytes in formerly glaciated terrain throughout much of the circumpolar, boreal forest zone. Male and female plants occur in roughly equal proportions, with only ca. 15% sterile, at five boreal forest sites in southeastern Manitoba, whereas at temperate sites in southern Britain female and sterile shoots are present in similar numbers, antheridia and sporophytes both being very rare. However, at geographically intermediate sites in northern Scotland shoots with archegonia outnumbered those with antheridia by 8—10:1, and archegonial shoots outnumbered the total of antheridial + sterile shoots by 2:1. Similar sex ratios were recorded at mild-arctic sites in northern Manitoba. In this case it is clear that male plants become less abundant than females at sites both to the north and to the south of the main area of distribution, but it is possible that failure in antheridial production also contributes to the high ratio of archegonial to antheridial shoots.[14,48]

Finally, *Plagiomnium undulatum* may be cited as an example where there is evidence that both a preponderance of female plants and failure in antheridial production by the males contribute to the observed rarity of antheridia. This species appears to be rare in fruit throughout much of its range in Europe and parts of Asia and North Africa. Most British populations contain sterile and archegonial plants, with antheridia and sporophytes being rare. Newton[76,77] determined the sex of sterile plants using a size difference between the heterochromatin body in interphase nuclei as a cytological marker. She found that females outnumbered males by ca. 6:1. However, males were widely distributed, and bisexual populations outnumbered female populations by 1.3:1. Thus, male plants are less abundant than females and commonly remain sterile for reasons discussed below (Sections III.A and III.B.2.a).

c. Other Factors

Spore production also may be inhibited by other factors, including abortion of gametangia, failure in fertilization through lack of water,[78] and abortion of developing sporophytes. In some cases these factors act in addition to those already discussed in restricting the spore output of a given species.

It is normal for a proportion of the gametangia initiated within a population to abort, but this proportion is sometimes so high that the reproductive process fails. Thus, gametangial phenology of *Pohlia cruda* in Britain is broadly similar to that outlined for *P. nutans* in Table 2. However, on the cool-antarctic island of South Georgia gametangial development is irregular, often with two crops initiated within an inflorescence; most gametangia abort, and sporophyte production is very rare.[17]

Frost appears to be a potent factor causing abortion of young moss sporophytes. A high incidence of abortion has been reported among some antarctic species.[32,46,79] It also occurs in *Pleurozium schreberi* in the mild-arctic, where the short growing season results in many sporophytes being in the immature, early calyptra intact stage at the onset of winter rather than the more advanced operculum intact stage as seen further south (Figure 2).[14] Autumn frosts can also cause high mortality among *Buxbaumia aphylla* sporophytes in Newfoundland.[80] Other factors may also be involved, for capsules of *Tortula standfordensis* fail to develop in Britain, where the species may not be native, because the parent gametophytes die at the beginning of summer when the sporophytes are immature.[81]

III. THE LABORATORY APPROACH

A. SEX DETERMINATION AND SEX RATIOS

Some 80 to 90% of hepatic species and ca. 60% of mosses are dioecious. Experimental studies reviewed by Ramsay and Berrie[82] suggest that sex expression is determined genetically in most dioecious bryophytes. Many species have pairs of sex-specific chromosomes differing in size or in heterochromatin content, as is probably the case in *Plagiomnium undulatum* (Section II.B.3.b). These are likely to function as sex chromosomes, although confirmation of this role is still required in most cases. Where sex expression is genetically determined meiosis would be expected to yield two male and two female spores in each tetrad. An explanation of the unequal ratios of male:female plants in many populations of such species as *Plagiomnium undulatum* and *Pleurozium schreberi* presumably lies in differential survival or growth of the male and female gametophytes at some stage from meiosis onward, although an imbalance, once established, could be maintained without further differential mortality if asexual reproduction predominates. Different responses of the two sexes to environmental factors are likely to be involved, particularly where sex ratios vary within the range of a species, as in *Pleurozium schreberi*.

An experimental investigation of this point in *Pleurozium schreberi* from Scotland proved inconclusive. Aborted spores were recorded in most capsules, but not in sufficient numbers to explain the minimum discrepancy of two females to one male plant (Section II.B.3.b), even assuming that all aborted spores were male. Rates of spore germination increased with temperature, independently of photoperiod, with no evidence of two populations of spores germinating at different times. Male and female gametophytes showed similar growth rates under a range of conditions and, contrary to expectation, regeneration from leaves occurred significantly more frequently in material from male than from female shoots.[83]

Different and rather more suggestive results were obtained with *Plagiomnium undulatum*. Newton[77] initiated single spore cultures from the most vigorous sporelings in multispore plates and found a 4:1 ratio of females to males among the resulting clones. A similar ratio was obtained when the first buds produced in the multispore plates were subcultured. No significant departure from a 1:1 sex ratio was observed in similar experiments with *Mnium hornum*, a related species lacking an obvious sexual imbalance in nature. In a single experiment with *P. undulatum*, regenerants from leaves proved more prone to desiccation damage in the case of males than females. Newton[77] concluded that the 1:1 sex ratio is altered between spore formation and early protonemal growth, with the possibility of further modification by environmental factors such as desiccation.

B. GAMETANGIAL INDUCTION

The past 25 years have witnessed considerable interest in the control of reproductive development in bryophytes by temperature, photoperiod, and other physical factors of the environment, with emphasis on the initiation of gametangia. The regular seasonal patterns of development observed in natural populations have prompted speculation concerning the influence of photoperiod and other factors on gametangial induction,[24] but experimental evidence is essential before firm conclusions can be drawn. To date, the results of laboratory studies suggest that photoperiodic control of this initial stage in the reproductive process is more prevalent among hepatics and anthocerotes than in mosses.[84] In the following account the term photoperiod, or day length, refers to the numbers of hours of light during a 24-h cycle, although by analogy with higher plants the length of the dark period may be a more critical factor.

1. Hepatics and Anthocerotes

a. Long-Day Responses

Photoperiodic control of gametangial induction in bryophytes was first reported in 1925 when Wann[85] demonstrated that *Marchantia polymorpha* responds as a long-day plant. This result was later confirmed by Voth and Hamner[86] in experiments under more rigorous control. They showed that *M. polymorpha* thalli grew larger under long days and produced few if any gametangia in short days. Induction was also enhanced by high relative humidity, while gemma production was stimulated under short days. Lockwood[87] demonstrated a similar relationship between day length and the production of gametangia and gemmae in *Cephalozia media* (Jungermanniales).

A major advance was made by Benson-Evans and her associates,[39,73,88] who showed that, in addition to *M. polymorpha*, British material of the following nine hepatics shows long-day stimulation of gametangial initiation: *Conocephalum conicum, Lunularia cruciata* (Marchantiales), *Pellia epiphylla, Riccardia multifida, R. pinguis* (Metzgeriales), *Diplophyllum albicans, Lophocolea cuspidata,* and *Lophocolea heterophylla* (Jungermanniales). In most cases the response appeared to be qualitative in that few, if any, gametangia formed under short days. However, in contrast to the findings of Voth and Hamner[86] with American plants, the British material of *Marchantia polymorpha* showed a quantitative response, as the percentage of plants producing gametangia at 18°C ranged from 5% at a photoperiod of 6 h to 40% at 8 h, 60% at 14 h, and 80 to 90% at 16 to 24 h.

In many hepatics the speed of response increases with rising light irradiance (e.g., *Conocephalum conicum, Marchantia polymorpha,* and *Pellia epiphylla*), and the photoperiodic response has been shown to interact with temperature in species where this point was investigated. Thus, *C. conicum* and *M. polymorpha* became fertile under long days at 21°C, but not at 10°C, while *Preissia guadrata* and *Pellia epiphylla* produced more gametangia at the higher temperature. Archegoniophore elongation in *Reboulia hemisphaerica,* which occurs in the field in late spring, shows a long-day response, but may be delayed by low temperature. Indoleacetic acid and gibberellic acid applied in lanolin to the archegoniophore heads failed to stimulate elongation in short days, possibly due to ineffective absorption.[89]

The requirement of high temperature combined with long days for gametangial induction in *Conocephalum conicum* is consistent with the appearance of gametangia during August in north-temperate populations of this species (Section II.A.5). All the species considered above are perennials, and in several it is known that young sporophytes develop during autumn. This could be explained by gametangial initiation under long days in summer, followed by rapid maturation and fertilization, a pattern differing substantially from that in *C. conicum.* Field studies are required to clarify this point.

The results of Benson-Evans and Hughes[73] on *Lunularia cruciata* are of particular interest in view of the rarity of gametangial formation in this species outside areas with Mediterranean climates (Section II.B.3.b). In British material, archegonia were found to develop only when thalli collected under cold conditions between December and March were transferred to warmer (15 to 21°C), long-day conditions. This result suggests that a cold treatment is required before the thalli respond to the stimulus of long days combined with a rise in temperature. However, no archegonia were formed by thalli that had regenerated from surviving apices following severe frost damage to mature parts of the plant. It was thus concluded that changes induced by the cold treatment occur in mature parts of the thalli and subsequently must be communicated to the developing apex, a hypothesis yet to be tested.

These findings are compatible with the appearance of archegonia during spring in British populations of *Lunularia cruciata* and with their more widespread occurrence in the south and west, where the winters are cool but severe frost is less frequent than in the north and east. Where antheridia occur they are first seen in British populations in late summer, subsequently overwintering in an immature stage, and none has been induced experimentally. It was suggested[73] that a high-temperature stimulus may be required for antheridial induction and that this might be provided more effectively in Mediterranean than in temperate climates.

It should be pointed out, however, that archegonial initiation in spring, at the beginning of the hot, dry summer, would not appear to be adaptive under Mediterranean conditions. Moreover, while British plants showed long-day stimulation of vegetative growth as well as archegonia induction,[39] plants from Israel studied by Schwabe[90] showed short-day stimulation of vegetative growth, which would promote development under relatively humid winter conditions, and became dormant under long days. These discrepancies suggest the occurrence of adaptive, genetic differentiation within *Lunularia cruciata,* and an investigation of the reproductive behavior of Mediterranean material would be of great interest. Confirmation of physiological differences between temperate and Mediterranean plants would favor the view that *L. cruciata* is native in temperate parts of Europe rather than a recent introduction from the Mediterranean region.

b. Short-Day Responses

Other members of the Marchantiales and several anthocerotes show short-day responses in terms of gametangial initiation. Bostic[91] reported that a clone of *Asterella tenella* from Arizona produced archegoniophores only under cool, short-day conditions, and a similar response was demonstrated in *Targionia hypophylla* from India.[92] British *Riccia glauca* formed gametangia only under short days at 10 and 18°C and failed to respond at 21°C regardless of day length. Vegetative growth was also most vigorous under cool, short-day conditions.[88] In the Anthocerotales, Ridgeway[93] found, for five different north temperate species, that antheridial induction occurred at day lengths of 4, 8, and 12 h, but not at 18 h, at 10°C. A temperature of 15°C proved optimal for antheridial induction at a 12-h day length, with a decline in fertility under warmer or cooler conditions. No comparable temperature response was shown by Benson-Evans[88] in another population of one of these species: in *Phaeoceros* (= *Anthoceros*) *laevis,* gametangial induction showed a short-day response at 10, 18, and 21°C, with vegetative growth also most vigorous under short days.

Riccia glauca and many temperate members of the Anthocerotales function as winter annuals, and the present results suggest that this phenological pattern may be controlled in part by photoperiod and temperature. In contrast to the low-temperature response in *R. glauca, R. gangetica* (which grows in India during the rainy season of July and August) shows optimal gametangial induction under warm conditions.[94] It also may be noted that Proskauer[95] has described a hybrid swarm of anthocerotes from western Himalaya, with a form of *Phaeoceros laevis* as a putative parent, which function as long-day plants in terms of both growth and gametangial induction. He attributed this response to selection pressure exerted by a monsoon climate with summer rain. The perennial species *Targionia hypophylla,* while widely distributed elsewhere, is particularly characteristic of rock crevices in deserts.[35] Stimulation of gametangial induction by cool, short-day conditions would be adaptive in deserts with rainfall concentrated in winter. However, rainfall patterns vary widely between desert regions,[96] and it would be of interest to determine whether this is reflected in intraspecific variation in physiological responses in species such as *T. hypophylla.*

c. *Quantitative Photoperiodic and Day-Neutral Responses*

Relatively few liverworts have been shown so far to be day-neutral or quantitatively photoperiodic in terms of gametangial induction. Initiation was stimulated by low temperature in Indian material of *Riccia crystallina,*[97] the numbers of gametangia increasing with increasing day length, and spore germination in this winter annual showed a similar response. Gametangial induction varied with light irradiance, 6500 lux being optimal. *Cryptothallus mirabilis* (Metzgeriales), a subterranean hepatic, proved to be day-neutral: gametangial induction was stimulated by a rise in temperature, independent of day length, and there was an indication that a cold pretreatment might be necessary as in *Lunularia cruciata.*[39]

d. *Inorganic Nutrition*

Nutrient concentration and pH can affect gametangial formation in hepatics growing *in vitro*. The addition of sucrose to the culture medium stimulates gametangial formation in several species.[98,99] In the monoecious species *Riccia crystallina,* pH 4.5 proved optimal for the formation of antheridia and pH 6.5 for archegonia, although some gametangia of both sexes were initiated over the range 4.5 to 7.5. Concentration of inorganic nutrients also influenced the ratio of male to female gametangia.[99]

Particularly interesting results were obtained by Selkirk[100] for *Riccia* spp. In *R. fluitans* and *R. rhenana* archegonia formed only at low nutrient concentration, whereas in *R. duplex* they developed over a wide range of concentrations. In the field *R. duplex* is commonly fertile, but gametangia and sporophytes have never been observed in *R. rhenana* and only one fertile population of *R. fluitans* has been described, with archegonia and sporophytes.[101] Antheridia have never been seen in either species, although they were presumably present in the fruiting population of *R. fluitans*; thus, it is not clear whether these taxa are monoecious or dioecious. *R. fluitans* is commonly found either floating or on mud beside ponds, and its normal infertility could be related to the somewhat eutrophic conditions typical of such habitats. Carbohydrate and mineral nutrient concentrations also influence gametangial induction in some mosses.[102]

2. Mosses

a. *Qualitative Photoperiodic Responses*

A qualitative effect of photoperiod on gametangial induction has, to the present author's knowledge, been recorded in only two moss species. British material of *Sphagnum subnitens* (= *plumulosum*) reacts as a short-day plant, a result in keeping with the appearance of gametangia from February to March in the field.[39,88] In contrast, *Plagiomnium undulatum* acts as a long-day plant, and it is of interest in view of the sex ratios discussed in Sections II.B.3.b and III.A that different responses have been recorded for males and females.[21] Archegonia formed only under long days at 10°C, but they developed under both long and short days in plants subjected to a rise in temperature from 10 to 20°C. The conditions required for antheridial induction appear to be more exacting, as antheridia formed only under long days with diurnal fluctuation in temperature. *P. undulatum* also needs a humid environment for active vegetative growth, and Newton[21] suggested that this combination of requirements limits the range of habitats in which fertile male plants are found.

b. Quantitative Photoperiodic Responses

In a rather greater number of mosses it has been shown that gametangial initiation occurs under a range of photoperiods, but either the response is accelerated or greater proportions of shoots respond under long days than short days. There is commonly interaction between the effects of day length, light irradiance, and temperature. Thus, Chopra and Rahbar[102] found that the time required for gametangial induction in *Bartramidula bartramioides* at 4000 lux and 25°C decreased with increasing photoperiod from 8 to 24 h, with a corresponding increase in percentage response. This species showed a relatively high temperature optimum (25°C) and a moderate optimum light irradiance (4000 lux) under otherwise favorable conditions. The responses of vegetative growth generally resembled those for gametangial induction, but with the differences between treatments proportionally less.

Indian material of *Bryum argenteum* studied by Chopra and Bhatla[103] showed a similar pattern of response, but here a lower light irradiance of 1800 to 2000 lux proved optimal. In this case the maximal induction of gametangia under warm, long-day conditions appears at variance with the appearance of gametangia in winter in British populations (Table 2). We subsequently tested the response of British plants, suspecting that they might be adapted differently than the Indian material, but our results[15] were broadly in agreement with those of Chopra and Bhatla. Gametangial initiation in the British populations occurs at the end of a period of vegetative growth, and its timing could thus be related to vegetative phenology or possibly to an internal rhythm of the type reported in *Mnium hornum* (Section III.B.2.d).

Indian material of *Barbula gregaria* and *Bryum coronatum* (male clones) also responded to temperature and photoperiod in a manner broadly similar to *Bartramidula* in terms of vegetative growth and antheridial production, except that while the proportion of fertile shoots increased with increasing photoperiod the speed of response was unaffected.[104] In British material of *Leptobryum pyriforme* fertility was again increased by increasing photoperiod and light irradiance, but the optimum temperature was only 10°C. No antheridial induction occurred at 25°C, but archegonia subsequently developed in inflorescences transferred to 25°C after antheridial induction under cooler conditions.[105]

c. Temperature Responses

Gametangial induction in *Funaria hygrometrica* occurs at relatively low temperature, with only a minor or inconsistent photoperiodic response beyond a minimum light requirement. Monroe[106] showed that continuous darkness or a 4-h photoperiod at a light irradiance of 1000 to 2000 lux failed to elicit gametangial formation in material from the U.S., and indeed gametangial initiation in the dark has not been reported yet in bryophytes. Induction in Monroe's plants was not influenced by photoperiod in the range 8 to 16 h in two populations, but a third exhibited a long-day response. All three populations showed a rather specific temperature requirement, forming gametangia at a day/night regime of 10/7°C, but not at 17/15°C or 5/3°C. However, isolates from other North American populations were less exacting in this respect, forming gametangia over the range 5 to 15°C and, exceptionally, up to 20°C.[107,108] Dietert's studies[107] agreed with those of Weitz and Heyn[37] in suggesting the occurrence of adaptive variation in this and other responses in provenances from different latitudes, but all populations that Dietert studied showed more rapid gametangial initiation and a less rigorous temperature requirement at a photoperiod of 12 h than under longer or shorter days.

Low temperature requirements for gametangial initiation (<20°C) have also been demonstrated in *Philonotis turneriana,*[104] *Physcomitrium pyriforme,* and *Physcomitrella patens,*[108] with the response independent of day length, at least in *Physcomitrella patens.*[109] In contrast, British material of *Pogonatum* (= *Polytrichum*) *aloides* showed archegonial initiation at 21°C, but not at 10°C, regardless of photoperiod.[39] An interesting result was reported in the autoecious species *Brachythecium rutabulum* by Moutschen,[110] who showed that the optimal temperature for antheridial induction is as low as 5°C: few archegonia are initiated at this temperature, but they are freely produced at laboratory temperature (20 to 25°C), under which conditions no antheridia are formed. This differential response could be partly responsible for the different time of appearance of antheridia and archegonia noted in Table 2.

d. Internal Rhythms

Most of the studies so far considered have involved plants grown axenically in liquid medium or on agar, with the observations commonly restricted to the initial phase of gametangial development. In contrast, Clarke and Greene[33] followed gametangial production and development in *Pohlia nutans* and *P. cruda* over a period of 80 weeks, using portions of intact colonies growing on soil, and they were able to demonstrate cyclic development of gametangia under constant conditions. Both species proved intolerant of continuous warmth, mortality occurring in cultures maintained at 20°C, irrespective of photoperiod. At 10°C seven successive gametangial cycles, from initiation to dehiscence, occurred in British material of *P. nutans.* The cycles overlapped in that juveniles of one crop appeared before all those of the previous cycle had dehisced. Temperature and quantitative photoperiodic responses were shown, as five, four, and two cycles were completed under 10°C short day, 5°C long day, and 5°C short day, respectively.

Only three cycles were completed within 80 weeks at 5°C long day, two at 5°C short day, and two under both photoperiods at 10°C in cool-antarctic *Pohlia nutans.*[33] This material was less able to respond to higher temperatures than the temperate plants, and gametangial initiation and development were less restricted by night frost (5/−2.5°C LD) than in British material. *P. cruda* appears to be even more cold adapted than *P. nutans* in that neither British nor cool-antarctic material formed gametangia at 10°C (long day or short day), with up to two cycles occurring in 80 weeks at 5°C and induction of juveniles under conditions of night frost. Given these results for *P. nutans* it is not clear how the field pattern of a single cycle of gametangial production annually is maintained (Table 2).

In similar experiments with *Mnium hornum,* Newton[21] found that the production of male and female inflorescences showed a peak at intervals of approximately 12 months under constant conditions. Neither temperature (10 and 20°C) nor day length (7.25, 12, and 15 h) had a major influence on gametangial initiation, although 7.25 h had a slight delaying effect. Newton therefore concluded that the annual periodicity in gametangial development in this species (Table 2) is determined by an endogenous rhythm. The timing of spore release and germination and the influence of short days in winter were suggested as factors that might be involved in the seasonal synchronization of the cycle.

C. GAMETANGIAL MATURATION

There have been few experimental studies of factors affecting the rate of gametangial maturation. However, it has been shown that antheridia and archegonia of *Pleurozium schreberi* collected in Britain during the winter resting phase (Section II.A.3) will resume development in response to a rise in temperature.[111] Of the temperatures tested, 10°C proved close to optimal, as development was very slow at 5°C, and while maturation also occurred at 20°C it was accompanied by high rates of abortion, particularly of antheridia. The response was not affected by photoperiod. As the experiments were not initiated until December, the possibility remains that a cold treatment is required before the gametangia are able to respond to increasing temperature.

In British *Pohlia nutans,* gametangial maturation was more rapid at 10°C than at 5°C, with a suggestion that long days accelerated development at either temperature, but development was arrested by night frost (5°C day/ − 2.5°C night). A different response was shown by cool-antarctic plants in which maturation was more rapid at 5°C than at 10°C, with some development occurring under conditions of night frost and high rates of abortion under short days. The difference in temperature response between the provenances can again be viewed as possibly reflecting adaptation to contrasting natural environments.[33]

D. SPERM RELEASE AND FERTILIZATION

It has been noted in many investigations involving both mosses and liverworts that fertilization in agar cultures occurs only when plants bearing mature gametangia are flooded with water.[97,103,107,112] It is not yet clear, however, whether the water is required to permit sperm motility or to stimulate gametangial dehiscence. Fertilization has occurred readily without flooding in our agar cultures of other species, e.g., *Physcomitrella patens* and *Brachythecium rutabulum.*

Detailed studies by Muggoch and Walton[113] showed that the mechanism of antherozoid release from the antheridium varies widely. Thus, spraying the thalli of *Conocephalum conicum* with water stimulated explosive dehiscence of the antheridia, with the antherocytes being ejected as spurts of white spray which rose vertically as much as 8 cm. Antherozoids were subsequently released from the antherocytes. In *Sphagnum* spp. the cap cells of the antheridium separate from each other, the mouth of the antheridium curves outward, and the antherozoids are immediately released from the antherocytes and become free-swimming.

In *Mnium hornum* the cap cells swell when mature antheridia are placed in water, an event attributed by Muggoch and Walton[113] to absorption of water by mucilage present in their walls, and these cells rupture. The mass of antherocytes is then extruded from each antheridium as a unit and remains intact while immersed in water. It was thought that the spermatocytes were ejected through pressure exerted by a fluid accumulating at the base of the antheridium as a result of water entering the antheridium through the living cells of the wall. When the antherocyte mass reaches an air-water interface it breaks up with great rapidity and spreads, forming a film over the surface of the water before the sperm become free-swimming. The spreading was attributed to the low surface tension of lipids shown to be present in the spermatocyte mass; no lipid has been recorded in *Conocephlum* or *Sphagnum,* where different mechanisms operate.[113]

As the response in *Mnium* was repeated in several other mosses, Muggoch and Walton[113] viewed rapid spreading of spermatocytes over the water surface as playing an important role in sperm dispersal. This would be especially true in species such as *Mnium hornum* in which discoid perigonia function as splash cups because any water droplet splashed out of a perigonium would tend to be covered by a layer of antherocytes. In laboratory experiments in which drops of dyed water have been allowed to fall onto discoid perigonia the resulting splashes have traveled distances from 50 cm in *Mnium ciliare* to 230 cm in *Dawsonia superba.*[62,64,65] The frequency of splashes declined with increasing distance from the perigonia in each case. Conversely, it is not clear that the falling drops had always reached their terminal velocity. Similarly, insect limbs and hairs dipping into the fluid in a perigonium would be likely to pick up antherocytes from the surface layer on being withdrawn, and there have been several reports that insects may be implicated in the dispersal of sperm to the perichaetia.[114]

This picture of sperm release in mosses has been refined by the more recent work of Paolillo.[115-120] While the details vary between species, his observations reveal that the antherocyte mass is commonly discharged in two phases, an initial rapid phase powered by contraction of the antheridial wall followed by a period of slower release due to the pressure of fluid in the base. Paolillo has confirmed that lipids are present between the spermatocytes, but he has also shown that the antheridium develops a cuticle as it matures. Thus, he believes that the water present within the antheridium, responsible for rupture of the cap cells and in part for expelling the antherocytes, is drawn from the gametophyte and not from the exterior as Muggoch and Walton[113] assumed. He suggests, however, that the mature antheridium is metastable and that dehiscence may be triggered by a mechanical disturbance which is probably supplied in the field by raindrops striking the perigonium.[119] If so, this would ensure that the antherocytes are normally ejected into a film of water on which surface spreading can occur.

After release from the antherocytes the antherozoids are reputedly capable of swimming as much as 1 to 2 m at speeds of 100 to 200 $\mu m\ sec^{-1}$.[121] However, they tend to move aimlessly, with little net change in position, unless appropriate chemotactic stimuli are present.[113] Sucrose excreted by receptive archegonia is effective in this regard, at least in *Funaria*,[122] but it is not clear how far from the archegonium this stimulus extends. In view of these observations, Paolillo[119] has suggested that the spreading of antherocytes on a water surface may be more important than sperm motility in dispersal between perigonia and perichaetia in many mosses and perhaps also in other land plants, with sperm motility more important as a fertilization mechanism than in dispersal.

E. SPOROPHYTIC DEVELOPMENT

1. Mosses

The factors controlling sporophytic development in bryophytes are not understood well. Belkengren[123] noted that formation of sporophytes in cultures of the moss *Amblystegium riparium* could be stimulated by subjecting gametophytes to a period unfavorable for vegetative growth, e.g., by growing the gametophytes in continuous light on a medium lacking a carbohydrate source and depriving them of CO_2 for 1 month. Sporophytes then developed if CO_2 was subsequently supplied. However, it is unclear whether the absence of sporophytes from vegetatively vigorous cultures resulted from failure in sporophyte development or from absence of mature gametangia. Many bryophytes develop sporophytes under conditions favorable for active vegetative growth of the gametophytes. Similarly, Engel[109] reported maximum sporophyte development in cultures of *Physcomitrella patens* at a temperature of 15 to 19°C, but this could be due to enhanced gametangial formation under these conditions.

One of the most penetrating investigations of sporophyte development in mosses remains that of Hughes[124] in the Polytrichaceae. He showed that during an early vegetative phase of development the seta is formed through the activity of a single apical cell. The derivatives form two histogens in the meristematic region immediately proximal to the apical cell, an endothecium in which the cells are large and seldom divide and a surrounding amphithecium comprising several layers of smaller, more actively dividing cells. Some distance behind the apex (<0.5 mm) the cells of the endothecium resume division, and the seta tissue differentiates from the products of division in the two histogens. At the onset of what Hughes[124] termed the reproductive phase the apical cell ceases to divide, no more cells are contributed to the seta from the meristematic region, endothecial as well as exothecial cells commence division throughout the meristematic region, and the latter elongates and expands to form the capsule following cell differentiation.

In *Pogonatum aloides* it was shown[124] that timing of the switch from the vegetative to the reproductive phase at 27°C is independent of photoperiod; the rate of capsule development is greatest in short days, but capsules are eventually formed under long days. Setae are longer in plants raised under long days than short days, but this is due to greater cell elongation and not to increased numbers of cells. In the field the capsules normally reach maturity in late summer as natural day length shortens, and Hughes[124] viewed the quantitative photoperiodic effect on capsule development in *P. aloides* as an adaptation facilitating rapid growth of late starters.

The switch from vegetative to reproductive phase of sporophyte development in *Polytrichum piliferum* occurs in the field during November, with development subsequently very slow during winter. Hughes[124] found that plants brought into cultivation at 27°C after the switch had occurred developed normal (although rather small) capsules, with no evident photoperiodic influence. No capsules were produced in cultivation by plants collected before the switch, and Hughes speculated that a low-temperature stimulus might be required. However, apical growth of sporophytes in the early samples ceased and segmentation of the endothecium, with some cell enlargement, took place, suggesting to Hughes that initiation of sporangial differentiation and ending of apical growth of the sporophyte are independent events.

While capsules of *Pogonatum aloides* may develop most rapidly in short days, Paolillo and his associates[125,126] have shown that light is essential for normal capsule maturation in *Polytrichum* spp. and in *Funaria hygrometrica.* These authors quote Haberlandt as stating that *Physcomitrium* capsules develop in the dark if carbohydrate reserves are adequate, but abnormal differentiation was reported in *Funaria* capsules grown in the dark with glucose in the medium.[125] The light requirement was satisfied at low irradiance in the polytricha and in *Funaria;* this requirement was shown by explants with and without a carbohydrate source in the medium, and red was the most effective wavelength. Paolillo therefore concluded that a photomorphogenetic effect rather than a photosynthetic effect was operating, probably involving phytochrome. Shading experiments with *Funaria* suggested that the photoreceptors are located in the precapsular region (the meristematic region of Hughes[124]) and possibly also in the seta. For reasons that were not fully understood, dry weight gain of the capsule increased in relation to the length of seta remaining on the explants in all the species investigated.[125,126]

In *Atrichum undulatum* it has been shown that plants cultured under warm conditions produce substantially longer setae than those maintained at low temperature,[127] as a result of both increased cell division and greater cell enlargement. It is not clear, however, whether the increased number of cells reflects a later switch to the reproductive phase of development at high temperature. Dietert[107] showed that setal elongation in cultures of *Funaria hygrometrica* is completed most rapidly under short-day conditions, with 15°C being the optimum temperature. No elongation occurred at 5°C. Different developmental patterns were also noted, for at 15 to 20°C the calyptra enlarged only after capsule expansion, whereas the reverse was true at lower temperatures, as is often the case in the field. In contrast to the results for *F. hygrometrica,* capsule formation in *Bryum argenteum* occurred in sporophytes maintained at 25°C, but not at 18°C.[103]

Finally, it has been shown that developing sporophytes of *Splachnum sphaericum* exhibit both positive phototropism and negative geotropism, thus ensuring vertical orientation. The response occurs through curvature in the actively elongating zone. In early stages of growth this elongation zone extends throughout the length of the seta. Later, the zone of growth, and thus of curvature, is restricted to the setal apex below the meristematic region. Shading experiments showed that light perception is restricted to the responsive area and that removal of the meristematic region does not affect either the phototropic or geotropic response.[128]

2. Hepatics

The only significant studies so far reported on environmental control of sporophytic development in hepatics concern the stage of setal elongation. In contrast to the pattern in mosses, hepatic setae elongate after maturation of the capsule and spores. The process involves an up to 50-fold increase in cell length, unaccompanied by cell division and with no cell differentiation. Thomas[129] has pointed out that these and other features render the hepatic seta a particularly favorable system for fundamental studies of cell elongation and its control, and he has presented a comprehensive review of the relevant literature. Studies to date have centered on *Pellia epiphylla*.

Setal elongation in liverworts occurs rapidly, at rates up to 5 mm h^{-1}.[129] Benson-Evans[39] showed that elongation in several species takes place during a short period immediately after sunset. Three such periods during successive evenings were required to complete the process in *Pellia epiphylla*, compared with two in *Cryptothallus mirabilis* and one in *Lophocolea cuspidata*.

Sporophytes of *Pellia epiphylla* in Britain become differentiated during the autumn, with meiosis and sporogenesis occurring in winter and setal elongation delayed until spring. Elongation can be stimulated during the autumn and winter, even before the spores are mature, by transferring plants to warm conditions. However, Crombie and Paton[130] showed that the delay between transfer and positive response in 50% of individuals decreased progressively in successive samples transferred from September through February. Provision of a cold treatment failed to accelerate setal elongation in the earlier samples. The authors concluded that either synthesis of a growth stimulator or removal of an inhibitor occurs in nature during the winter, and it has been confirmed that exogenous application of auxins during winter will stimulate elongation.[131]

The influence of physical factors on setal growth in *Pellia epiphylla* collected during January has been investigated by Slade.[132] He showed that setal elongation was stimulated most rapidly at ca. 20°C, although final seta length was greatest at the lower temperature of 5°C. Variation in photoperiod from 6 to 24 h had little effect, but elongation was inhibited by continuous darkness. It was also shown that the response was more sensitive to day temperature than to night temperature despite the normal occurrence of elongation after sunset. Elongation was also inhibited by low relative humidity and by high soil water tension,[132] the latter possibly by restricting osmotic uptake of water.[129]

Sporophytes of *Pellia epiphylla* are positively phototropic during setal elongation, due to faster extension on the side shaded from a undirectional light source. As in *Splachnum,* the mechanism differs from that in flowering plants in that it is shown by decapitated setae, there is no associated polar auxin transport, and curvature occurs only at the site of illumination without evidence of vertical transmission of the light stimulus.[133] Thomas et al.[134] have shown that curvature of setae exposed to unilateral light occurs along the entire length of the organ, with growth along the shaded side increased and on the illuminated side reduced compared with dark controls. The response was greatest to blue light and was visible within 10 to 15 min.[134] One assumes that low light irradiance is adequate to cause the response, as setal elongation normally occurs at dusk. Although other explanations were possible, Thomas et al.[134] considered the results consistent with the hypothesis that curvature resulted from light-induced, lateral auxin transport. A similar response was shown by archegoniophores of *Conocephalum conicum,* although here growth involves cell division as well as cell elongation. Moreover, growth is reduced by decapitating the archegoniophore, although the phototropic response persists.[135]

IV. CONCLUSION

It is now abundantly clear that many bryophyte species show distinct seasonal periodicity in gametangial and sporophytic development. The seasonal cycles show infinite variety between species, although common trends appear to be established in certain environments.

The cycles of given species may be modified to some extent by variation in environmental conditions, but they tend to remain fundamentally similar throughout the range of widely distributed taxa.

These comments apply particularly to species with long-lived, perennial gametophytes occupying stable habitats. Greater flexibility is shown by many annuals or short-lived perennials functioning as colonists or particularly as shuttle species or fugitives in unstable habitats. Furthermore, the perennials tend to show dioecism, often with a low frequency of sporophyte production in parts of their range, whereas the shuttle species and fugitives are predominantly monoecious with abundant sporophyte production. There is also some evidence that spores of long-lived perennials are less effective at germinating and giving rise to new gametophytes under field conditions than those of colonists and fugitives.[136]

Experimental studies have shown that gametangial initiation in bryophytes is regulated by a variety of responses to temperature, photoperiod, and light irradiance, much as in flowering plants. Habitat-correlated, intraspecific variation in response patterns has been reported between provenances in several taxa. Despite the inevitably unnatural conditions in laboratory experiments, many of which have been conducted at constant temperature, the experimentally determined responses of a given species normally show reasonable agreement with climatic conditions prevailing at the time of gametangial appearance in the field, where this is known. However, the situation in *Bryum argenteum* and *Mnium hornum* suggests that more complex mechanisms of controlling gametangial initiation occur in some species, with gametangial induction linked to growth patterns in the gametophyte or to endogenous rhythms.

Qualitative photoperiodic responses in gametangial induction appear to be more widespread among hepatics than in mosses. It may be significant, however, that the two mosses showing this type of response are both long-lived perennials. In contrast, many of the mosses where photoperiodic control is quantitative or lacking are annuals or short-lived perennials. In these latter species, which are often particularly convenient for cultural studies, the absence of obligate photoperiodic control is consistent with the reduced level of seasonal periodicity in reproductive development. While a number of winter annual hepatics and anthocerotes do show short-day stimulation of gametangial induction, it is nevertheless possible that the present picture regarding mosses is biased and that qualitative photoperiodic control of gametangial induction may be widespread among perennial species.

It should be noted that most of the experiments investigating photoperiodic responses in bryophytes have been carried out under broad-spectrum lighting. Thus, it is not always feasible to determine whether the effects of photoperiod represent photomorphogenetic or photosynthetic responses, especially where long days result in greater vegetative growth as well as stimulation of gametangial induction. However, during vegetative growth in a strain of *Lunularia cruciata* from Israel a short-day response was shown to be controlled by red light, with a light break effect, far-red reversal, and other manifestations of the phytochrome system,[137] while gametangial induction in *Marchantia polymorpha* is most prolific under incandescent lighting.[138]

Gametangial initiation also may be influenced by pH and levels of inorganic nutrients and, as shown particularly by Chopra and his associates,[84,99,105,139,140] by auxins, cAMP, and other growth regulators. In some cases changes in endogenous concentrations of growth regulators have been shown to accompany gametangial induction,[141] and it would be of great interest to determine how such concentrations are influenced by climatic factors as a step toward elucidating the mechanisms of environmental control. Work on the control of gametangial induction has so far provided only limited insight into factors underlying the unbalanced sex ratios that characterize many dioecious taxa in the field, and this problem merits further study.

The processes of gametangial maturation and, following fertilization, of sporophytic development also appear to be controlled by a variety of responses to temperature and photoperiod. However, little work has been attempted, and these areas remain open for investigation. It is clear that bryophytes offer ideal systems for addressing fundamental problems concerning the control of plant growth and development at most stages of their life history.

REFERENCES

1. **Arnell, H. W.,** Phenological observations on mosses, *Bryologist,* 8, 41, 1905.
2. **Gilbert, A. E.,** The fruiting season of the hair cap moss, *Bryologist,* 7, 35, 1904.
3. **Grimme, A.,** Über die Blüthezeit deutscher Laubmoose und die Entwicklungsdauer ihrer Sporogone, *Hedwigia,* 42, 1, 1903.
4. **Hagerup, O.,** Zur Periodizität im Laubwechsel der Moose, *K. Danske Vidensk. Selsk. Biol. Medd.,* 11, 1, 1935.
5. **Jendralski, U.,** Die Jahresperiodizität in der Entwicklung der Laubmoose im Rheinlande, *Decheniana,* 108, 105, 1955.
6. **Lackner, J.,** Über die Jahresperiodizität in der Entwicklung de Laubmoose, *Planta,* 29, 534, 1939.
7. **Towle, P. M.,** Notes on the fruiting season of *Catharinea, Bryologist,* 8, 44, 1905.
8. **Towle, P. M.,** Notes on the life history of *Mniums, Bryologist,* 9, 54, 1906.
9. **Zander, R. H.,** Patterns of sporophyte maturation dates in the Pottiaceae, *Bryologist,* 82, 538, 1979.
10. **Greene, S. W.,** The maturation cycle, of the stages of development of gametangia and capsules in mosses, *Trans. Br. Bryol. Soc.,* 3, 736, 1960.
11. **Forman, R. T.,** A system for studying moss phenology, *Bryologist,* 68, 289, 1965.
12. **Longton, R. E. and Greene, S. W.,** The growth and reproduction of *Polytrichum alpestre* Hoppe on South Georgia, *Philos. Trans. R. Soc. London Ser. B,* 252, 295, 1967.
13. **Stark, L. R.,** Reproductive biology of *Entodon cladorrhizans* (Bryopsida, Entodontaceae). I. Reproductive cycle and frequency of fertilization, *Syst. Bot.,* 8, 381, 1983.
14. **Longton, R. E.,** Reproductive biology and susceptibility to air pollution in *Pleurozium schreberi* (Brid.) Mitt. (Musci) with particular reference to Manitoba, Canada, *Monogr. Syst. Bot. Mo. Bot. Gard.,* 11, 51, 1985.
15. **Miles, C. J., Odu, E. A., and Longton, R. E.,** Phenological studies on British mosses, *J. Bryol.,* 15, 607, 1989.
16. **Longton, R. E. and Greene, S. W.,** The growth and reproductive cycle of *Pleurozium schreberi* (Brid.) Mitt., *Ann. Bot. (London),* 33, 83, 1969.
17. **Clarke, G. C. S. and Greene, S. W.,** Reproductive performance of two species of *Pohlia* at widely separated stations, *Trans. Br. Bryol. Soc.,* 6, 114, 1970.
18. **Stark, L. R.,** Phenology and species concepts: a case study, *Bryologist,* 88, 190, 1985.
19. **Stark, L. R.,** The life history of *Forsstroemia trichomitria* (Hedw.) Lindb., an epiphytic moss, *Lindbergia,* 12, 20, 1986.
20. **Longton, R. E.,** Studies on growth, reproduction and population ecology in relation to microclimate in the bipolar moss *Polytrichum alpestre* Hoppe, *Bryologist,* 82, 325, 1979.
21. **Newton, M. E.,** An investigation of photoperiod and temperature in relation to the life cycles of *Mnium hornum* Hedw. and *M. undulatum* Sw. (Musci) with reference to their histology, *Bot. J. Linn. Soc.,* 65, 189, 1972.
22. **van der Wijk, R.,** De periodiciteit in de Ontwikkeling der Bladmossen, *Buxbaumia,* 14, 25, 1960.
23. **Benson-Evans, K. and Brough, M. C.,** The maturation cycles of some mosses from Fforest Ganol, Glamorgan, *Trans. Cardiff Nat. Soc.,* 92, 4, 1966.
24. **Zehr, D. R.,** Phenology of selected bryophytes in southern Illinois *Bryologist,* 82, 29, 1979.
25. **Glime, J. M.,** Physio-ecological factors relating to reproduction and phenology in *Fontinalis dalecarlica, Bryologist,* 87, 17, 1984.
26. **Glime, J. M. and Vitt, D. H.,** The physiological adaptations of aquatic Musci, *Lindbergia,* 10, 41, 1984.
27. **Dixon, H. N.,** *The Student's Handbook of British Mosses,* 3rd ed., Sumfield and Day, Eastbourne, England, 1924.
28. **Johnsen, A. B.,** Phenological and environmental observations on stands of *Orthotrichum, Bryologist,* 72, 397, 1969.
29. **Egunyomi, A.,** Autecology of *Octoblepharum albidum* in western Nigeria. II. Phenology and water relations, *Nova Hedwigia Z. Kryptogamenkol.,* 31, 377, 1979.

30. **Odu, E. A.,** Reproductive phenology of some tropical African mosses, *Cryptogamie Bryol. Lichénol.*, 2, 91, 1981.
31. **Longton, R. E.,** *Biology of Polar Bryophytes and Lichens,* Cambridge University Press, London, 1988.
32. **Longton, R. E.,** Reproduction of Antarctic mosses in the genera *Polytrichum* and *Psilopilum* with particular reference to temperature, *Br. Antarct. Surv. Bull.*, 27, 51, 1972.
33. **Clarke, G. C. S. and Greene, S. W.,** Reproductive performance of two species of *Pohlia* from temperate and sub-Antarctic stations under controlled conditions, *Trans. Br. Bryol. Soc.*, 6, 278, 1971.
34. **During, H. J.,** Life strategies of bryophytes: a preliminary review, *Lindbergia,* 5, 2, 1979.
35. **Longton, R. E.,** Life history strategies among bryophytes of arid regions, *J. Hattori Bot. Lab.*, 64, 15, 1988.
36. **MacArthur, R. H. and Wilson, E. D.,** *The Theory of Island Biogeography,* Princeton University Press, Princeton, NJ, 1967.
37. **Weitz, S. and Heyn, C. C.,** Intra-specific differentiation within the cosmopolitan moss species *Funaria hygrometrica, Bryologist,* 84, 315, 1981.
38. **Taylor, J. and Hollensen, R. H.,** Sexual reproductive cycle of the liverwort *Conocephalum conicum, Mich. Bot.*, 23, 77, 1984.
39. **Benson-Evans, K.,** Environmental factors and bryophytes, *Nature (London),* 191, 255, 1961.
40. **Smith, A. J. E.,** *The Moss Flora of Britain and Ireland,* Cambridge University Press, London, 1978.
41. **Longton, R. E. and Miles, C. J.,** Studies on the reproductive biology of mosses, *J. Hattori Bot. Lab.*, 52, 219, 1982.
42. **Longton, R. E. and Schuster, R. M.,** Reproductive biology, in *New Manual of Bryology,* Vol. 1, Schuster, R. M., Ed., Hattori Botanical Laboratory, Nichinan, Japan, 1983, 386.
43. **Brassard, G. R.,** The mosses of northern Ellesmere Island, Arctic Canada. I. Ecology and phytogeography, with an analysis for the Queen Elizabeth Islands, *Bryologist,* 74, 233, 1971.
44. **Holmen, K.,** The mosses of Peary Land, north Greenland, *Medd. Grøenl.*, 162(2), 1, 1960.
45. **Selkirk, P. M.,** Vegetative reproduction and dispersal of bryophytes on subantarctic Macquarie Island and in Antarctica, *J. Hattori Bot. Lab.*, 55, 105, 1984.
46. **Webb, R.,** Reproductive behaviour of mosses on Signy Island, South Orkney Islands, *Br. Antarct. Surv. Bull.*, 36, 61, 1973.
47. **Rastorfer, J. R.,** Vegetative regeneration and sporophyte development of *Bryum antarcticum* in an artificial environment, *J. Hattori Bot. Lab.*, 34, 391, 1971.
48. **Longton, R. E. and Greene, S. W.,** Relationship between sex distribution and sporophyte production in *Pleurozium schreberi* (Brid.) Mitt., *Ann. Bot. (London),* 33, 107, 1969.
49. **Stoneburner, A.,** Fruiting in relation to sex ratios in colonies of *Pleurozium schreberi* in northern Michigan, *Mich. Bot.*, 18, 73, 1979.
50. **Gemmel, A. R.,** Studies in the bryophyta. I. The influence of sexual mechanism on varietal production and distribution in British Musci, *New Phytol.*, 49, 64, 1950.
51. **Smith, A. J. E. and Ramsay, H. P.,** Sex, cytology and frequency of bryophytes in the British Isles, *J. Hattori Bot. Lab.*, 52, 275, 1982.
52. **Iwatsuki, Z.,** Geographical isolation and speciation of bryophytes in some islands in eastern Asia, *J. Hattori Bot. Lab.*, 35, 126, 1972.
53. **Longton, R. E.,** Reproductive biology and evolutionary potential in bryophytes, *J. Hattori Bot. Lab.*, 41, 205, 1976.
54. **Une, K., Higuchi, M., and Nishimura, N.,** Bryophytes of the Hiruzen Highlands. III. Sexual conditions of acrocarpous mosses, *Bull. Hiruzen Res. Inst. Okayama Univ. Sci.*, 8, 33, 1983.
55. **Rohrer, J. R.,** Sporophyte production and sexuality of mosses in two northern Michigan habitats, *Bryologist,* 85, 394, 1982.
56. **Inoue, H.,** Some taxonomic problems in the genus *Plagiochila, J. Hattori Bot. Lab.*, 38, 105, 1974.
57. **Lefebvre, J.,** Fertilité et souplesse adaptive chez les Plagiotheciaceae de Belgique, *Rev. Bryol. Lichénol.*, 36, 162, 1969.
58. **Proctor, V. W.,** The genus *Riella* in North and South America: distribution, culture and reproductive isolation, *Bryologist,* 75, 281, 1972.
59. **Anderson, L. E. and Lemmon, B. E.,** Gene flow distances in the moss *Weissia controversa* Hedw., *J. Hattori Bot. Lab.*, 38, 76, 1974.
60. **Bedford, T. H. B.,** Sex distribution in colonies of *Climacium dendroides* W. and M. and its relation to fruit bearing, *Northwest. Nat.*, 13, 312, 1938.
61. **McQueen, C. B.,** Spatial pattern and gene flow distances in *Sphagnum subtile, Bryologist,* 88, 333, 1985.
62. **Brodie, H. J.,** The splash-cup mechanism in plants, *Can. J. Bot.*, 29, 224, 1951.
63. **Bedford, T. H. B.,** The fruiting of *Breutelia arcuata* Schp., *Naturalist,* p. 113, 1940.
64. **Clayton-Greene, K. A., Green, T. G. A., and Staples, B.,** Studies of *Dawsonia superba.* I. Antherozoid dispersal, *Bryologist,* 80, 439, 1977.

65. **Reynolds, D. N.,** Gamete dispersal in *Mnium ciliare, Bryologist,* 83, 73, 1980.
66. **Wyatt, R.,** Spatial pattern and gamete dispersal distances in *Atrichum angustatum,* a dioecious moss, *Bryologist,* 80, 284, 1977.
67. **Schuster, R. M.,** *The Hepaticae and Anthocerotae of North America,* Vol. 4, Columbia University Press, New York, 1980.
68. **Udar, R., Srivastava, S. C., and Srivastava, G.,** Observations on endemic liverwort taxa from India. I. Reproductive biology and SEM details of spores in *Stephensoniella brevipedunculata* Kash., *J. Hattori Bot. Lab.,* 54, 321, 1983.
69. **Hague, S. and Welch, W. H.,** Observations regarding the scarcity of sporophytes in *Bryoxiphium norvegicum, Bryologist,* 54, 214, 1951.
70. **Löve, A. and Löve, D.,** Studies on *Bryoxiphium, Bryologist,* 56, 73 and 183, 1953.
71. **Crum, H. A. and Anderson, L. E.,** *Mosses of Eastern North America,* Vol. 1, Columbia University Press, New York, 1981.
72. **Reese, W. D.,** Reproductivity, fertility and range of *Syrrhopodon texanus* Sull. (Musci), a North American endemic, *Bryologist,* 87, 217, 1984.
73. **Benson-Evans, K. and Hughes, J. G.,** The physiology of sexual reproduction in *Lunularia cruciata* (L.) Dum., *Trans. Br. Bryol. Soc.,* 2, 513, 1955.
74. **Whitmore, A.,** The occurrence of *Lunularia cruciata* in California, *Bryologist,* 85, 320, 1982.
75. **Goodman, G. T.,** Sexual *Lunularia cruciata* (L.) Dum. in south Wales, *Trans. Br. Bryol. Soc.,* 3, 98, 1956.
76. **Newton, M. E.,** A cytological distinction between male and female plants in *Mnium undulatum* Hedw., *Trans. Br. Bryol. Soc.,* 6, 230, 1971.
77. **Newton, M. E.,** Sex-ratio differences in *Mnium hornum* Hedw. and *M. undulatum* Sw. in relation to spore germination and vegetative regeneration, *Ann. Bot. (London),* 36, 163, 1972.
78. **Schofield, W. B.,** Bryology in Arctic and boreal North America and Greenland, *Can. J. Bot.,* 50, 1111, 1972.
79. **Greene, S. W.,** Bryophyte distribution, in *Terrestrial Life of Antarctica,* Bushnell, V. C., Ed., American Geographical Society, New York, 1967, 11.
80. **Hancock, J. A. and Brassard, G. R.,** Phenology, sporophyte production and life history of *Buxbaumia aphylla* in Newfoundland, Canada, *Bryologist,* 77, 501, 1974.
81. **Smith, A. J. E. and Whitehouse, H. L. K.,** The sporophyte and male plants of *Tortula standfordensis* Steere and the taxonomic position of this and *T. khartoumensis* Pettet and *T. rhizophylla* (Sak.) Iwats. & Saito, *J. Bryol.,* 8, 9, 1974.
82. **Ramsay, H. P. and Berrie, G. K.,** Sex determination in bryophytes, *J. Hattori Bot. Lab.,* 52, 255, 1982.
83. **Longton, R. E. and Greene, S. W.,** Experimental studies on growth and reproduction in the moss *Pleurozium schreberi* (Brid.) Mitt., *J. Bryol.,* 10, 321, 1979.
84. **Chopra, R. N. and Bhatla, S. C.,** Regulation of gametangial formation in bryophytes, *Bot. Rev.,* 49, 29, 1983.
85. **Wann, F. B.,** Some of the factors involved in the sexual reproduction of *Marchantia polymorpha, Am. J. Bot.,* 12, 307, 1925.
86. **Voth, P. D. and Hamner, K. C.,** Responses of *Marchantia polymorpha* to nutrient supply and photoperiod, *Bot. Gaz. (Chicago),* 102, 169, 1940.
87. **Lockwood, L. G.,** The influence of photoperiod and exogenous nitrogen-containing compounds on the reproductive cycles of the liverwort *Cephalozia media, Am. J. Bot.,* 62, 893, 1975.
88. **Benson-Evans, K.,** Physiology of the reproduction of bryophytes, *Bryologist,* 67, 431, 1964.
89. **Koevenig, J. L.,** Effect of photoperiod, temperature and plant growth hormones on initiation of archegoniophore elongation in the thallose liverwort *Reboulia hemisphaerica, Bryologist,* 76, 501, 1973.
90. **Schwabe, W. W.,** Photoperiodism in liverworts, in *Light and Plant Development,* Smith, H., Ed., Butterworths, Boston, 1976, 371.
91. **Bostic, S. R.,** Laboratory induction of sexuality in *Asterella tenella* (L.) Beauv. (Aytonaceae), *Bryologist,* 84, 89, 1981.
92. **Bapna, K. R., Singh, R. P., and Chaudhary, B. L.,** Induction of sex organs in *Targionia hypophylla* L., *Bryologist,* 87, 340, 1984.
93. **Ridgeway, J. E.,** Factors initiating antheridial formation in six Anthocerotales, *Bryologist,* 70, 203, 1967.
94. **Dua, S., Singal, N., and Chopra, R. N.,** Studies on growth and sexuality in *Riccia gangetica* Ahmad grown in vitro, *Cryptogamie Bryol. Lichénol.,* 3, 189, 1982.
95. **Proskauer, J.,** Studies on the Anthocerotales.VII.13. On day length and the western Himalayan hornwort flora, and on some problems in cytology, *Phytomorphology,* 17, 61, 1967.
96. **Walter, H.,** *Vegetation of the Earth,* English Universities Press, London, 1973.
97. **Chopra, R. N. and Sood, S.,** *In vitro* studies on the reproductive biology of *Riccia crystallina, Bryologist,* 76, 278, 1973.

98. **Mehra, P. N. and Pental, D.**, Induction of apospory, callus and correlated morphogenetic studies in *Athalamia pusilla* Kash., *J. Hattori, Bot. Lab.*, 40, 151, 1976.
99. **Chopra, R. N. and Sood, S.**, *In vitro* studies in Marchantiales. I. Effects of some carbohydrates, agar, pH, light and growth regulators on the growth and sexualuty in *Riccia crystallina, Phytomorphology*, 23, 230, 1973.
100. **Selkirk, P. M.**, Effect of nutritional conditions on sexual reproduction in *Riccia, Bryologist*, 82, 37, 1979.
101. **Paton, J. A.**, *Riccia fluitans* L. with sporophytes, *J. Bryol.*, 7, 253, 1973.
102. **Chopra, R. N. and Rahbar, K.**, Temperature, light and nutritional requirements for gametangial induction in the moss *Bartramidula bartramioides, New Phytol.*, 92, 251, 1972.
103. **Chopra, R. N. and Bhatla, S. C.**, Effect of physical factors on gametangial induction, fertilization and sporophyte development in the moss *Bryum argenteum* grown *in vitro, New Phytol.*, 89, 439, 1981.
104. **Kumra, P. K. and Chopra, R. N.**, Effect of some physical factors on growth and gametangial induction in male clones of three mosses grown in vitro, *Bot. Gaz. (Chicago)*, 144, 533, 1983.
105. **Chopra, R. N. and Rawat, M. S.**, Studies on the induction of sexual phase in the moss *Leptobryum pyriforme, Beitr. Biol. Pflanz.*, 53, 353, 1977.
106. **Monroe, J. H.**, Some factors evoking formation of sex organs in *Funaria, Bryologist*, 68, 337, 1965.
107. **Dietert, M. F.**, The effect of temperature and photoperiod on the development of geographically isolated populations of *Funaria hygrometrica* and *Weissia controversa, Am. J. Bot.*, 67, 369, 1980.
108. **Nakosteen, P. C. and Hughes, K. W.**, Seasonal life cycles of three species of Funariaceae in culture, *Bryologist*, 81, 307, 1978.
109. **Engel, P. P.**, The induction of biochemical and morphological mutants in the moss *Physcomitrella patens, Am. J. Bot.*, 55, 438, 1968.
110. **Moutschen, J.**, L'Hérédeté des caractères gamétophytiques chez les mousses, *Rev. Bryol. Lichénol.*, 36, 617, 1969.
111. **Longton, R. E.**, An Experimental Study of Reproduction in the Bryophyta, Ph.D. thesis, University of Birmingham, Birmingham, England, 1963.
112. **Hughes, J. G. and Wiggin, A. J. A.**, Light intensity and sexual reproduction in *Phascum cuspidatum, Trans. Br. Bryol. Soc.*, 5, 823, 1969.
113. **Muggoch, H. and Walton, J.**, On the dehiscence of the antheridium and the part played by surface tension in the dispersal of spermatocytes in Bryophyta, *Proc. R. Soc. London Ser. B*, 130, 448, 1942.
114. **Harvey-Gibson, R. J. and Miller-Brown, D.**, Fertilization of Bryophyta, *Ann. Bot. (London)*, 41, 190, 1927.
115. **Paolillo, D. J.**, The release of sperms from the antheridia of *Polytrichum juniperinum, New Phytol.*, 74, 287, 1975.
116. **Paolillo, D. J.**, On the release of sperms in *Atrichum, Am. J. Bot.*, 64, 81, 1977.
117. **Paolillo, D. J.**, Release of sperms in *Funaria hygrometrica, Bryologist*, 80, 619, 1977.
118. **Paolillo, D. J.**, On the lipids of the sperm masses of three mosses, *Bryologist*, 82, 93, 1979.
119. **Paolillo, D. J.**, The swimming sperms of land plants, *BioScience*, 31, 367, 1981.
120. **Hausmann, M. K. and Paolillo, D. J.**, On the development and maturation of antheridia in *Polytrichum, Bryologist*, 80, 143, 1977.
121. **Richards, P. W.**, The taxonomy of bryophytes, in *Essays in Plant Taxonomy*, Street, H. E., Ed., Academic Press, London, 1978, 177.
122. **Watson, E. V.**, *The Structure and Life of Bryophytes*, 3rd ed., Hutchinson, London, 1971.
123. **Belkengren, R. O.**, Growth and sexual reproduction of the moss *Amblystegium riparium* under sterile conditions, *Am. J. Bot.*, 49, 567, 1962.
124. **Hughes, J. G.**, The effects of day-length on the development of the sporophytes of *Polytrichum aloides* Hedw. and *P. piliferum* Hedw., *New Phytol.*, 61, 266, 1962.
125. **French, J. C. and Paolillo, D. J.**, Effect of light and other factors on capsule expansion in *Funaria hygrometrica, Bryologist*, 79, 457, 1976.
126. **Krisko, M. E. and Paolillo, D. J.**, Capsule expansion in the hairy-cap moss, *Polytrichum, Bryologist*, 75, 509, 1975.
127. **Stevenson, D. W., Rastorfer, J. R., and Showman, R. E.**, Effects of temperature on seta elongation in *Atrichum undulatum, Ohio J. Sci.*, 72, 146, 1972.
128. **Walsh, W.**, Geotropism and phototropism in the sporophyte of *Splachnum sphaericum* Linn. fil., *Naturalist*, 823, 137, 1947.
129. **Thomas, R. J.**, Cell elongation in hepatics: the seta system, *Bull. Torrey Bot. Club*, 107, 339, 1980.
130. **Crombie, W. M. and Paton, J. A.**, An age effect on seta elongation in *Pellia epiphylla, Nature (London)*, 182, 541, 1958.
131. **Asprey, G. F., Benson-Evans, K., and Lyon, A. G.**, Effect of gibberellin and indoleacetic acid on seta elongation in *Pellia epiphylla, Nature (London)*, 4619, 1351, 1958.
132. **Slade, J. S.**, Physical factors influencing seta elongation in *Pellia epiphylla, Bryologist*, 68, 440, 1965.

133. **Garjeanne, A. J. M.,** Physiology, in *Manual of Bryology,* Verdoorn, F., Ed., Nijhoff, The Hague, 1932, 207.
134. **Thomas, R. J., Caron, P. J., and Wall, R. S.,** Time-lapse photography of phototropism in *Pellia epiphylla, Bryologist,* 90, 390, 1987.
135. **Newton, M. E.,** Probing cytological and reproductive phenomena by means of bryophytes, *J. Biol. Educ.,* 19, 189, 1985.
136. **Miles, C. J. and Longton, R. E.,** The role of spores in reproduction in mosses, *J. Linn. Soc.,* in press.
137. **Wilson, J. R. and Schwabe, W. W.,** Growth and dormancy in *Lunularia cruciata* (L.) Dum. III. The wavelengths of light effective in photoperiodic control, *J. Exp. Bot.,* 15, 368, 1964.
138. **Courtoy, R.,** Contribution a l'etude du role de la lumiere dans la sexualization du gemetophyte de *Marchantia polymorpha* L., *Phytochem. Photobiol.,* 5, 441, 1966.
139. **Bhatla, S. C.,** Onset of reproductive phase in the moss *Bryum argenteum* Hedw.: involvement of cyclic AMP, *J. Hattori Bot. Lab.,* 56, 167, 1984.
140. **Bhatla, S. C. and Chopra, R. N.,** Hormonal regulation of gametangial formation in the moss *Bryum argenteum* Hedw., *J. Exp. Bot.,* 32, 1243, 1981.
141. **Rao, M. P. and Das, V. S. R.,** Metabolic changes during reproductive development in liverworts, *Z. Plazenphysiol.,* 59, 87, 1968.

Chapter 9

IN VITRO STUDIES ON CHEMICAL REGULATION OF GAMETANGIAL FORMATION IN BRYOPHYTES

R. N. Chopra and Sarla

TABLE OF CONTENTS

I. INTRODUCTION

In recent years considerable work has been done on chemical regulation of sexual reproduction in bryophytes grown *in vitro*. Although the conditions of growth in axenic cultures are in some ways far from natural, it is hoped that ultimately the data obtained in artificial conditions will be helpful in understanding the way sexual reproduction is regulated *in vivo*. The significance of responses to applied chemicals has its basis in the fact that all the major growth regulators have been reported in bryophytes.

Since sexual reproduction is a culmination of normal vegetative growth, it has been thought appropriate to include in this chapter the relevant data on the effects of chemical additives on gametophytic growth. An attempt has also been made to compare the responses of bryophytes with those of pteriodophytes and angiosperms.

Among the chief chemical factors whose effect has been investigated on the induction of gametangia are growth regulators, sugars, chelating agents, and nitrogenous substances. The parameters taken into consideration are the time of gametangial formation, percentage of fertile gametophytes, and number of antheridia and archegonia produced per thallus/gametophore.

II. AUXINS

High auxin levels inhibit growth in the majority of investigated liverworts and mosses.[1-7] In some bryophytes auxins prove inhibitory at all levels,[8] whereas in others they stimulate growth at low concentrations.[6,9-12]

Auxins increase femaleness in liverworts, but in mosses they are either ineffective or increase maleness. Auxin-stimulated archegonial production has been reported in three species of *Riccia*. In *R. crystallina* maximum archegonia were noticed in response to indole-3-acetic acid (IAA), followed by 1-naphthaleneacetic acid (NAA), *o*-chlorophenoxyacetic acid (*o*-CPA), and 2,4,5-trichlorophenoxyacetic acid (2,4,5-T).[13] In *R. gangetica* enhancement of archegonial production by IAA, 2,4-dichlorophenoxyacetic acid (2,4-D), NAA, and β-naphthoxyacetic acid (NOA) at their optimal levels was approximately 340, 300, 300, and 185% of the control, respectively.[9] In *R. frostii,* of the four auxins tried, NAA initiated maximum archegonia and was followed in order of effectiveness by IAA, 2,4-D, and NOA (Figure 1).[11]

Among mosses IAA caused substantial stimulation of antheridial production in the male clone of *Bryum argenteum* (140% of the control).[6] In *Barbula gregaria* and *Bryum coronatum*

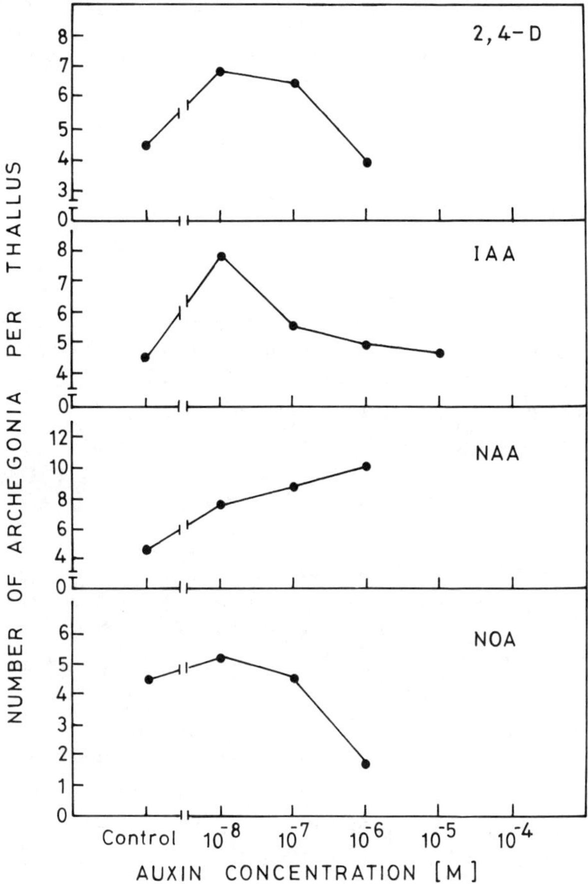

FIGURE 1. Effect of auxins on archegonial formation in the female clone of *Riccia frostii*.[11] Each datum indicates an average of 12 cultures. Data are from 60-day-old cultures.

antheridial production was enhanced at low concentrations of IAA (the optimum being 10^{-7} *M*).[7] In *Microdus brasiliensis* IAA proved inhibitory for antheridial formation, but NAA, 2,4-D, and NOA were promotive, and relative enhancement of antheridial production at their optimal levels was 189, 155, and 145% of the control, respectively.[12] However, in *R. gangetica* antheridial formation was not affected appreciably with auxins.[9] In *R. crystallina* antheridial formation was enhanced with *o*-CPA and IAA, but was inhibited with NAA.[13] Benson-Evans[14] observed that application of auxins (2,4-D and NAA) to *Marchantia polymorpha* induced structures morphologically comparable to receptacles. In *Conocephalum conicum* 2,4-D induced receptacles, but these did not bear any gametangia.

A promotive effect of auxins on archegonial formation has also been reported in some ferns like *Microlepia speluncae* and *Lygodium flexuosum*.[15,16] Among higher plants auxin treatment is reported to induce female flowers on male plants of *Cannabis*.[17]

III. ANTIAUXINS

Antiauxins generally inhibit vegetative growth,[1,3,5,18-20] but in some bryophytes they are even reported to enhance it.[10,12,21]

With maleic hydrazide (MH) and 2,3,5-triidobenzoic acid (TIBA) archegonial number decreased in *R. gangetica,* but antheridial production was stimulated.[20] In the male clone

TABLE 1
Effectiveness of Different Cytokinins in Promoting Vegetative Growth in *Riccia* Species

Taxon	Order of effectiveness	Ref.
Riccia frostii	BAP > kinetin = 2iP	11
Riccia gangetica	BAP > kinetin > zeatin	9
Riccia discolor	Zeatin > kinetin > 2iP > BAP	10

of *Microdus brasiliensis* the percentage of fertile gametophores increased at all the tried concentrations of MH and also at low concentrations of TIBA. Relative enhancement in the number of fertile gametophores at the optimal levels of MH and TIBA was 144 and 133% of the control, respectively.[12] In *R. discolor* MH and TIBA failed to induce gametangia,[10] and they inhibited the formation of gametangia in *R. crystallina.*[13]

IV. CYTOKININS

Cytokinins, in general, enhanced vegetative growth in three species of *Riccia,* although their relative effectiveness varied with the species (Table 1). There are also reports of an inhibitory effect of kinetin on growth of some liverworts.[13,19] Among mosses cytokinins may stimulate growth[6,7,12] or may result in the production of stunted gametophores, especially at higher concentrations.[22-25]

Cytokinins usually enhance femaleness in bryophytes. In the monoecious *R. gangetica* the relative enhancement in archegonial number with N^6-[2-isopentenyl] adenine (2iP), 6-[4-hydroxy-3-methyl-but-2-enylamino] purine (zeatin), 6-furfurylaminopurine (kinetin, Kn), and 6-benzylaminopurine (BAP) was approximately 380, 325, 245, and 230% of the control, respectively. On the other hand, antheridial production either was not affected or it decreased in response to cytokinins (Figure 2).[9] In the male clone of *R. discolor* the thalli remained sterile on cytokinin-supplemented media.[10] In female gametophytes of *R. frostii* 2iP proved best for the production of archegonia, and it was followed in order of effectiveness by kinetin and BAP.[11] In the monoecious *R. crystallina* kinetin enhanced archegonial production, whereas antheridial number was not significantly affected.[13]

Among mosses, antheridial formation was inhibited with kinetin in the male clone of *Bryum argenteum,* whereas in the female clone the percentage response per culture was enhanced at lower concentrations (10^{-8} and 10^{-7} *M*).[6] In the male clone of *Microdus brasiliensis* the percentage of fertile gametophores increased in response to 2iP.[12] In the male clones of *Barbula gregaria* and *Bryum coronatum* kinetin did not have any appreciable effect on antheridial production at lower levels (10^{-7} and 10^{-6} *M*), but at higher concentrations it proved inhibitory (10^{-5} and 10^{-4} *M*).[7] In the monoecious *Leptobryum pyriforme* cytokinins did not alter the ratio of antheridia to archegonia.[26]

In *R. gangetica* antheridia were noticed after 15 to 20 days of growth and archegonia appeared about 5 days later on cytokinin-supplemented media, as in control cultures.[9] Bryokinin (2iP) hastened the initiation of archegonia in the female clone of *R. frostii.*[11] In the male clone of *Microdus brasiliensis* the time of initiation of antheridia was not affected at lower concentrations of 2iP, but they appeared 2 to 3 days earlier at higher concentrations.[12]

Cytokinins also enhance femaleness in higher plants. Treatment of hermaphrodite flowers of *Bryophyllum crenatum* resulted in a shift of the balance toward femaleness. Kinetin promoted femaleness in the monoecious *Luffa acutangula* under short days and day-neutral conditions.[17]

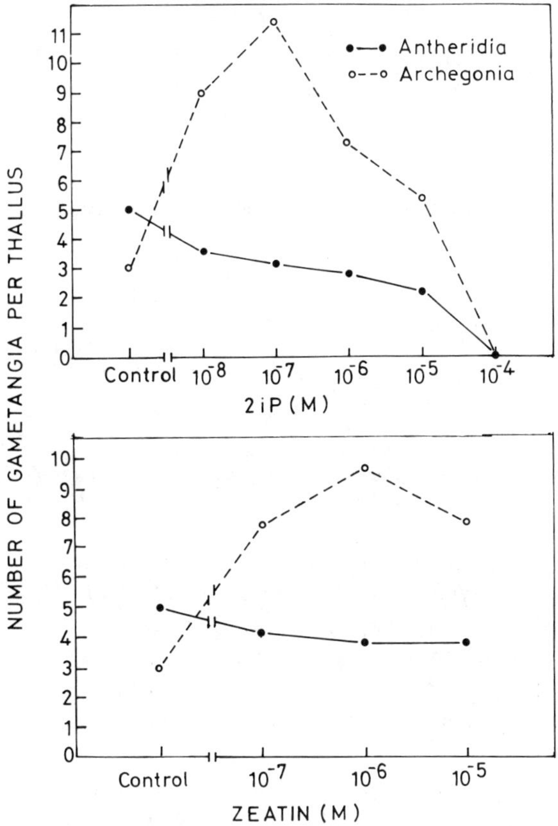

FIGURE 2. Effect of 2iP and zeatin on gametangial formation in the monoecious *Riccia gangetica.*[9] Each datum indicates an average of 12 cultures. Data are from 45-day-old cultures.

V. GIBBERELLINS

The effect of gibberellic acid (GA_3) on the growth of bryophytes is quite variable. In several liverworts growth was stimulated at lower concentrations and inhibited at higher levels.[3,10,19,22] In some bryophytes growth was stimulated at all the tried concentrations of GA_3.[7,12,20] Gibberellic acid may not have any appreciable effect on growth.[1,5,6,27]

In some systems, like *Bryum argenteum,* GA_3 enhanced antheridial formation in the male clone, and 10^{-5} *M* proved to be optimal (169% of control). In the female clone GA_3 had no marked effect at lower concentrations (10^{-8} and 10^{-7} *M*), but at higher levels it resulted in inhibition of archegonial formation.[6] Chopra and Kumra[7] observed that in male clones of *Barbula gregaria* and *Bryum coronatum* the percentage of fertile gametophores increased markedly at lower concentrations (10^{-7} to 10^{-5} *M*) of GA_3, 10^{-6} *M* being optimal. At optimal concentrations the relative enhancement of antheridial production and gametophore formation was 138 and 155% of the controls in *Barbula* and 130 and 134% in *Bryum coronatum,* respectively. In *Microdus brasiliensis* and *Bartramidula bartramioides* the percentage of fertile gametophores increased at lower concentrations, 10^{-7} *M* being optimal.[12,28] At the optimal level, relative enhancement of gametangial induction in *Microdus* was 189% of the control (Table 2).[12]

Chopra and Kumra[20] reported that in the monoecious *R. gangetica* the number of antheridia increased at all levels of GA_3, and 10^{-4} *M* proved to be the best. Relative enhancement of antheridial production was approximately 242% of the control. On the other hand,

TABLE 2
Effect of GA_3 on Gametangial Formation in *Microdus brasiliensis* Maintained in Controlled Conditions of Irradiance (10 to 11.5 W m^{-2}) at 25 ± 2°C[12]

Treatment (mol dm^{-3})	% of fertile gametophores
Control	18 ± 2
10^{-8} GA_3	20 ± 2
10^{-7} GA_3	34 ± 3
10^{-6} GA_3	24 ± 2
10^{-5} GA_3	15 ± 2
10^{-4} GA_3	—

Note: —: antheridia are not formed. Mean values and standard error are from 12 replicates. Data are from 60-day-old cultures.

archegonial production decreased with increasing concentration of GA_3. In another monoecious species, *R. crystallina,* gibberellins (GA_3, GA_{13}, and $GA_{4/7}$) greatly enhanced antheridial production. GA_3 was the most effective, and maximum antheridia were produced at 10^{-4} *M*. At lower levels of GA_3 (10^{-8} to 10^{-6} *M*) a slight increase in archegonial production was observed, but GA_{13} and $GA_{4/7}$ proved inhibitory at all concentrations.[13] In *R. frostii,* a dioecious species, GA_3 suppressed archegonial formation in the female clone and also delayed their appearance.[22]

Gibberellic acid is reported to induce antheridia in bryophytes which remain sterile on basal medium. As examples, *Philonotis turneriana*[7] and *R. discolor*[10] may be mentioned. Antheridia were induced in *R. discolor* at 10^{-8} and 10^{-7} *M* GA_3, and growth of thalli was also stimulated only at lower levels.

Some endogenous substances (antheridiogens) are known to regulate antheridial formation in ferns like *Anemia phyllitidis, A. rotundifolia, Lygodium japonicum, Dryopteris filix-mas, Onoclea sensibilis,* and *Ceratopteris.*[15,29,30] These compounds are structurally similar to gibberellins.[31] Takeno and Furuya[32] reported that in *L. japonicum* archegonial differentiation is inhibited by the application of GA_3. Gibberellic acid is reported to promote flowering in some higher plants,[33] but its effect on sex expression is variable. It increased femaleness in some plants, whereas in others it enhanced maleness.[34] Mohan Ram and Jaiswal[35] demonstrated that gibberellins induce male flowers in genetically female plants of *Cannabis sativa.* The female plants, however, reverted to the production of female flowers when the applied gibberellins were depleted.

VI. ANTIGIBBERELLIN

The effect of cycocel (CCC) on growth varies with the system. It may prove stimulatory,[10,12] may not have any appreciable effect,[13] or may retard growth.[20,36]

In *R. crystallina* and *R. gangetica* CCC retarded antheridial production. Archegonial number increased in the former species, whereas there was no appreciable effect in the latter.[13,20] On the contrary, in *Microdus brasiliensis* antheridial formation was markedly enhanced with CCC, 10^{-8} *M* being optimal (178% of the control).[12]

The effect of CCC on liverworts seems to be similar to that on vascular plants. Treatment of the sunflower with CCC reduced the levels of endogenous gibberellins and retarded shoot elongation, as also occurs in *Silene.*[37,38] Halevy et al.[39] reported that CCC changed the orthogeotropic growth of erect-type peanut plants to diageotropic growth.

TABLE 3
Comparison of Effects of Gibberellic Acid and Cycocel

Taxon	Gibberellic acid	Cycocel	Ref.
Riccia crystallina (monoecious)	Enhanced antheridial production	Antheridial production retarded with increase in concentration	13
	Archegonial production reduced	Archegonial production stimulated	
Riccia gangetica (monoecious)	Vegetative growth stimulated at all levels	Vegetative growth retarded, with maximum inhibition at 10^{-4} *M*	20
	Archegonial production decreased with increase in concentration	Archegonial production not affected appreciably	
	Antheridial production enhanced	Antheridial production inhibited	
Riccia discolor (male clone)	Vegetative growth stimulated at lower levels and inhibited at higher levels	Growth stimulated at 10^{-6} to 10^{-4} *M*, with a maximum at 10^{-4} *M*	10
	Antheridia induced at lower levels	Gametangia failed to appear	
Microdus brasiliensis (male clone)	Vegetative growth stimulated at all levels	Growth stimulated, except at 10^{-4} *M*	12
	Percentage of fertile gametophores increased at lower levels, higher levels proved inhibitory	Antheridial formation markedly enhanced	

Judging from the overall effects of gibberellic acid and CCC on vegetative growth and gametangial production in the investigated liverworts and mosses, it becomes clear that in the majority of systems the effects of CCC are quite opposite those of gibberellic acid (Table 3). It can thus be concluded that in bryophytes CCC acts as an antigibberellin.

VII. ANIMAL SEX HORMONES

Several steroid hormones are known to exist in plants, and they seem to play a prominent role in their growth and development.[39A] In the male clones of *Barbula gregaria* and *Bryum coronatum* progesterone and testosterone stimulated antheridial production only at lower levels (10^{-7} and 10^{-6} *M*).[7] In *R. gangetica* testosterone significantly reduced the time required for gametangial initiation and markedly increased their number, especially that of antheridia. Progesterone did not affect the time of initiation of gametangia, but enhanced their number at lower levels.[20] In *Microdus brasiliensis* both the hormones promoted antheridial formation and enhanced growth at lower concentrations. Testosterone was more effective than progesterone, and it also reduced the time of antheridial initiation.[12] In *R. discolor* regeneration and growth were stimulated by testosterone, whereas progesterone inhibited growth. Gametangia failed to appear in response to both the hormones.[10]

VIII. ABSCISIC ACID

In bryophytes abscisic acid (ABA) inhibits vegetative growth[6,10,12,40] as well as gametangial induction.[40]

In *Bryum argenteum* the gametangial induction response was reduced to 32 and 21% of the controls at 10^{-4} *M* ABA in male and female clones, respectively.[6] In *Microdus brasiliensis* the relative response of antheridial induction came down to 17% of the control at 10^{-5} *M* ABA.[12]

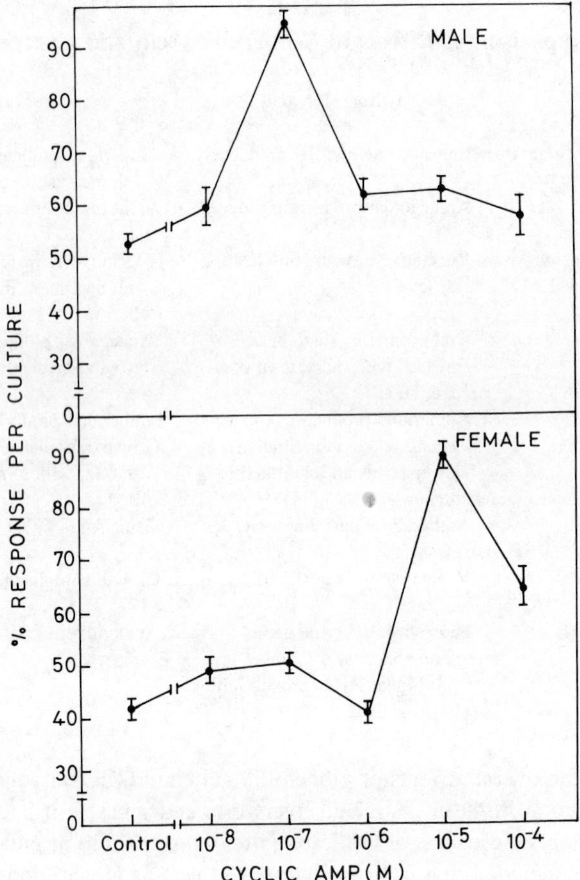

FIGURE 3. Effect of cAMP on gametangial formation in male and female clones of *Bryum argenteum*.[42] Each datum indicates the mean and standard error from 12 replicates. Data are from 40-day-old cultures.

IX. CYCLIC ADENOSINE 3′,5′-MONOPHOSPHATE (cAMP) AND OTHER PURINE DERIVATIVES

Gametangial formation is remarkably enhanced in *Bryum argenteum* by cAMP.[41] For optimal effect the male clone requires a lower dose than the female clone (Figure 3). The extent of stimulation also differs in the two sexes: it is greater in the female (214% of the control) than in the male clone (177% of the control). In this moss the effect on gametangial formation is not accompanied by any significant increase in vegetative growth. In order to observe if the response to cAMP was specific, the effect of other purine derivatives was studied.[42] Gametangial formation was enhanced by 3′-AMP at 10^{-6} *M*. Adenosine and ADP also evoked some response, but adenine and ATP had practically no effect. Growth was, however, favored by all these purine derivatives. Cyclic 2′,3′-AMP slightly stimulated gametangial formation in the female clone without appreciably affecting vegetative growth. It remained ineffective in the male clone. Among these purine derivatives maximum response was elicited by 3′-AMP, but even this was insignificant compared to the response to cAMP. cAMP has also been observed to bring about some stimulation of antheridial formation in *Bryum coronatum* and *Barbula gregaria*.[43]

Administration of inhibitors of phosphodiesterase (PDE, the enzyme catalyzing the

TABLE 4
Effect of IAA-GA_3 Interaction on Gametangial Formation in *Riccia gangetica*[40]

Treatment (mol dm^{-3})	Average number of gametangia per thallus	
	Antheridia	Archegonia
Control	5.0 ± 0.5	3.0 ± 1.2
10^{-6} IAA	5.0 ± 0.7	8.6 ± 0.6
10^{-5} IAA	5.0 ± 0.6	10.2 ± 1.5
10^{-5} GA_3	9.0 ± 1.2	1.6 ± 0.3
10^{-4} GA_3	12.1 ± 0.8	1.6 ± 0.6
10^{-5} IAA + 10^{-5} GA_3	7.3 ± 1.9	6.5 ± 1.5
10^{-5} IAA + 10^{-4} GA_3	10.2 ± 2.0	6.3 ± 1.0
10^{-6} IAA + 10^{-5} GA_3	7.1 ± 1.3	6.4 ± 1.6
10^{-6} IAA + 10^{-4} GA_3	9.9 ± 1.9	6.0 ± 1.2

Note: Each datum indicates the mean and standard error from 12 thalli.

hydrolysis of cAMP) is known to increase the intracellular levels of cAMP in many biological systems.[44] Therefore, these inhibitors should be able to mimic the response evoked by cAMP. It has indeed been observed that I.C.I. 58,301 [3-acetamido-6-methyl-8-*n*-propyl-*s*-triazolo(4,3a)pyrazine] and theophylline (PDE inhibitors) enhance gametangial formation in *Bryum argenteum*.[42] These inhibitors also may be acting by blocking the leaching of cAMP into the medium.[45]

The response evoked by cyclic 3′,5′-guanosine monophosphate (cGMP) is unique. In contrast to cAMP and cyclic 2′,3′-AMP, cGMP inhibits gametangial formation in *B. argenteum*, the female clone being more sensitive than the male clone. At lower concentrations (10^{-8} and 10^{-7} *M*) cGMP retards only gametangial formation, but at higher levels (up to 10^{-4} *M*) it inhibits vegetative growth as well.[42] It is probable that cGMP affects plant processes by selectively stimulating the activity of cAMP nucleotide phosphodiesterase, which catalyzes the hydrolysis of cAMP.[46]

X. INTERACTION OF GROWTH REGULATORS

Some combinations of growth regulators have also been tried, and the findings are presented below.

A. AUXINS AND GIBBERELLIC ACID

In higher plants the favorable effect of NAA on production of female flowers is inhibited by treating cucumber with a mixture of NAA and GA_3, and the number of male flowers is enhanced.[47] In *R. crystallina*, interaction of GA_3 with auxins (IAA, NAA, 2,4,5-T, and *o*-CPA) resulted in decreased archegonial production, possibly because of the dominating influence of GA_3.[13] However, in *R. gangetica* a combination of GA_3 and IAA decreased the number of archegonia as well as antheridia as compared to those produced in response to IAA or GA_3, respectively (Table 4).[40] Asprey et al.[48] demonstrated that when isolated sporophytes of *Pellia epiphylla* were treated simultaneously with IAA and potassium gibberellate, full seta elongation occurred, whereas it was incomplete when treated with either hormone alone. They suggested that gibberellin may promote auxin-mediated elongation through inactivation of auxin oxidase or that of an auxin oxidase cofactor.

B. CYTOKININS AND AUXINS

In *Bryum argenteum* kinetin and IAA synergistically stimulated gametangial formation

TABLE 5
Effect of Kinetin-IAA Interaction on Gametangial Formation in *Microdus brasiliensis* Maintained in Controlled Conditions of Irradiance (10 to 11.5 W m^{-2}) at 25 ± 2°C[12]

Treatment (mol dm^{-3})	% of fertile gametophores
Control	18 ± 2
10^{-8} IAA	12 ± 2
10^{-7} IAA	13 ± 2
10^{-6} IAA	17 ± 1
10^{-5} IAA	15 ± 2
10^{-4} IAA	—
10^{-6} Kn	—
10^{-6} Kn + 10^{-8} IAA	26 ± 2
10^{-6} Kn + 10^{-7} IAA	28 ± 3
10^{-6} Kn + 10^{-6} IAA	22 ± 4
10^{-6} Kn + 10^{-5} IAA	18 ± 2

Note: —: gametophores are abnormal, antheridia not formed. Mean values and standard error are from 12 replicates. Data are from 60-day-old cultures. Kn = kinetin.

in both male and female clones. Their interaction nullified the inhibitory effect of kinetin and IAA on antheridial and archegonial production, respectively. Thus, at optimal combination the gametangial formation response in male and female clones was 180 and 250% of the controls (−IAA and −Kn), respectively.[6] In *Microdus brasiliensis* kinetin induced moruloid buds which later developed into leafless, sterile gametophores. On IAA-supplemented media antheridia were observed, but the response was less than that in control cultures. Interaction of IAA and kinetin (10^{-6} *M*) resulted in the formation of normal gametophores, and the percentage of fertile plants increased as compared to that with IAA alone (Table 5). Thus, interaction of kinetin and IAA promotes antheridial production in this moss. Maximum response was observed at 10^{-7} *M* IAA + 10^{-6} *M* kinetin, and this combination was also optimal for vegetative growth.[12,49] In *R. gangetica* IAA is inhibitory for growth at higher concentrations, but this effect is overcome with the coaddition of kinetin. This interaction also had an additive effect on archegonial formation, but antheridial induction was not appreciably affected.[40] In *R. discolor,* when the optimal concentration of BAP (10^{-5} *M*) was combined with 2,4-D (10^{-8} to 10^{-4} *M*), fresh and dry weight yields decreased and inhibition increased with an increase in the concentration of 2,4-D. Individually, BAP stimulated growth, the optimum being at 10^{-5} *M*, whereas growth was enhanced with 2,4-D only at lower levels and was inhibited at higher concentrations.[10]

C. CYTOKININS AND GIBBERELLIC ACID

In *R. discolor* a combination of BAP (10^{-5} *M*) with GA_3 (10^{-8} to 10^{-4} *M*) supported excellent regeneration and growth. Thalli were exceptionally broad and dark green. GA_3 at higher concentrations or BAP at all concentrations remained ineffective in bringing about fertility, whereas their combination induced a large number of antheridia per thallus at all concentrations of GA_3 (Table 6).[10]

TABLE 6
Effect of Interaction of BAP with Various Concentrations of GA_3 on Gametangial Induction in *Riccia discolor*[10]

Treatment (*M*)	Number of antheridia per gametophyte
Control	—
10^{-5} BAP	—
10^{-8} GA_3	2.0
10^{-7} GA_3	4.0
10^{-6} GA_3	—
10^{-5} GA_3	—
10^{-4} GA_3	—
10^{-5} BAP + 10^{-8} GA_3	16.5
10^{-5} BAP + 10^{-7} GA_3	24.5
10^{-5} BAP + 10^{-6} GA_3	20.0
10^{-5} BAP + 10^{-5} GA_3	15.0
10^{-5} BAP + 10^{-4} GA_3	10.0

Note: —: indicates absence of antheridia. Cultures maintained in 3500 lux of continuous light at 25 ± 2°C. Each datum indicates mean of 12 replicates. Data are from 120-day-old cultures.

In *Microdus brasiliensis* the interaction of kinetin (10^{-6} *M*) and GA_3 led to an increase in the number of gametophores per culture, but the percentage of fertile gametophores decreased with this combination.[50]

D. KINETIN AND CYCLIC AMP

In *Bryum argenteum* cAMP prevented the inhibitory effect of kinetin in the male clone, whereas in the female clone it stimulated the response elicited by kinetin.[6] In *Microdus brasiliensis* kinetin enhances the number of buds, but it inhibits the formation of normal gametophores. With cAMP alone the number of buds is reduced and the gametophores are stunted. Coaddition of kinetin and cAMP increases the number of buds, improves the morphology of gametophores, and in some combinations brings about considerable enhancement of antheridial production.[49]

E. GIBBERELLIC ACID AND ABSCISIC ACID

Gibberellin-induced responses like stem elongation and seed germination are inhibited by ABA.[51] ABA also influences sex expression in several plants, probably by antagonizing the effect of gibberellin.[52] Induction of male flowers by GA_3 in *Cannabis sativa* could be partially or completely inhibited by the simultaneous application of ABA.[35]

In *R. gangetica* ABA reduced the stimulatory effect of GA_3 on vegetative growth and antheridial formation. The interaction of GA_3 and ABA suppressed antheridial formation, and the inhibition was directly proportional to the concentration of ABA.[40]

F. COCONUT MILK AND MALEIC HYDRAZIDE

In *R. discolor* regeneration and growth were vigorous at 5 and 15% coconut milk. At these levels senescence was delayed and antheridia were induced. A higher concentration (25%) proved inhibitory for growth and the thalli remained sterile. The morphology of thalli

A

FIGURE 4. *Riccia discolor*.[10] (A) Thin and long thalli growing on Nitsch's basal medium (NB) supplemented with coconut milk (5%). Antheridia appear in such cultures after 60 days. 47-day-old culture. (Magnification × 5.6.) (B) Broader and thicker thalli with distinct median grooves. These were grown on NB medium supplemented with 25% coconut milk. Globular callus masses have developed from the basal cut end of the inoculum and from the apex of a regenerant. A few rhizoids have arisen on the dorsal surface. The gametophytes remain sterile. 47-day-old culture. (Magnification × 3.9.)

was altered considerably: they became broader and thicker, the median grooves were well marked, callus masses were initiated, and some rhizoids developed on the dorsal surface (Figures 4A and B).[10] In the female clone of *R. frostii* the formation of archegonia was promoted by coconut milk at 10 and 15%, the former being more effective.[22]

Combination of coconut milk with MH enhanced vegetative growth of *R. discolor* and induced antheridia in large numbers (Figure 5B), whereas MH alone remained ineffective in eliciting this response.[10]

XI. SUGARS

The effect of several sugars on growth and reproduction of bryophytes has been studied. In general, sucrose and/or glucose prove best for vegetative growth, and others are effective to a lesser degree (Table 7). Curiously, there are also some reports of inhibition of growth by sugars.[10,13,22,53,54]

Onset of the reproductive phase is also affected by sugars. Wann[55] noticed that a relatively high carbohydrate/nitrogen (C/N) ratio in the medium resulted in the appearance of gametangiophores in *Marchantia polymorpha*. In some systems, like *Bryum argenteum*, *Bryum coronatum*, *Barbula gregaria*, and *Microdus brasiliensis*, gametangia are formed on media without sugar, but with the addition of sucrose (0.5 to 1%) the response increases.[43,50,56] Other taxa like *R. crystallina*, *Athalamia pusilla*, *Bartramidula bartramioides*, and *R. frostii* develop gametangia only in the presence of sugar.[5,13,22,57]

FIGURE 4B.

It is interesting to note that the two sexes may have different requirements for carbohydrates. Thus, in *Reboulia hemisphaerica* male gametangiophores appeared on media with 1% glucose, whereas female gametangiophores developed on sugar-free medium.[58] In *Riccia gangetica* archegonia did not appear on sugar-free medium, but some antheridia were noticed on medium without sugar.[54] In *Fimbriaria angusta* almost a double C/N ratio was necessary for initiation of female gametangiophores as compared to male gametangiophores.[59] In *Bryum argenteum* the best response by male and female clones was shown at 0.5 and 1% sucrose, respectively.[56,60]

The relative effectiveness of sugars in supporting gametangial formation also differs in different bryophytes. In *Riccia crystallina, Riccia gangetica,* and *Microdus brasiliensis* sucrose was most effective, whereas in *Riccia frostii* sucrose and glucose were almost equally potent.[13,22,50,54] In general, higher concentrations of sucrose suppress gametangial formation.[13,41,54] In higher plants as well there are reports of inhibition of the reproductive phase in response to high levels of sucrose.[61]

Judging from the responses of bryophytes to sucrose and glucose it seems that the effect of sugars on gametangial production is not specific, and it may be a consequence of good vegetative growth.

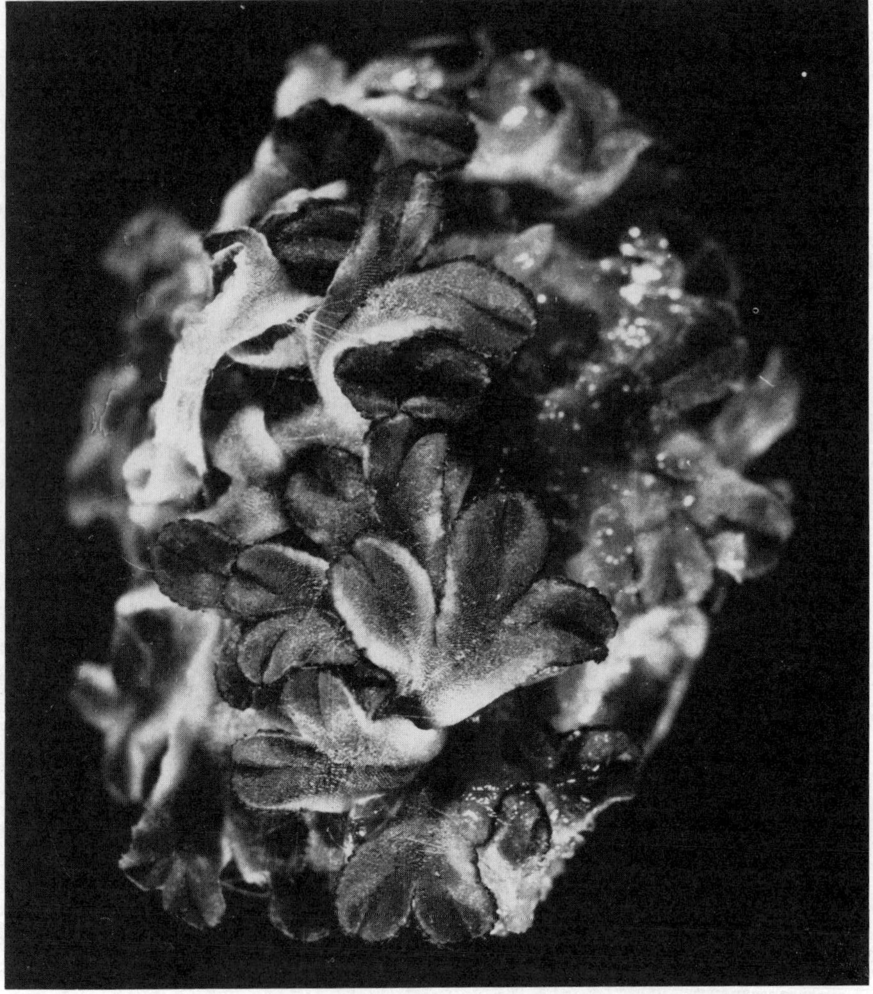

A

FIGURE 5. *Riccia discolor.*[10] (A) Large and robust thalli growing compactly on Murashige and Skoog's medium (MS) supplemented with casein hydrolysate (5 ppm). Note antheridial papillae and pores of antheridial chambers on dorsal surface of thalli. The marginal scales are well developed. 120-day-old culture. (Magnification × 7.3.) (B) A thallus enlarged from Nitsch's basal medium (NB) supplemented with coconut milk (15%) and maleic hydrazide (10^{-8} *M*). The dorsal surface shows prominent antheridial papillae and pores. 120-day-old culture. (Magnification × 33.)

XII. CHELATING AGENTS

Chelating agents ethylenediamine-di-(*o*-hydroxyphenylacetic acid) (EDDHA) and ethylenediaminetetraacetic acid (EDTA) and their iron salts, salicylic acid, and ferric citrate, in general, stimulate vegetative growth of liverworts and mosses. In the three species of *Riccia* their relative effectiveness varies (Table 8).

Chelates also affect gametangial production in bryophytes. In the monoecious *Riccia crystallina* they bring about an increase in gametangial number, especially archegonia.[63] In *R. gangetica* Fe-EDDHA and Fe-EDTA significantly enhanced antheridial production and, to a lesser extent, archegonial formation. EDDHA and EDTA were more effective than their iron salts. In this species salicylic acid increased antheridial formation, but archegonial production was only stimulated slightly.[64]

FIGURE 5B.

TABLE 7
Effectiveness of Different Sugars in Promoting Vegetative Growth in Bryophytes

Taxon	Order of effectiveness	Ref.
Sphaerocarpos texanus	Glucose > sucrose > mannose > fructose > cellobiose > maltose > xylose > ribose > galactose > arabinose	62
Athalamia pusilla	Sucrose > glucose > mannose	5
Fossombronia himalayensis	Sucrose = glucose	27
Riccia crystallina	Sucrose > glucose > fructose	13
Riccia gangetica	Sucrose > fructose > glucose > mannose	54
Riccia frostii	Glucose > sucrose > fructose > galactose > mannose	22
Riccia discolor	Sucrose > glucose > fructose > mannose	10
Microdus brasiliensis	Sucrose > glucose > fructose	50

Among mosses also chelating agents have a profound effect on growth and gametantial formation. In the male clones of *Barbula gregaria* and *Bryum coronatum* EDDHA and EDTA enhanced the percentage of fertile gametophores. Fe-EDDHA, Fe-EDTA, and ferric citrate increased antheridial production as well as vegetative growth, and the former effect was more striking. Salicylic acid inhibited both responses.[65] Bhatla and Chopra[66] reported that in the dioecious moss *Bryum argenteum* EDTA and EDDHA did not affect the time of gametangial formation, but markedly increased the percentage of fertile gametophores. The response to EDTA was significant, with an optimum at 10^{-7} *M* (167% of the control) and at 10^{-6} *M* (129% of the control) in male and female clones, respectively. In the male clone the response at the optimal level of EDDHA was 164% of the control, whereas in the female clone it was 155% of the control. In the male clone of *Microdus brasiliensis* EDDHA increased gametophore number as well as percentage of fertile gametophores at lower levels

TABLE 8
Relative Effectiveness of Chelating Agents in Promoting Vegetative Growth in *Riccia* Species

Taxon	Order of effectiveness	Ref.
Riccia crystallina	Ferric citrate > Fe-EDTA > Fe-EDDHA	63
Riccia gangetica	EDDHA = Fe-EDTA > Fe-EDDHA > EDTA	64
Riccia discolor	EDDHA > SA > EDTA > Fe-EDDHA > Fe-EDTA > Ferric citrate	10

(10^{-8} to 10^{-6} *M*), whereas EDTA enhanced both responses at all levels (10^{-8} to 10^{-4} *M*). The iron salts of these chelates increased gametophore number and percentage of fertile gametophores, but the latter response was more striking. The number of antheridia per head also increased with Fe-EDTA (12 to 14 as compared to 8 to 10 in controls), and at higher concentrations antheridia were induced 4 days earlier. Salicylic acid increased the number of gametophores at all concentrations. The percentage of fertile gametophores increased only at lower concentrations (10^{-8} to 10^{-6} *M*), and antheridia were induced 3 days earlier.[67]

Estimations of endogenous levels of iron and copper have been made in plants treated with chelates. While iron content increases with an increase in gametangial induction response in *Barbula gregaria* and *Microdus brasiliensis,* no definite correlation was observed for copper.[43,67] On the other hand, in *Bryum argenteum* the content of neither metal showed definite correlation with the morphogenetic response.[66]

Since exogenously supplied ferric citrate also increases the percentage of fertile plants in *Bryum argenteum, Microdus brasiliensis, Barbula gregaria, Bryum coronatum,* and *Riccia gangetica,* it is possible that iron is involved in the promotion of gametangial formation.[64-67] In contrast to these observations, ferric citrate does not affect sexuality in *Riccia crystallina.*[63] EDTA remains ineffective in inducing buds in *Bryum pallescens,* but its iron salt (Fe-EDTA) elicits this response.[68] This might also be due to bioavailability of metal ions such as iron.

Thus, even though the chelates may enhance endogenous iron and copper content in bryophytes, the correlation between their uptake and the morphogenetic response is not conclusive.

Investigations on the role of chelating agents in higher plants indicate their significance in the physiological processes leading to flowering.[69-71] Seth et al.[69] also reported that the promotion of flowering in *Wolffia microscopica* by EDDHA is accompanied by a several-fold increase in the endogenous iron content. Hillman[72] had demonstrated earlier that iron plays a specific role in the flowering of *Lemna gibba.* However, subsequent studies on duckweeds have cast doubt on the chelation hypothesis.[71]

XIII. AMINO ACIDS

Studies on amino acid-deficient mutants of *Marchantia polymorpha* have proven that endogenous amino acids are involved in regulation of growth and development.[73] In this liverwort amino acids elicit these responses by influencing protein synthesis.[74-76] Margaris[77] studied five mosses and recorded significantly higher amounts of aspartic acid as compared to other amino acids, especially in *Tortula princeps* and *Platyhypnidium riparioides.*

In *Haplomitrium hookeri* amino acids in various combinations resulted in excellent growth of plants.[78] The relative effectiveness of amino acids for improving growth varies in different taxa (Table 9).

Amino acids also influence gametangial production. In *Riccia gangetica* aspartic acid and glutamic acid favor antheridial production. In contrast, asparagine, serine, and tryptophan enhance archegonial formation, and among these asparagine elicits the best response. Glycine

TABLE 9
Effectiveness of Different Amino Acids in Enhancing Growth in Some Bryophytes

Taxon	Order of effectiveness	Ref.
Riccia crystallina	Glutamic acid > threonine > valine > aspartic acid > hydroxyproline > leucine > serine > tryptophan > asparagine > glycine > alanine	63
Athalamia pusilla	Tryptophan > lycine > histidine > methionine > glycine > arginine	5
Riccia gangetica	Glutamic acid > aspartic acid > glycine > serine > asparagine > tryptophan	79
Riccia discolor	Threonine > lysine > aspartic acid > methionine > asparagine	10
Riccia frostii	Aspartic acid > tryptophan > threonine	80
Microdus brasiliensis	Methionine > isoleucine	50

not only proves best for antheridial production, but also increases the number of archegonia.[79] In *R. frostii* aspartic acid and threonine stimulate archegonial production at all levels, whereas tryptophan does so only at lower levels. At optimal levels threonine initiated maximum archegonia and was followed in order of effectiveness by aspartic acid and tryptophan.[80] In *R. crystallina* archegonial production is favored by hydroxyproline, serine, threonine, asparagine, glutamic acid, alanine, and leucine. On the other hand, valine, tryptophan, glycine, and aspartic acid enhance antheridial formation.[63,81,82] In *Cephalozia media* amino acids in general inhibit gametangial formation. The inhibitory effect of cysteine, arginine, and proline is greater in female than in male gametophytes.[83] In *Microdus brasiliensis* aspartic acid and tryptophan inhibited vegetative growth as well as antheridial production.[50]

XIV. CASEIN HYDROLYSATE

In general, gametophytes of liverworts show normal growth at lower concentrations of casein hydrolysate, but at higher levels growth is retarded and callus/callus-like cell masses develop.[27,63,80,84]

In *Riccia discolor* addition of casein hydrolysate to Murashige and Skoog's (MS) medium markedly stimulated normal growth of thalli, 5 ppm being optimal (Figure 5A). Deletion experiments indicated that major salts of MS medium in combination with casein hydrolysate are mainly responsible for stimulation of vegetative growth.[85]

Experimental studies on gametangial induction in response to casein hydrolysate are meager. In the monoecious *Riccia crystallina* there was no significant increase in antheridial production, but archegonial number was enhanced on casein hydrolysate-supplemented medium.[63] In the dioecious *R. frostii* casein hydrolysate reduced archegonial formation except at the lowest concentration tried (50 ppm).[80] The male clone of *R. discolor* remained sterile on basal medium. Antheridia were induced only on complete MS medium + casein hydrolysate, 20 ppm being optimum for this response. It may be argued that fertility on MS medium + casein hydrolysate is a consequence of stimulated vegetative growth. However, it does not appear to be so because the optimal concentrations of casein hydrolysate for vegetative growth and antheridial induction are different. Furthermore, antheridia should have been induced on MS medium without hormones, minor salts, or vitamins, since their removal did not bring about any appreciable reduction in growth. Hence, antheridial induction in *R. discolor* seems to be a specific response to casein hydrolysate-supplemented MS medium.[85]

XV. UREA

In *Riccia crystallina* very robust, dark-green thalli were produced in response to urea, and archegonial and antheridial production increased up to 10^{-3} *M*.[86] In *R. discolor* urea

TABLE 10
Effect of Varying Concentration of Minor Salts Solution of MS Medium (Containing One Tenth Strength Major Salts Solution) on Antheridial Induction in *Riccia discolor*[10]

Concentration of minor salts solution (ppm)	Number of antheridia per gametophyte
Control (1 ppm)	—
2.0	3.5
4.0	6.0
6.0	13.2
8.0	11.5

Note: —: antheridia are not formed. Cultures were maintained in 3500 lux of continuous light at 25 ± 2°C. Each datum indicates mean of 12 replicates. Data are from 120-day-old cultures.

supported good regeneration and growth and also delayed senescence. Antheridia were induced at all concentrations of urea in the male clone of this liverwort.[10] In *Pohlia nutans* archegonia were induced with urea at 5 to 50 mg/l.[87]

In *Microdus brasiliensis* the time taken for antheridial formation was not affected, but the percentage of fertile plants increased at lower concentrations of urea. Relative enhancement of gametangial production and vegetative growth at the optimal concentration (10^{-7} *M*) was 172 and 156% of controls, respectively.[50]

Experiments on carrot roots indicated that urea furnishes a more complete nitrogen nutrition.[88]

XVI. YEAST EXTRACT

In *Riccia crystallina* the number of antheridia increased in the presence of up to 1000 ppm yeast extract (YE), whereas archegonial production was optimal at 1500 ppm.[63] In *R. frostii* YE retarded vegetative growth as well as archegonial formation except at the lowest concentration tried (0.5%).[80] In *R. discolor* regeneration and growth were enhanced at all levels of YE. None of the concentrations tried induced gametangia in this liverwort.[10]

XVII. CHLORAL HYDRATE

In *Riccia frostii* chloral hydrate stimulated archegonial formation at 10^{-8} *M*, but it inhibited this response at all other levels.[22] When chloral hydrate was added to MS medium, growth of *R. discolor* was vigorous and thalli remained dark green up to $3^1/_2$ months. Antheridia were induced in this species on complete MS medium supplemented with chloral hydrate, with its optimal concentration being 20 ppm. When chloral hydrate was added to Nitsch's basal medium growth and regeneration were normal, but the thalli remained sterile and senesced in about 2 months.[89]

XVIII. MINOR SALTS

Normally, 1 ppm of minor salts solution is used in MS medium. Regeneration and growth of *Riccia discolor* were promoted at increased levels (2, 4, 6, and 8 ppm) of minor salts. Antheridia were induced at all enhanced concentrations of minor salts, while control cultures (1 ppm) remained sterile (Table 10).[10]

XIX. METABOLIC CHANGES ACCOMPANYING GAMETANGIAL FORMATION

Rao and Das[59] studied the changes in respiratory rate, C/N ratio, endogenous levels of auxin, nucleic acid, and protein during the reproductive development in five liverworts: *Exormotheca tuberifera, Plagiochasma articulatum, Reboulia hemisphaerica, Fimbriaria angusta,* and *Pallavicinia canarus.* It was observed that a transition from vegetative to reproductive phase in female gametophytes was characterized by a sharp rise in respiratory rate and invariably a doubling of the C/N ratio. Closely associated with the production of archegoniophores were also the enhanced levels of endogenous IAA, RNA, and protein. Male plants exhibited a reverse trend during antheridial formation, and there was no appreciable rise in the C/N ratio.

Maravolo[90] observed that three phenolic compounds are present in the antheridial disc tissue of *Marchantia,* and these are not detected in the uninduced thalli. Phenols were present in the lowest concentration in the intermediate thallus region, in higher concentration in basal vegetative areas, and in the highest concentration in the sexually differentiated apex of *M. polymorpha.*[91] Phenolic substances have been demonstrated to act as growth regulators through IAA oxidase interactions.[92]

Hilgenberg et al.[93] observed variations in the activity of tryptophan synthase, IAA oxidase, and peroxidase during the development of *Marchantia polymorpha.* The specific activity of tryptophan synthase and peroxidase was reduced in the antheridiophores and archegoniophores and that of IAA oxidase increased sharply as compared to that in the thallus. Maravolo et al.[94] observed changes in esterase, peroxidase, and phosphatase activity in the uninduced and induced thalli, antheridiophores, and archegoniophores of *M. polymorpha.* There are some common sites of activity of these enzymes in all the structures, but additional sites of esterase and peroxidase activity were found in antheridiophore discs. The antheridia provided the additional esterases. No additional sites of activity were observed in archegoniophores, owing probably to relatively fewer archegonia.

XX. CONCLUDING REMARKS

Several bryophytes have been cultured successfully, but induction of gametangia appears somewhat difficult. The life history (spore to spore) of very few bryophytes has been completed so far *in vitro.* In view of this it is correct to state that we are far from a complete understanding of the mechanisms underlying the onset of the reproductive phase.

There are no reports of sex reversal in bryophytes, but it has been possible to induce gametangia in plants which remain sterile on basal medium and to increase the number of antheridia or archegonia by adding specific chemicals.

The *in vitro* studies on reproductive biology are comparatively recent, and in order to derive any meaningful generalizations such work should be extended to many more taxa. Multifactorial experiments and biochemical aspects of the mode of action of additives should also be stressed in order to understand why diverse factors elicit similar responses.

ACKNOWLEDGMENTS

We are thankful to the Council of Scientific and Industrial Research for financial assistance in the form of a Research Associateship to Sarla. Thanks are also due to Dr. B. D. Vashistha for help in finalizing the chapter and to Mr. Krishan Lal and Mr. Sumar Kumar Dass for preparing the illustrations.

REFERENCES

1. **Allsopp, A., Pearman, C., and Rao, A. N.,** The effects of some growth substances and inhibitors on the development of *Marchantia* gemmae, *Phytomorphology,* 18, 84, 1968.

1a. **Chopra, R. N. and Kumra, P. K.,** *Biology of Bryophytes,* Wiley Eastern Ltd., New Delhi, 1988, 178.

2. **Ilahi, I. and Allsopp, A.,** Studies in the Metzgeriales (Hepaticae). II. The effects of certain auxins on the growth and morphology of some thalloid species, *Phytomorphology,* 19, 381, 1969.

3. **Allsopp, A. and Ilahi, I.,** Studies in the Metzgeriales (Hepaticae). III. The effects of sugars and auxins on *Noteroclada confluens, Phytomorphology,* 20, 9, 1970.

4. **Allsopp, A. and Ilahi, I.,** Studies in the Metzgeriales (Hepaticae). VII. Regeneration in *Noteroclada confluens* and *Blasia pusilla, Phytomorphology,* 20, 173, 1970.

5. **Mehra, P. N. and Pental, D.,** Induction of apospory, callus, and correlated morphogenetic studies in *Athalamia pusilla* Kash., *J. Hattori Bot. Lab.,* 40, 151, 1976.

6. **Bhatla, S. C. and Chopra, R. N.,** Hormonal regulation of gametangial formation in the moss *Bryum argenteum* Hedw., *J. Exp. Bot.,* 32, 1243, 1981.

7. **Chopra, R. N. and Kumra, P. K.,** Hormonal regulation of growth and antheridial production in three mosses grown in vitro, *J. Bryol.,* 12, 491, 1983.

8. **Kapur, A.,** In Vitro Studies on Some Bryophytes, Ph.D. thesis, University of Delhi, Delhi, India, 1983.

9. **Kumra, S. and Chopra, R. N.,** Effects of some auxins and cytokinins on growth and gametangial formation in the liverwort *Riccia gangetica* Ahmad, *Ann. Bot. (London),* 54, 605, 1984.

10. **Sarla,** Morphogenetic Studies on some Bryophytes, Ph.D. thesis, University of Delhi, Delhi, India, 1986.

11. **Vashistha, B. D.,** Effects of some auxins and cytokinins on growth and archegonial formation in the liverwort *Riccia frostii* Aust., *Biochem. Physiol. Pflanz.,* 182, 309, 1987.

12. **Chopra, R. N. and Mehta, P.,** Effect of some known growth regulators on growth and fertility in male clones of the moss *Microdus brasiliensis* (Dub.) Ther., *J. Exp. Bot.,* 38, 331, 1987.

13. **Chopra, R. N. and Sood, S.,** In vitro studies in Marchantiales. I. Effects of some carbohydrates, agar, pH, light, and growth regulators on the growth and sexuality in *Riccia crystallina, Phytomorphology,* 23, 230, 1973.

14. **Benson-Evans, K.,** Environmental factors and bryophytes, *Nature (London),* 191, 255, 1961.

15. **Fellenberg-Kressel, M.,** Untersuchungen über die Archegonien-und Antheridienbildung bei *Microlepia speluncae* (L.) Moore in Abhängigkeit von inneren und äusseren Faktoren, *Flora (Jena),* 160, 14, 1969.

16. **Rashid, A.,** Spore germination, protenema development and bud formation in *Anoectangium thomsonii, Phytomorphology,* 20, 49, 1970.

17. **Mohan Ram, H. Y.,** Hormones and flower sex, *Plant Biochem. J.,* S. M. Sircar Memorial Volume, 77, 1980.

18. **Maravolo, N. C. and Voth, P. D.,** Morphogenic effects of three growth substances on *Marchantia* gemmalings, *Bot. Gaz. (Chicago),* 127, 79, 1966.

19. **Ilahi, I. and Allsopp, A.,** Studies in the Metzgeriales (Hepaticae). IV. Further investigations on the effects of various physiologically active substances on *Noteroclada confluens, Phytomorphology,* 20, 68, 1970.

20. **Chopra, R. N. and Kumra, S.,** Hormonal regulation of growth and gametangial formation in *Riccia gangetica* Ahmad, *Beitr. Biol. Pflanz.,* 61, 99, 1986.

21. **Sarla and Chopra, R. N.,** Effect of some auxins and antiauxins on protonemal growth and bud formation in *Bryum pallescens* Schleich. ex Schwaegr. grown in vitro, *Plant Sci.,* 51, 251, 1987.

22. **Vashistha, B. D.,** In Vitro Investigations on Some Indian Bryophytes, Ph.D. thesis, University of Delhi, Delhi, India, 1985.

23. **Chopra, R. N. and Rashid, A.,** Auxin-cytokinin interaction in shoot-bud formation of a moss: *Anoectangium thomsonii* Mitt., *Z. Pflanzenphysiol.,* 61, 192, 1969.

24. **Sharma, P. and Chopra, R. N.,** Effect of chelating agents and metal ions on growth and fertility in the male clones of the moss *Microdus brasiliensis* (Dub.) Ther., *J. Exp. Bot.,* 36, 494, 1985.

25. **Chopra, R. N. and Sarla,** Bud formation in the moss *Garckea phascoides* (Hook.) C. Muell. I. Effects of auxins, cytokinins and their interaction, *J. Hattori Bot. Lab.,* 61, 75, 1986.

26. **Chopra, R. N. and Rawat, M. S.,** Studies on the initiation of sexual phase in the moss *Leptobryum pyriforme, Beitr. Biol. Pflanz.,* 53, 353, 1977.

27. **Mehra, P. N. and Pahwa, M. S.,** Phenotypic variations in *Fossombronia himalayensis* Kash. In vitro effect of sugars, auxins, growth substances, and growth inhibitors, *J. Hattori Bot. Lab.,* 40, 371, 1976.

28. **Mehra, K.,** Studies on the reproductive biology of the moss *Bartramidula bartramioides.* Effect of hormones and chelates, *J. Hattori Bot. Lab.,* 56, 175, 1984.

29. **Schraudolf, H.,** Die Wirkung von Phytohormonen auf Keimung und Entwicklung von Farnprothallien. I. Auslösung der Antheridienbildung und Dunkelkeimung bei Schizaeaceen durch Gibberellinsäure, *Biol. Zentralbl.,* 81, 731, 1962.

30. **Warne, T. R., Hickok, L. G., and Scott, R. J.,** Characterization and genetic analysis of antheridiogen-insensitive mutants in the fern *Ceratopteris, J. Linn. Soc. Bot. (London),* 96, 371, 1988.

31. **Brandes, H.,** Gametophyte development in ferns and bryophytes, *Annu. Rev. Plant Physiol.,* 24, 115, 1973.
32. **Takeno, K. and Furuya, M.,** Inhibitory effects of gibberellins on archegonial differentiation in *Lygodium japonicum, Physiol. Plant.,* 39, 135, 1977.
33. **Nanda, K. K., Kumar, S., and Sood, V.,** Effect of adenosine monophosphates on the flowering of *Impatiens balsamina* L., a qualitative short-day plant, *Planta,* 130, 93, 1976.
34. **Vince-Prue, D.,** *Photoperiodism in Plants,* McGraw-Hill, New York, 1975.
35. **Mohan Ram, H. Y. and Jaiswal, V. S.,** Induction of male flowers on female plants of *Cannabis sativa* by gibberellins and its inhibition by abscisic acid, *Planta,* 105, 263, 1972.
36. **Sarla,** Bud formation in the moss *Garckea phascoides* (Hook.) C. Muell. II. Effect of some growth regulators, *J. Hattori Bot. Lab.,* 62, 151, 1987.
37. **Jones, R. L. and Phillips, I. D. J.,** Organs of gibberellin synthesis in light-grown sunflower plants, *Plant Physiol.,* 41, 1381, 1966.
38. **Cleland, C. F. and Zeevaart, J. A. D.,** Gibberellins in relation to flowering and stem elongation in the long-day plant *Silene armeria, Plant Physiol.,* 46, 392, 1970.
39. **Halevy, A. H., Ashri, A., and Ben-Tal, Y.,** Peanut: gibberellin antagonists and genetically controlled differences in growth habit, *Science,* 164, 1397, 1969.

39a. **Genus, J. M. C.,** Steroid hormones and plant growth and development, *Phytochemistry,* 17, 1, 1978.

40. **Kumra, S. and Chopra, R. N.,** Combined effect of some growth regulators on growth and gametangial formation in the liverwort *Riccia gangetica* Ahmad, *J. Exp. Bot.,* 37, 1552, 1986.
41. **Bhatla, S. C. and Chopra, R. N.,** Inhibition of sex induction in *Bryum argenteum* due to high concentration of sucrose and its reversal by cyclic 3′,5′-adenosine monophosphate, *Z. Pflanzenphysiol.,* 92, 375, 1979.
42. **Chopra, R. N. and Bhatla, S. C.,** Involvement of cyclic 3′,5′-adenosine monophosphate and other purine derivatives in sex induction in the moss *Bryum argenteum, Z. Pflanzenphysiol.,* 103, 393, 1981.
43. **Kumra, P. K.,** Morphogenetic and Physiological Studies on Some Mosses, Ph.D. thesis, University of Delhi, Delhi, India, 1981.
44. **Amrhein, N. and Filner, P.,** Adenosine 3′,5′-cyclic monophosphate in *Chlamydomonas reinhardtii:* isolation and characterization, *Proc. Natl. Acad. Sci. U.S.A.,* 70, 1099, 1973.
45. **Chlapowski, F. J., Kelly, J. A., and Butcher, R. W.,** Cyclic nucleotides in cultured cells, in *Advances in Cyclic Nucleotide Research,* Vol. 6, Greengard, P. and Robison, G. A., Eds., Raven Press, New York, 1975.
46. **Goldberg, N. G. and Haddox, M. K.,** Cyclic GMP metabolism and involvement in biological regulation, *Annu. Rev. Biochem.,* 46, 823, 1977.
47. **Galun, E.,** Effects of gibberellic acid and naphthaleneacetic acid on sex expression and some morphological characters in cucumber plant, *Phyton,* 13, 1, 1959.
48. **Asprey, G. F., Benson-Evans, K., and Lyon, A. G.,** Effect of gibberellin and indoleacetic acid on seta elongation in *Pellia epiphylla, Nature (London),* 181, 1351, 1958.
49. **Chopra, R. N. and Sharma, P.,** Effect of cyclic 3′,5′-adenosine monophosphate and kinetin on growth and fertility in the male clones of the moss *Microdus brasiliensis* (Dub.) Ther., *J. Plant Physiol.,* 117, 293, 1985.
50. **Mehta, P.,** Morphogenetic Studies on Some Indian Mosses, Ph.D. thesis, University of Delhi, Delhi, India, 1986.
51. **Evin, W. H. and Varner, J. E.,** Hormone-controlled synthesis of endoplasmic reticulum in barley aleurone cells, *Proc. Natl. Acad. Sci., U.S.A.,* 68, 1631, 1971.
52. **Cúlafić, A. and Neŝković, M.,** Effect of growth substances on flowering and sex expression in isolated apical buds of *Spinacia oleracea, Physiol. Plant,* 48, 588, 1980.
53. **Woodfin, C. M.,** Physiological studies on selected species of the liverwort family Ricciaceae, *J. Hattori Bot. Lab.,* 41, 179, 1976.
54. **Dua, S., Singal, N., and Chopra, R. N.,** Studies on growth and sexuality in *Riccia gangetica* Ahmad grown in vitro, *Cryptogamie Bryol. Lichenol.,* 3, 189, 1982.
55. **Wann, F.,** Some of the factors involved in the sexual reproduction of *Marchantia polymorpha, Am. J. Bot.,* 12, 307, 1925.
56. **Bhatla, S. C.,** Physiology of Sexual Reproduction in *Bryum argenteum* Hedw., Ph.D. thesis, University of Delhi, Delhi, India, 1980.
57. **Chopra, R. N. and Rahbar, K.,** Temperature, light, and nutritional requirements for gametangial induction in the moss *Bartramidula bartramioides, New Phytol.,* 92, 251, 1982.
58. **Allsopp, A.,** The metabolic status and morphogenesis, *Phytomorphology,* 14, 1, 1964.
59. **Rao, M. P. and Das, V. S. R.,** Metabolic changes during reproductive development in liverworts, *Z. Pflanzenphysiol.,* 59, 87, 1968.
60. **Chopra, R. N. and Bhatla, S. C.,** Chemical control of reproduction in some bryophytes, *Phytomorphology,* 33, 1, 1983.

61. **Sotta, B.,** Interaction du photopériodisme et des effects de la zéatine, du saccharose et de l'eau dans la floraison du *Chenopodium polyspermum, Physiol. Plant.,* 43, 337, 1978.
62. **Diller, V. M., Fulford, M., and Kersten, H. J.,** Culture studies on *Sphaerocarpos.* II. The effect of various sugars on the growth and form of *S. texanus, Am. J. Bot.,* 42, 819, 1955.
63. **Sood, S.,** In vitro studies in Marchantiales. II. Effect of mineral nutrients, chelates, and organic nitrogenous sources on growth and sexuality in *Riccia crystallina, Phytomorphology,* 24, 186, 1974.
64. **Chopra, R. N. and Kumra, S.,** Effect of some chelating agents on vegetative growth and sexuality in the liverwort *Riccia gangetica* Ahmad, *Beitr. Biol. Pflanz.,* 60, 367, 1986.
65. **Kumra, P. K.,** Effect of some chelating agents on growth and antheridial production in male clones of three mosses grown in vitro, *Ann. Bot. (London),* 50, 771, 1982.
66. **Bhatla, S. C. and Chopra, R. N.,** Effect of chelating agents and metal ions on gametangial formation in the moss *Bryum argenteum* Hedw., *Ann. Bot. (London),* 52, 755, 1983.
67. **Sharma, P. and Chopra, R. N.,** Effect of chelating agents and metal ions on growth and fertility in the male clones of the moss *Microdus brasiliensis* (Dub.) Ther., *J. Exp. Bot.,* 36, 494, 1985.
68. **Chopra, R. N. and Sarla,** Induction of buds in the moss *Bryum pallescens* Schleich. ex Schwaegr. by a metal chelate, Fe-EDTA, *J. Bryol.,* 13, 423, 1985.
69. **Seth, P. N., Venkataraman, R., and Maheshwari, S. C.,** Studies on the growth and flowering of a short-day plant, *Wolffia microscopica.* II. Role of metal ions and chelates, *Planta,* 90, 349, 1970.
70. **Pieterse, A. H.,** Experimental control of flowering in *Pistia stratiotes* L., *Plant Cell Physiol.,* 19, 1091, 1978.
71. **Khurana, J. P. and Maheshwari, S. C.,** Floral induction in *Wolffia microscopica* by salicylic acid and related compounds under non-inductive long days, *Plant Cell Physiol.,* 24, 907, 1983.
72. **Hillman, W. S.,** Photoperiodism, chelating agents and flowering in *Lemna perpusilla* and *L. gibba* in aseptic culture, in *Light and Life,* McElroy, W. D. and Glass, B., Eds., Johns Hopkins Press, Baltimore, MD, 1961, 673.
73. **Miller, M. W., Garber, E. D., and Voth, P. D.,** Nutritionally deficient mutants of *Marchantia polymorpha* induced by x-rays, *Bot. Gaz. (Chicago),* 124, 94, 1962.
74. **Dunham, V. L. and Bryan, J. K.,** Effects of exogenous amino acids on the development of *Marchantia polymorpha* gemmalings, *Am. J. Bot.,* 55, 745, 1968.
75. **Dunham, V. L. and Bryan, J. K.,** Synergistic effects of metabolically related amino acids on the growth of a multicellular plant, *Plant Physiol.,* 44, 1601, 1969.
76. **Dunham, V. L. and Bryan, J. K.,** Synergistic effects of metabolically related amino acids on the growth of a multicellular plant. II. Studies of ^{14}C-amino acid incorporation, *Plant Physiol.,* 47, 91, 1971.
77. **Margaris, N. S.,** Free amino acid pools of certain mosses from Greece, *Bryologist,* 77, 246, 1974.
78. **Sharma, K. K., Diller, V. M., and Fulford, M.,** Studies on the growth of *Haplomitrium.* II. Media containing amino acids, *Bryologist,* 63, 203, 1960.
79. **Chopra, R. N. and Kumra, S.,** Regulation of growth and gametangial formation in *Riccia gangetica* Ahmad by some amino acids, *J. Exp. Bot.,* 35, 1537, 1984.
80. **Vashistha, B. D. and Chopra, R. N.,** Effect of some amino acids and complex organic nitrogenous substances on vegetative growth and gametangial formation in the female clone of *Riccia frostii* Aust., *Plant Sci.,* 48, 175, 1987.
81. **Chopra, R. N.,** Environmental factors affecting gametangial induction in bryophytes, *J. Hattori Bot. Lab.,* 55, 99, 1984.
82. **Chopra, R. N. and Bhatla, S. C.,** Regulation of gametangial formation in bryophytes, *Bot. Rev.,* 49, 29, 1983.
83. **Lockwood, L. G.,** The influence of photoperiod and exogenous nitrogen-containing compounds on the reproductive cycles of the liverwort *Cephalozia media, Am. J. Bot.,* 62, 893, 1975.
84. **Chopra, R. N. and Dhingra-Babbar, S.,** Studies on callus induction, its growth and differentiation in *Marchantia palmata* Nees. I. Effect of some amino acids, complex organic substances and activated charcoal, *J. Hattori Bot. Lab.,* 60, 193, 1986.
85. **Sarla and Chopra, R. N.,** In vitro studies on growth and antheridial induction in the male clone of *Riccia discolor* Lehm. et Lindb., *Biochem. Physiol. Pflanz.,* 183, 67, 1988.
86. **Sood, S.,** Experimental Studies on Some Indian Bryophytes, Ph.D. thesis, University of Delhi, Delhi, India, 1972.
87. **Mitra, G. C.,** Induction of archegonia in *Pohlia nutans, Curr. Sci.,* 36, 134, 1967.
88. **Shantz, E. M. and Steward, F. C.,** Investigations on growth and metabolism of plant cells. VII. Sources of nitrogen for tissue cultures under optimal conditions for their growth, *Ann. Bot. (London),* 23, 371, 1959.
89. **Sarla,** In vitro studies on growth and antheridial induction in the male clone of *Riccia discolor* Lehm. et Lindb., 14th Int. Bot. Congr., Berlin, July 24 to August 1, 1987, Abstract 2-110-8.
90. **Maravolo, N. C.,** Biochemical Changes during Sexual Development in *Marchantia polymorpha,* Ph.D. dissertation, University of Chicago, Chicago, IL, 1966.

91. **Reynolds, A. C. and Maravolo, N. C.,** Phenolic compounds associated with development in the liverwort *Marchantia polymorpha, Am. J. Bot.*, 60, 406, 1973.
92. **Stonier, T., Singer, R. W., and Yang, H. M.,** Studies on auxin protectors. X. Inactivation of certain protectors by polyphenol oxidase, *Plant Physiol.*, 46, 454, 1970.
93. **Hilgenberg, W., Baumann, G., and Knab, R.,** Veränderungen von Tryptophan-Synthase-, Indole-3-essigsäure-Oxidase- und Peroxidase-Aktivität im Verlauf der Entwicklung von *Marchantia polymorpha* L., *Z. Pflanzenphysiol.*, 87, 103, 1978.
94. **Maravolo, N. C., Garber, E. D., and Voth, P. D.,** Biochemical changes during sexual development in *Marchantia polymorpha.* I. Esterases, *Am. J. Bot.*, 54, 1113, 1967.

Chapter 10

INDUCTION OF CONTROLLED DIFFERENTIATION OF CALLUS IN MOSSES

M. K. C. Menon and M. Lal

TABLE OF CONTENTS

I. INTRODUCTION

Mosses exhibit a concrete periodicity in their growth pattern by perceiving stimuli in the environment. They show manifestation of gametophytic and sporophytic phases. Regeneration products of varying nature can be obtained from the same organ, such as the sporophyte.[1-4] When a young sporogonium is placed on a culture medium, the apical cell and its derivatives are generally meristematic and differ in regeneration potentialities from "seta" cells and haustorium-like "foot" cells. Moreover, the cultures derived from spores of mosses give an opportunity to discover the regenerational potentialities of such gametophytic organs as the protonema, stem, leaf, and gametangia.[5-7] The frequency of vegetative development of sporogonia in the diploid race of *Phascum cuspidatum* shows that physiological comparison between haploid sexual gametophytes and diploid apogamous plants lies in their metabolic reaction to daylight, which plays a significant role in the reproduction of these plants. The fact that leafy parts of the haploid and diploid plants differ in their reaction to light revealed that metabolic products played a crucial role in determination of the gametophytic and sporophytic phases.[8,9] Interestingly, Ashton and Cove[10] have shown that in auxotrophs self-sterility segregates as a pleotropic effect of mutations and can cause developmental abnormalities in *Physcomitrella patens*.

Compared to the complexity of the two phases in responding to environmental stimuli, a simpler intermediate stage between gametophytes and sporophytes in the form of a callus growth has been isolated from various mosses under different cultural conditions. Callus cells grow very rapidly and can be maintained in a constant physiological condition for experimentation. They can be induced to differentiate when transferred to different culture media. They readily disintegrate into single cells or groups of cells and can easily be spread out and manipulated in cultures. The calli obtained from gametophytes and sporophytes have evoked much interest, since fundamental physiological problems can be tackled using these undifferentiated tissue systems.

II. INDUCTION AND MAINTENANCE OF CALLUS

Wettstein[11] had predicted that hybrids produced by reciprocal crosses among the members of Funariaceae would be excellent materials for understanding developmental physiology in mosses. Bauer[1,12] induced callus from the sporophyte cultures of *Physcomitrium pyriforme* and also from the hybrid sporogonia obtained from a cross between *P. pyriforme* and *Funaria hygrometrica*. On the basis of his experiments he concluded that the morphological nature of the tissue regenerating from the hybrid sporogonia depends on the physiological age of the regenerating zone. Attempts were made to induce mutations using "univalent" and "bivalent" protonema regenerated from *P. pyriforme*.[13] The mutants showed variations in their phenotypic characters. In reciprocal crosses between *F. hygrometrica* and *P. acuminatum*[14] the F_1 generation did show phenotypic differences from the parents. However, it was recorded that the protonema regenerated from the hybrid sporogonia was remarkably different in its osmotic potentialities. Bauer[12] concluded that embryonic regions of the developing hybrid sporogonium possess a specific state of proliferative capability which is preserved in subsequent subcultures due to changes in their developmental physiology. Thus, the tissue regenerated from the extreme tip retains embryonic features and comprises undifferentiated apolar callus cells.

Quiescent cells from the gametophyte also may be activated to divide and form a callus tissue. Callus was induced from gametophytes originated from spore platings of *Polytrichum commune* and *Atrichum undulatum*.[15,16] Similar callus tissues have been induced from various sites of the gametophytic and sporophytic phases (Table 1).

It can be seen from Table 1 that supplementing the medium with sucrose[15,17-19,21] has

TABLE 1
Induction and Maintenance of Callus of Various Moss Species in Gametophytic Phase and Sporophytic Phase

Moss species	Site	Medium for induction	Medium for maintenance[a]	Ref.
		Gametophytes derived from spores		
Polytrichum commune	Gametophore	Knudson's + sucrose	Knudson's + sucrose + casamino acids	15
Atrichum undulatum				
Physcomitrium coorgense	Stem axis and gametangia	Knop's + sucrose	Knop's + sucrose + coconut milk	17
Funaria hygrometrica	Stem axis	Knop's + sucrose (aging and drying cultures)	Knop's + coconut milk + casein hydrolysate	18
Physcomitrium pyriforme	Stem axis, protonema, and leaf	Moore's + sucrose	Moore's + sucrose + yeast extract + low light	19
Pylaisiella selwynii	Protonema	Inorganic nutrients + cytokinins/*Agrobacterium tumefaciens*	—	20,54
Hyophila involuta	Gametophore	Murashige & Skoog's (ammonium nitrate + chelated iron + vitamins + sucrose + 2,4-D)	Murashige & Skoog's (ammonium nitrate + chelated iron + vitamins + sucrose + 2,4-D + casein hydrolysate)	21
Funaria hygrometrica	Gametophytic buds, protonema	Murashige & Skoog's (kinetin + IAA + vitamins) or Murashige & Skoog's + UV irradiation	Stable on induction medium	22
Bryum coronatum	Gametophore, protonema	Murashige & Skoog's (kinetin + IAA + vitamins) or Murashige & Skoog's + UV irradiation	Stable on induction medium	23
Anoectangium thomsonii	Protonema	Murashige & Skoog's (half-strength macronutrients + sucrose)	—	24
		Sporophytes and Seta Cuttings		
Physcomitrium pyriforme	Sporophyte	Knop's + glucose (dry culture conditions; 3—4% agar)	Knop's + yeast extract	1
Pottia intermedia	Capsule, seta regenerated, gametophytic leaf	Inorganic nutrient medium	—	25
Funaria hygrometrica (n = 14) × *Physcomitrium pyriforme* (n = 36)	Sporophyte	Knop's + chloral hydrate + glucose	Knop's + yeast extract + glucose	12,34

[a] — = not maintained

afforded a unique opportunity for obtaining calli from the cells of the gametophyte (Figure 1A,B) and sporophyte. Subculture of the callus on media containing yeast extract[12,19] and/or casamino acids/coconut milk[17,18,21] helps in maintaining the proliferative capacity of these undifferentiated tissues for further metabolic and physiological experimentation (Figure 1C, D). Also of interest is the induction of callus in *Pylaisiella selwynii* by *Agrobacterium tumefaciens*[20] through the ability of the bacteria to attach to the cell wall of this moss and through enzymes involved in octopine and nopaline utilization.

III. DIFFERENTIATION OF CALLUS ON AGAR CULTURES

In 1956 Bauer[26] observed that apogamy could be induced at the leaf apices under aseptic cultural conditions. Since then Bauer,[12] Lal,[17] Lazarenko,[25] and Ward[15] have proposed that the determination of gametophytes and sporophytes takes place as a consequence of nutritional changes in the meristematic cells. It is generally agreed that gametophore differentiation is preceded by an early establishment of an apical cell with three cutting faces and sporophytic differentiation by determining an apical cell with two cutting faces.[4,7] It has been possible to trace the sequence of laying of the cell walls up to a crucial three- or four-celled stage, after which one or the other type of morphogenesis takes place in the protonema. It was observed that an apical cell with three cutting faces was formed by a passage through an apical cell with two cutting faces. If the factors conducive to the advent of sporophytic development are available, then the apical cell with two cutting faces becomes stabilized. On the other hand, if the factors for the normal development of the apical cell with three cutting faces are available then a gametophore develops from the secondary protonema or callus. The pattern of further development of these young buds is reflected in the planes of division of their respective apical cells. Control of these morphogenetic events has been attempted successfully using callus tissues of mosses.

Both in *Physcomitrium coorgense* and in *P. pyriforme*[5,19] induction of callus was possible on a medium containing sucrose (2%). The apogamous sporophytes and callus arose on the linearly growing protonema by first differentiating a special small intercalary cell among the ordinary long cylindrical cells (Figure 2A,C). This specialized cell divided anticlinally and organized a sporophyte with an apical cell having two cutting faces or sometimes divided irregularly to form a callus. A sporophyte or a callus could also arise from a seemingly unmodified cell from which a gametophyte would normally arise (Figure 2B,D). When the callus derived from the secondary protonema of *P. pyriforme* was subcultured on a kinetin-supplemented liquid medium it formed apogamous sporophytic buds with virtual exclusion of gametophytes. Therefore, it was suggested that, rather than having a morphoregulatory role, kinetin may be responsible merely for enhancing cell proliferation, and the determination of an apical cell with two cutting faces or one with three cutting faces is under the control of other factors, both external and internal.[27] It was proposed that accumulation of a sporophytic factor in the gametophytic callus cells is diluted during the process of differentiation. Abscisic acid (ABA) did not suppress the differentiation of the quiescent callus cells. Conceivably, ABA regulates cell growth in mosses before the factors for apical cell determination accumulate in them.[28]

On the basal medium supplemented with 2% sucrose the callus cells of *Physcomitrium pyriforme* differentiated only partially. The remainder turned brown and became quiescent. Such callus masses, when subcultured on fresh medium of the same composition, turned green and differentiated readily after showing slight multiplication. Under high light intensity predominantly gametophytic differentiation and in low light intensity predominantly sporophytic differentiation was observed. Primary differentiation (i.e., directly from callus) in low light intensity was sporophytic, while secondary differentiation (through the intermediate stage of secondary protonema) was largely gametophytic. Some sporophytes did appear even

FIGURE 1. Induction and maintenance of callus in *Physcomitrium pyriforme* Brid. (A) $4^1/_2$-month-old gametophores from basal medium supplemented with 2% sucrose upon subculture on the same medium produce callus and apogamous sporophytes from their leaves (arrows); 8-week-old culture. (Magnification × 1.6.) (B) Gametophores cultured on basal medium containing NAA and sucrose have formed secondary protonemata which produce callus masses in addition to buds; 8-week-old culture. (Magnification × 3.) (C) Protonemal callus subcultured on basal medium + coconut milk (15% v/v) + sucrose (2%) in low light intensity show predominantly sporophytic differentiation after an initial spell of proliferation. Note that the sporophytes have differentiated (arrow) directly from the callus. 6-week-old culture. (Magnification × 2.8.) (D) Protonemal callus subcultured on basal medium supplemented with yeast extract (0.03%) shows only proliferation. (Magnification × 3.5.)

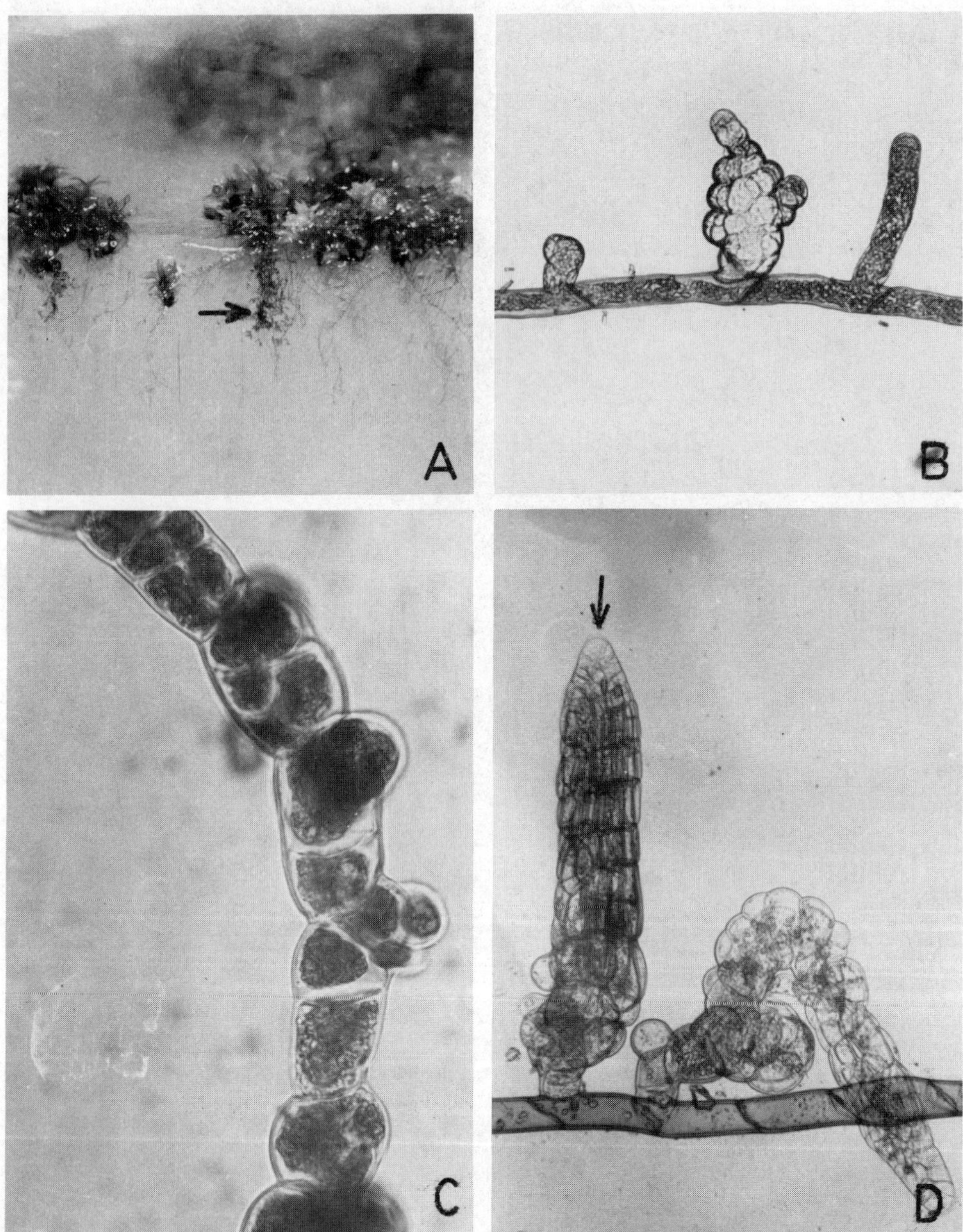

FIGURE 2. Origin of callus from secondary protonema of *Physcomitrium pyriforme* Brid. (A) Callus masses formed on the secondary protonema (arrow) in basal medium supplemented with 2% sucrose and exposed to low light intensity; 8-week-old culture. (Magnification × 3.) (B) Callus from a bud/branch initial originating as lateral outgrowth of the protonema. (Magnification × 188.) (C) Formation of callus from unusual small cells by internal divisions. (Magnification × 626.) (D) Protonema growing on basal medium + sucrose (2%) under low light intensity and producing callus and stunted sporophytes with clear-cut apical cell having two cutting faces (arrow). (Magnification × 333.)

in high light as a primary differentiation product. These were green and quite stunted. However, the gametophytes outnumbered the sporophytes and grew to such an extent that they covered the stunted sporophytes.[29]

The callus was also subcultured on the basal medium devoid of sucrose. In high light intensity the differentiation was mainly gametophytic, and in low light intensity the callus remained largely undifferentiated except for the appearance of some protonemal filaments. On the basal medium supplemented with 6% sucrose no differentiation of any kind was observed, but the callus multiplied profusely. The cultures were subjected to both high and low light intensity. The calli became pale green and slowly turned brown. However, to maintain the callus without differentiation, low light intensity and basal medium supplemented with 6% sucrose were found to be effective. Another method of maintaining the callus in a proliferative state is to subculture it on a medium supplemented with yeast extract in high or low light intensity. This method had already been applied by Bauer[12] to obtain and maintain large quantities of callus for biochemical experiments. However, when the callus derived from the above cultures was subcultured on basal medium supplemented with 2% sucrose, it proliferated for about 4 weeks and then began to differentiate, contrary to the behavior of callus in the inductive medium. Long-term subculturing on basal medium supplemented with 6% sucrose or yeast extract (0.3%) made the callus lose its capacity to differentiate, at least immediately on first subculture on any other medium.

The callus grown on basal medium supplemented with 2% sucrose was subcultured on basal medium supplemented with 2% sucrose and 15% coconut milk (CM) after it had become quiescent. Cultures were subjected to high as well as low light intensity. In low light intensity, after an initial proliferation of the inoculated mass, there was direct differentiation of sporophytes from the callus (Figure 1C). The sporophytes were well formed and had long setae and globular capsules. Some of these capsules contained viable spores (about 1% of the capsules). Although sporophytic differentiation prevailed, some gametophytes also appeared on the periphery of the cultures. On the other hand, when the cultures on CM medium were subjected to high light intensity, gametophytes as well as stunted sporophytes were formed. In some cultures the setae developed orange-red pigmentation.[29]

Bauer[30] reported the formation of an aposporous protonema from the culture of a hybrid sporogonium. This protonema produced apogamous sporophytes as long as it was in organic union with the parent sporophyte. When isolated from the parent sporophyte it did not show any apogamous tendency, and Bauer[31] concluded that a factor emanating from the diploid sporophyte is translocated to the aposporous protonema and induces it to differentiate apogamous sporophytes or callus. Using a callus derived from the secondary protonema of *Physcomitrium pyriforme* an experiment was conducted to determine whether the differentiation of callus, freshly formed on the protonema, would be affected if the latter was not in organic communion with the parent gametophyte. Therefore, an isolated protonema, together with the induced callus masses, was inoculated on the basal medium supplemented with 2% sucrose and subjected to high and low light intensities. It was observed that predominantly sporophytic differentiation occurred in low light intensity and predominantly gametophytic differentiation in high light intensity. In this system, therefore, continued connection with the parent gametophyte is not essential for sporophytic structures to be produced. Hence, the possibility of a gametophytic influence traveling through the protonema and controlling the differentiation of callus is also ruled out.[29]

IV. DIFFERENTIATION OF CALLUS IN SUSPENSION CULTURES

Ward[15] isolated proliferous callus tissues from the mosses *Atrichum* and *Polytrichum*. In suspension cultures this tissue dissociated into single cells and cell aggregates which later gave rise only to gametophytes.[16] Surprisingly, no sporophytes were ever recorded as developing from this diploid tissue.

When green callus masses of *Physcomitrium pyriforme* from semisolid medium were suspended in liquid medium in 100-ml flasks and shaken reciprocally or in rotatory shakers they readily dispersed into single cells and aggregates of two to ten cells. This indicated

FIGURE 3. Differentiation of callus in suspension cultures of *Physcomitrium pyriforme* Brid. (A) Callus from liquid basal medium + coconut milk (15% v/v)) + sucrose (2%) subcultured in suspension cultures containing basal medium supplemented with kinetin (4 ppm) showed differentiation of stunted sporophytes; 8-week-old culture. (Magnification × 26.) (B) Same as in A. Note the pyriform capsules (arrow). 10-week-old culture. (Magnification × 28.) (C) Metaphase plate of a callus cell with 72 chromosomes. (Magnification × 2650.)

that the callus masses were very fragile, and it was easy to obtain single cells for further experimentation within a period of about 48 h.

Callus masses dispersed in liquid basal medium supplemented with CM (15% v/v) and subjected to shaking revealed the presence of single cells and tiny proliferating masses, many of which were distinctly polarized and each having an apical cell with two cutting faces, typical of young sporophytes. These polarized masses stayed as such in this medium. However, when these masses from liquid CM medium were transferred to liquid basal medium + kinetin (4 ppm) lacking sucrose, they differentiated into very stunted sporophytes. These sporophytes lacked setae and were composed of just minute pyriform capsules. Thus, further growth and differentiation of polarized callus masses into sporophytes was possible even in the liquid medium containing kinetin. Some of these sporophytes were whitish and contained brown sporogenous tissue (Figure 3A,B), as revealed by squash preparations.[27]

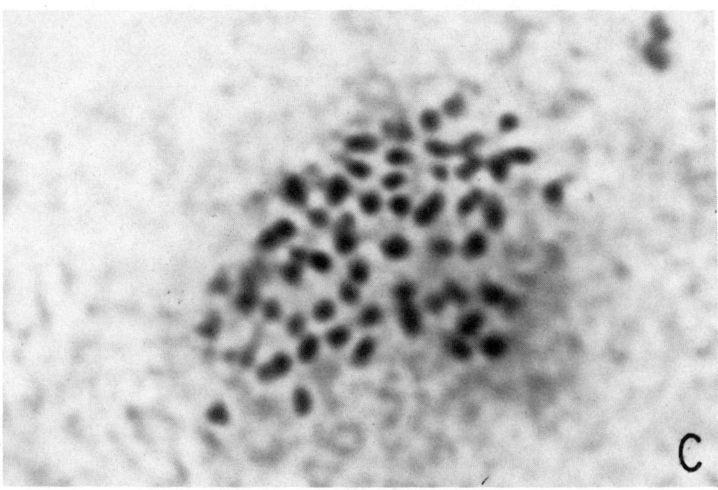

FIGURE 3C.

Also, these polarized masses from liquid CM medium, upon subculture in liquid basal medium devoid of sucrose, formed numerous sporophytes without organized capsules and under high light intensity produced protonemal filaments on which leafy gametophores eventually formed. In all these subcultures of callus in liquid medium it was worth noting that, while differentiation of callus directly into a sporophyte was quite common, the formation of a leafy gametophore always occurred through an intervening protonemal stage. When the callus aggregates were plated on an agar surface in a petri plate carrying basal medium supplemented with sucrose (2%) and CM (15% v/v), some of the callus aggregates differentiated into sporophytes while others just multiplied (Figures 4A to E). However, in the medium devoid of CM some gametophytes were also produced by these callus masses (Figure 4F). Differentiation responses were not prevalent throughout the cell population. Also, the diverse morphogenetic potentialities were retained even after passage through the liquid media.[29] Although Ward[16] and Ward and Frederick[32] have reported the origin of gametophores from single callus cells of *Polytrichum,* no ontogenetic data are presented in the results. However, the photographs indicate that calli had put forth filaments on which gametophores might have differentiated. However, it is certain that the gametophytic buds can be organized only on a filamentous protonemal system arising from the callus.

A significant observation about callus suspension cultures was that kinetin promoted sporophytic differentiation. If we agree that the action of sucrose in the secondary protonema is to promote synthesis of substances for sporophytic differentiation, then we will be ready to accept that the callus cells readily differentiated into sporophytes in the presence of kinetin because there was a general enhancement of growth in these cells leading to sporophytic differentiation by the interaction of kinetin and sucrose. It is noteworthy that kinetin brought about only filamentous growth and proliferation of the suspended cells in the absence of sucrose. This goes to prove that because of its absolute dependence on sucrose for the synthesis of sporophytic factor the callus reverted to protonemal development with the general activation of growth by kinetin. The primary role of kinetin appeared to be the general promotion of growth.

V. CYTOLOGY OF CALLUS CELLS

Bauer[31] reported that *Funaria hygrometrica* having chromosome numbers $n = 14$ produced only protonema when sporogonia were cultured. Sporogonia of a natural polyploid

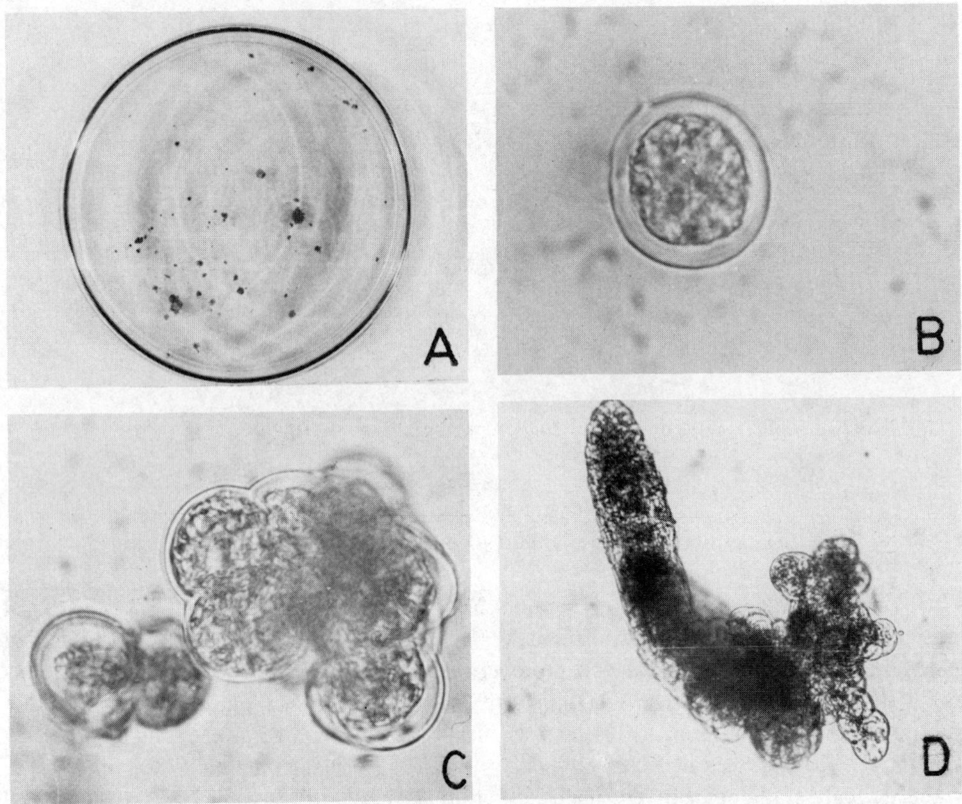

FIGURE 4. Plating and differentiation of callus cells of *Physcomitrium pyriforme* Brid. (A) Growth of callus cells on transfer from liquid to agar medium (basal medium + sucrose 2% + CM 15% v/v). A view of petri plate with undifferentiated callus aggregates; 3-week-old culture. (Magnification × 0.4.) (B) An isolated cell from the liquid basal medium containing sucrose (2%) + coconut milk (15% v/v). (Magnification × 625.) (C) Cell aggregates from the above medium; 6-week-old culture. (Magnification × 416.) (D) Young sporophytes differentiated from callus aggregates in the same medium; 6-week-old culture. (Magnification × 166.) (E,F) Callus aggregates differentiating into sporophytes, secondary protonema and gametophores (arrow) in agar medium (same as A). (Magnification of E × 10; F × 22.)

race (n = 28) could be induced to produce sporophytic tissue instead of or in addition to protonemata. Bauer[12,34] also obtained a callus from a hybrid sporogonium which was derived from a cross between *Funaria hygrometrica* (n = 14) and *Physcomitrium pyriforme* (n = 36), which also produced apogamous sporophytes. Therefore, it was concluded that spontaneous polyploids produced apogamous sporophytes through callus. However, Mehra[33] put forward a "gene block hypothesis", according to which expression of various phenotypic characters of a genome is divisible into definite gene blocks. These are complementary facets of the same genome, and irrespective of its duplicity any of these facets can be triggered to activity.

The chromosome number of the gametophytic callus cells of *Physcomitrium coorgense* from which apogamous sporophytes arose was found to be 51.[17] Three chromosome numbers have been reported for *P. pyriforme* in nature, i.e., n = 9, n = 18, and n = 36.[35-38] Most of the callus cells derived from the secondary protonema had 72 chromosomes each (Figure 3C). Not many cells could be observed in the divisional phase in the squashes. Some cells had two conspicuous nuclei, of which only one was preparing for the next division. Generally, chromosomes had a strong tendency to clump together, and it was difficult to separate them to enable an exact count. The difficulty of catching all the cell masses in the divisional

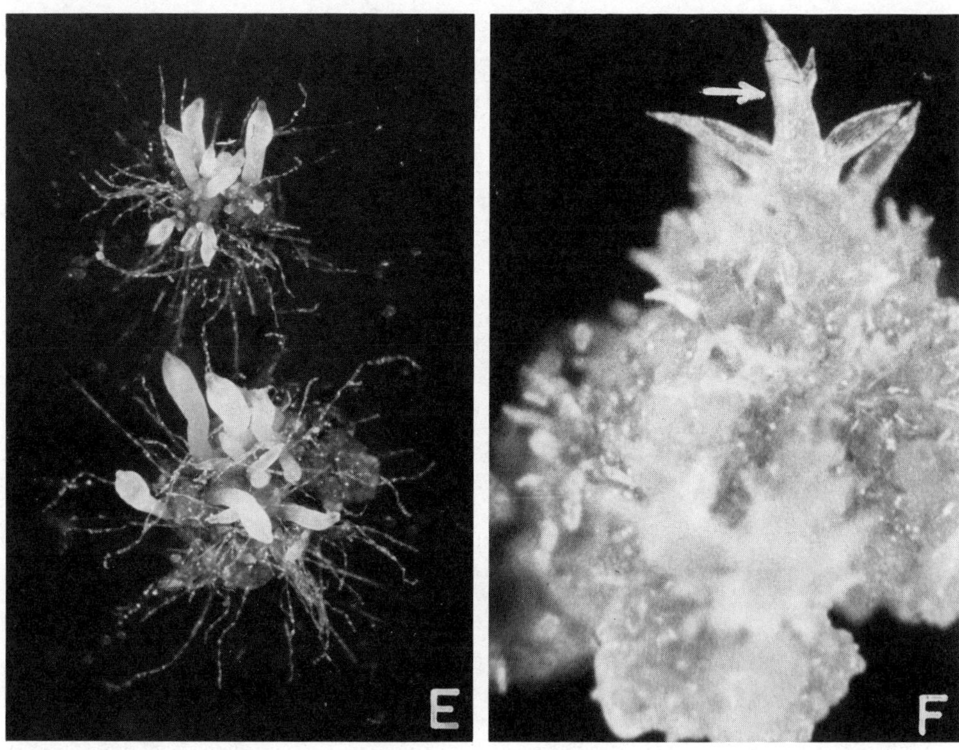

FIGURE 4E,F.

phase does not warrant any safe correlation to be established between the chromosomal number and the morphogenetic fate of a given cell.[29] Microdensitometry following staining with the Feulgen reagent gave no evidence of variation in chromosome number associated with the initiation of the sporophyte from callus cells. Fluorescence microscopy showed that the callus cells initiating the sporophyte were callused over the entire surface. Also, there was intense RNA fluorescence from the callus cells following staining with acridine orange.[39] Endoduplication of nuclear DNA during protonemal growth has been reported[40] and, therefore, chromosomal instability during callus growth cannot be ruled out. Formation of polyploid cell lines cannot be prevented in conventional tissue culture systems. However, the effects of the chromosomal change could be masked by the predominating cell lines and their differentiation products. Therefore, both of the hypotheses[33,34] remain to be tested, and no conclusions are warranted at this stage.

Ultrastructural study of the intercalary isodiametric cells of the filamentous protonema in *Physcomitrium pyriforme,* which give rise to callus or apogamous sporophyte, showed that the cells initiating the sporophyte develop labyrinthine outgrowths and do not form any recognizable placenta. Labyrinthine walls of the cells initiating the apogamous sporophyte arise in two ways. In the first, the inner surface of the wall remains more or less smooth, but an internal labyrinth is formed by the inclusion of narrow extensions of the protoplast (Figure 5A). In the second type, a labyrinth develops from anastomosing fingers or flanges growing into the protoplast from the inner surface of the wall (Figure 5B). Ultimately, the labyrinth formed by the ingrowths is as extensive as that found in gametophytic cells adjacent to the foot of the sexual sporophyte, and plasmodesmata are obliterated. Plastids in the callus cells dedifferentiate, and the disappearance of starch in the plastid coincides with the secretion of mucilage in the periphery of the callus cell.[39,41] The substantial incorporation of radioactive sugars into the noncellulosic fraction of the callus wall points to a preponderance of hem-

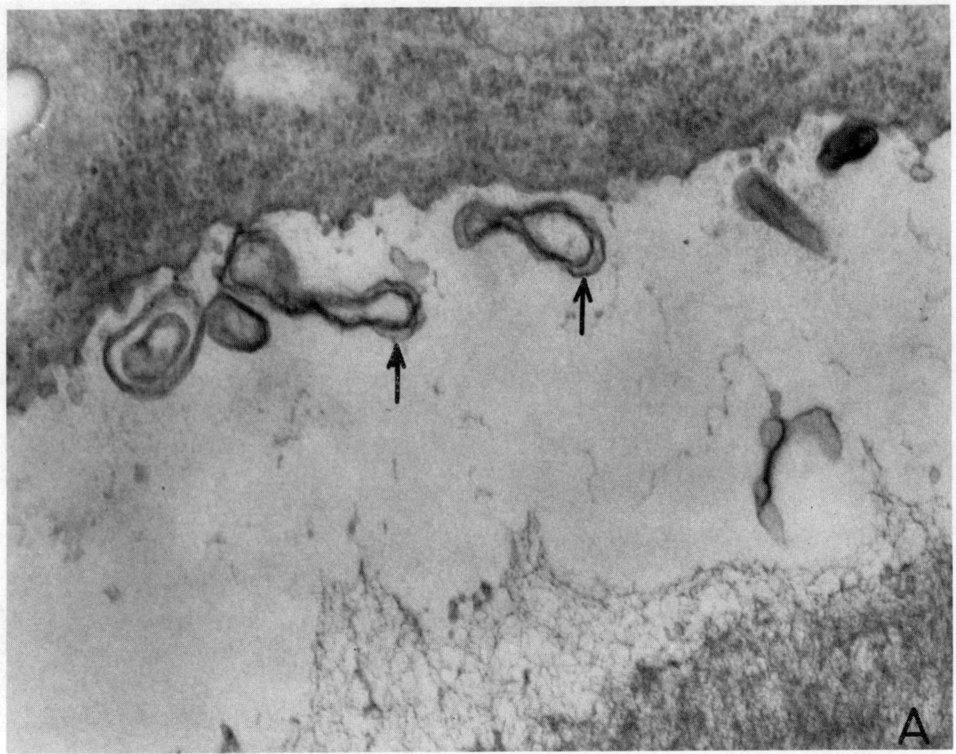

A

FIGURE 5. Ultrastructure of callus cells differentiating apogamous sporophytes in *Physcomitrium pyriforme* Brid. (A) A callus cell formed from the intercalary cell of the secondary protonema showing internal labyrinth formed by the inclusion of narrow extensions of the protoplast (arrow) into the wall. (Magnification × 35,000.) (B) A callus cell showing the second type of labyrinth, developed from anastamosing fingers or flanges growing into the protoplast from the inner surface. Large vesicles (about 0.5 μm in diameter) containing electron-transparent material (V), mitochondria (M), and Golgi bodies (G) often occur adjacent to them. (Magnification × 35,000.)

icelluloses and pectins in these labyrinthine walls.[42] Mucilagenous polysaccharides probably account for the ease with which single cells, each of which behaves like a zygote,[27] can be detached from the callus in shake cultures. However, in *P. coorgense* gametophytic callus cells destined to form apogamous sporophytes accumulated starch and developed wall ingrowths. Plasmodesmatal connections were retained between the apogamous outgrowth and the subtending callus.[43]

VI. METABOLISM IN CALLUS CULTURES

The use of tissue culture material for biosynthetic investigations and physiological studies carries the assumption that the results can be viably extrapolated to the whole gametophytic or sporophytic phase. Some of the early metabolic investigations on the callus cultures of moss sporophytes were carried out using a tissue derived from the hybrid sporogonium obtained via a cross between *Physcomitrium pyriforme* and *F. hygrometrica*.[44] The callus cells produced relatively large quantities of ''bryokinin'', which was secreted into the substrate. Its concentration in the cell sap was estimated to be ca. 10^{-5} *M*. This bryokinin had a characteristic ''kinetin-like'' effect on the protonema of *F. hygrometrica* in the production of buds. A methanolic solution of bryokinin showed an absorption maximum at 270 nm and an absorption minimum at 240 nm. If the callus cells were grown in 8-^{14}C-adenine about

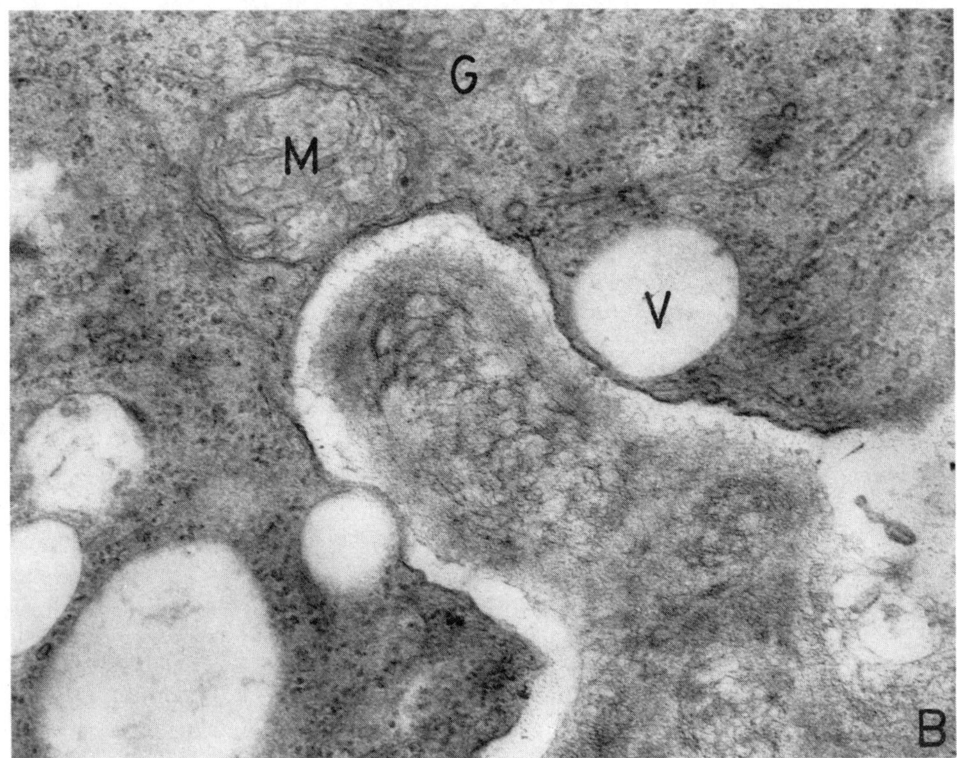

FIGURE 5B.

2% of the radioactivity was detected in bryokinin, and it was observed that on hydrolysis bryokinin produced adenine. Bauer[44] was successful in crystallizing 10 mg of bryokinin from 50 l of cell sap. Bryokinin was also found to replace kinetin as a growth factor for the growth of callus cultures of *Nicotiana tabacum.* It is also known to (1) delay senescence in the leaves of *N. rustica,* (2) cancel the apical dominance in shoots of *Microphyllum brasiliense,* and (3) induce apogamous sporophytes during the sexual maturation phase of *Splachnum ovatum.* The hybrid callus growth form also can be stabilized by bryokinin in interaction with auxin by inhibiting the growth form of the protonema.[44] It was further shown by chromatography that adenine precursors give rise to a cytokinin in the hybrid moss callus. This finding led to the view that the free cytokinin in the moss callus is not a degradation product of tRNA, but a product of adenine metabolism.[45] The structure of bryokinin was further confirmed as N^6-(Δ^2-isopentenyl)adenine by means of gas chromatography and mass spectrometry. The concentration of the compound that leached out into the medium was also found to be very low.[46]

Moss callus cells derived from the hybrid sporogonium were also used to study the developmental physiology as influenced by light and its various spectral components.[47] In the heterogeneous nature of the morphogenetic effects of light it was shown that under red light the RNA and protein contents increased, whereas under blue light the highest chlorophyll content was observed in relation to plastid development. Blue light enhanced the amino acid metabolism, whereas red light was found to affect carbon metabolism. This had been demonstrated by the existence of heterotrophy with regard to nitrogen and sugar. As an accumulation product of nitrogen metabolism allantoin was formed, and this is a typical expression of moss callus heterotrophy. This metabolic product could be distinctly connected with purine metabolism; this was proved by feeding ^{14}C-adenine to the culture. In darkness tryptophan content is significantly higher, whereas in white light α-aminobutyric acid and proline attain very high concentrations. In red light tyrosine was observed to be present in higher amounts in these hybrid callus cells.[47] In the protonemal callus of *Physcomitrium pyriforme* it could be demonstrated that ^{14}C-proline is rapidly converted to hydroxyproline. This was found to be noncovalently linked to the cell wall and could be extracted with chaotropic salts. This showed the presence of a glycoprotein rich in hydroxyproline in the callus cell wall.[48]

The metabolism of the moss callus is also influenced through the phytochrome system. Not only P_{730} but also P_{660} plays an active physiological role in the moss callus cells. P_{660} and darkness can be differentiated from each other during development of the protonema from callus cells. The effects of P_{660} cannot be shown immediately after the reversion with P_{730}, but only after a certain amount of time has elapsed. Red light enhances bryokinin synthesis, whereas blue light inhibits this process. The stimulation of bryokinin synthesis is connected with an increase in growth.[47] Red light brings about the differentiation of callus into cell chains and small sporophytes. In blue light the growth form of callus becomes stable. It has been shown that 5-bromouracil induces protonemal formation and 8-azaguanine prevents sporophytic differentiation.[47]

The presence of a neurohormone in the moss callus has been demonstrated by means of pharmacological tests and by chromatography. The concentration of the hormone in the callus cells increases after red light treatment and decreases after red and far-red light illumination.[49] Acetylcholine concentration in moss callus was also measured using gas-liquid chromatography after irradiation for 10 days in different spectral regimes. Moss callus irradiated with red light contained 56 times as much acetylcholine per gram fresh weight as compared to that irradiated with red/far-red light. Moss callus cells growing in the dark did not show any acetylcholine activity. This clearly revealed that the acetylcholine concentration in moss callus is regulated by phytochrome and that under short time irradiations very little acetylcholine is synthesized.[50]

The exposure of dark-grown callus tissue to red light induced a rapid increase in O_2 uptake after switching off the light. The respiratory change depends on the exposure duration and the intensity of light. This showed that exchange of oxygen is a good parameter for the determination of rapid responses of green moss callus to light. Acetylcholine was physiologically active only when the cofactor ascorbic acid was present. ATP is an additional cofactor for acetylcholine action. When the callus is subjected to quick alternating of red and far-red illumination acetylcholine promotes oxygen output. This effect is inhibited by serine or red light irradiance. Acetylcholine is activated only after irradiation with red light. Therefore, Hartmann[51,52] concluded that light is necessary for acetylcholine effect. Furthermore, in *Physcomitrium pyriforme* it was shown that gametophytes have more ATP than callus cells when grown under white light irradiance. However, the callus was observed to store more ATP in the dark than in light. The O_2 uptake of the gametophytes was more than that of the callus cells both in light- and in dark-grown cultures. Addition of sucrose to the substrate decreased the O_2 uptake in dark-grown cultures of callus and in gametophytes growing in light. However, the callus cells in light showed an increased uptake of oxygen.[42] Studies conducted on four different morphological stages of the hybrid sporogonium regenerants (i.e., protonema, gametophores, sporophytes, and callus) through gel electrophoresis of their soluble proteins for 14 different enzymes did not reveal any correlation between enzyme synthesis and morphogenesis.[53]

Light was shown to influence the developmental pattern of callus derived from the protonema of *Physcomitrium pyriforme*. On subculturing under low light intensity the callus differentiated predominantly into apogamous sporophytes, whereas under higher light intensities the callus differentiated into protonema and gametophores. It was shown that there is a changing pattern of polysaccharide synthesis in light and dark periods which probably influences the changes in the microfibrillar orientation of the wall, leading to labyrinthine outgrowths as shown in Section V. During redifferentiation of the callus cells more matrix wall material is synthesized in addition to cellulose, indicating that a morphogenetic change of the wall texture can be induced by light. Noncellulosic wall polymers can be secreted into the medium, and this process is controlled by the provision of sugars in the medium.[42]

VII. FUTURE EXPERIMENTAL APPROACHES

It is clear from the above account that induction and maintenance of moss callus has become a routine laboratory procedure. Among the unrealized projects is the understanding of the mechanism of determination of an apical cell with two and with three cutting faces. Thus far, we have only been able to conjecture how this determination can take place. Why the callus cells begin to form a specific apical cell at some places is shrouded in mystery. Likewise, it is still not clear how a circular callus cell changes its morphogenesis to become a filamentous protonemal cell under the influence of light. The chances of finding a specific cortical microtubular system operating to change the microfibrillar orientation for morphogenesis of an apical cell, however, have now improved considerably with the advent of new techniques in various laboratories. Moreover, the use of fermentors to obtain mass cultures of callus cells has just begun, and one can look forward in the near future to finding some specific secondary moss metabolic products.

REFERENCES

1. **Bauer, L.,** Regenerationsversuche am Sporogon von *Physcomitrium piriforme* Brid., *Ber. Dtsch. Bot. Ges.,* 70, 424, 1957.
2. **Bauer, L.,** Über die Stabilität von Teilfaktoren aus dem Faktorenkomplex der Sporogonbildung von Laubmoosen, *Biol. Zentralbl.,* 80, 353, 1961.
3. **Bauer, L.,** Über die Kalluswuchsform des Laubmoossporophyten, *Naturwissenschaften,* 48, 507, 1961.
4. **Bauer, L.,** Determination von Gametophyt und Sporophyt, in *Encyclopedia of Plant Physiology,* Vol. 18, Ruhland, W., Ed., Springer-Verlag, Berlin, 1967, 235.
5. **Lal, M.,** Experimental Investigations on the Moss *Physcomitrium,* Ph.D. thesis, University of Delhi, Delhi, India, 1961.
6. **Lal, M.,** The culture of bryophytes, including apogamy, apospory, parthenogenesis and protoplasts, in *Experimental Biology of Bryophytes,* Ducket, J. G. and Dyer, A. F., Eds., Academic Press, London, 1984, 97.
7. **Menon, M. K. C. and Lal, M.,** Problems of development in mosses and moss-allies, *Proc. Indian Natl. Sci. Acad. B,* 47, 115, 1981.
8. **Hughes, J. G.,** Factors conditioning development of sexual and apogamous races of *Phascum cuspidatum* Hedw., *New Phytol.,* 68, 883, 1969.
9. **Hughes, J. G.,** A Study of Apogamy in Mosses, Ph.D. thesis, University of Cambridge, Cambridge, England, 1958.
10. **Ashton, N. W. and Cove, D. J.,** The isolation and preliminary characterisation of auxotrophic and analogue resistant mutants of the moss *Physcomitrella patens, Mol. Gen. Genet.,* 154, 87, 1977.
11. **Wettstein, F. V.,** Morphologie und Physiologie des Formwechsels der Moose auf genetischer Grundlage. I, *Z. Vererbungsl.,* 33, 1, 1924.
12. **Bauer, L.,** On the physiology of sporogonium differentiation in mosses, *J. Linn. Soc. London Bot.,* 58, 343, 1963.
13. **Barthelmess, A.,** Mutationsversuche mit einem Laubmoos *Physcomitrium piriforme.* II. Morphologische und physiologische Analyse der univalenten und bivalenten Protonemen einiger Mutanten, *Z. Vereblungsl.,* 79, 153, 1941.
14. **Bauer, L. and Brosig, M.,** Zur Kenntnis Reziproker Kreuzungen von Funariaceen. I. Die Bastarde *Funaria hygrometrica* × *Physcomitrium acuminatum* und Reziprok, *Z. Vererblungsl.,* 90, 400, 1959.
15. **Ward, M.,** Callus tissue from the mosses *Polytrichum* and *Atrichum, Science,* 132, 1401, 1960.
16. **Ward, M.,** Gametophytic plants induced from single cells of moss callus, *Nature (London),* 204, 400, 1964.

17. **Lal, M.,** Experimental induction of apogamy and the control of differentiation in gametophytic callus of the moss *Physcomitrium coorgense,* in *Plant Tissue and Organ Culture — A Symposium,* Maheshwari, P. and Rangaswamy, N. S., Eds., International Society of Plant Morphologists, Delhi, India, 1963, 363.
18. **Chopra, R. N. and Rashid, A.,** Apogamy in *Funaria hygrometrica* Hedw., *Bryologist,* 70, 206, 1967.
19. **Menon, M. K. C. and Lal, M.,** Influence of sucrose on the differentiation of cells with zygote-like potentialities in a moss, *Naturwissenschaften,* 59, 514, 1972.
20. **Spiess, L. D., Lippincott, B. B., and Lippincott, J. A.,** Comparative response of *Pylaisiella selwynii* to *Agrobacterium* and *Rhizobium* species, *Bot. Gaz. (Chicago),* 138, 35, 1977.
21. **Rahbar, K. and Chopra, R. N.,** Preliminary studies on callus induction in the moss *Hyophila involuta, Z. Pflanzenphysiol.,* 99, 199, 1980.
22. **Kumra, P. K.,** Morphogenetic and Physiological Studies on Some Mosses, Ph.D. thesis, University of Delhi, Delhi, India, 1981.
23. **Kumra, P. K. and Chopra, R. N.,** Effect of some growth substances, vitamins and ultraviolet radiation on callus induction in the moss *Bryum coronatum* Schwaegr., *Z. Pflanzenphysiol.,* 108, 143, 1982.
24. **Saxena, P. K. and Rashid, A.,** Induction of callus from protonemal cells of the moss *Anoectangium thomsonii* Mitt., *Beitr. Biol. Pflanz.,* 57, 301, 1982.
25. **Lazarenko, A. S.,** Apogamous sporogonia in the haplophase of *Pottia intermedia* (Turn.) Fürnr., *Dopo. Akad. Nauk Ukr. RSR,* 11, 1524, 1963.
26. **Bauer, L.,** Über vegetative Sporogonbildung bei einer diploiden Sippe von *Georgia pellucida, Planta,* 46, 604, 1956.
27. **Menon, M. K. C. and Lal, M.,** Morphogenetic role of kinetin and abscisic acid in the moss *Physcomitrium, Planta,* 115, 319, 1974.
28. **Menon, M. K. C. and Lal, M.,** Regulation of a sub-sexual life cycle in a moss: evidence for the occurrence of a factor for apogamy in *Physomitrium, Ann. Bot. (London),* 41, 1179, 1977.
29. **Menon, M. K. C.,** Morphogenetic Studies on Apogamy in the Moss *Physcomitrium,* Ph.D. thesis, University of Delhi, Delhi, India, 1974.
30. **Bauer, L.,** Auslösung apogamer Sporogonbildung am Regenerationsprotonema von Laubmoosen durch einen von Muttersporogon abgegebenen Faktor, *Naturwissenschaften,* 46, 154, 1959.
31. **Bauer, L.,** Zur Frage der Qualität der Sporogonregenerate bei *Funaria hygrometrica, Naturwissenschaften,* 46, 154, 1959.
32. **Ward, M. and Frederick, S. E.,** Propagation of aberrant gametophytes from aggregates and from single cell of moss callus, *Phytomorphology,* 17, 371, 1967.
33. **Mehra, P. N.,** Some aspects of differentiation in cryptogams, *Res. Bull. Panjab Univ. Sci.,* 23, 221, 1975.
34. **Bauer, L.,** Über die Induktion apogamer Sporogonbildung bei Laubmoosen durch chloralhydrat, *Beitr. Biol. Pflanz.,* 42, 113, 1966.
35. **Bryan, V. S.,** Cytotaxonomic studies in the Ephemeraceae and Funariaceae, *Bryologist,* 60, 103, 1957.
36. **Crum, H. and Anderson, L. E.,** Taxonomic studies in the Funariaceae, *Bryologist,* 58, 1, 1955.
37. **Khanna, K. R.,** Cytological studies in some Himalayan mosses, *Caryologia,* 13, 559, 1960.
38. **Pande, S. K. and Chopra, N.,** Cytological studies in Indian mosses, *J. Indian Bot. Soc.,* 36, 241, 1957.
39. **Menon, M. K. C. and Bell, P. R.,** Ultrastructural and cytochemical aspects of induced apogamy following abscisic acid pre-treatment of secondary moss protonema, *Planta,* 151, 427, 1981.
40. **Knoop, B.,** Multiple DNA contents in the haploid protonema of the moss *Funaria hygrometrica* Sibth., *Protoplasma,* 94, 307, 1978.
41. **Menon, M. K. C. and Hartmann, E.,** Secreton of polysaccharides from moss callus cells, in *Methods in Bryology,* Glime, J. M., Ed., Hattori Botanical Laboratory, Nichinan, Japan, 1988, 107.
42. **Menon, M. K. C., Grasmück, I., and Hartmann, E.,** Effect of light on the callus cells of the moss *Physcomitrium, in Photoreceptors and Plant Development,* De Greef, J., Ed., Antwerp University Press, Antwerp, Belgium, 1980, 557.
43. **Lal, M. and Narang, A.,** Ultrastructural and histochemical studies of transfer cells in the callus and apogamous sporophytes of *Physcomitrium coorgense* Broth., *New Phytol.,* 100, 225, 1985.
44. **Bauer, L.,** Isolierung und Testung einer KinetinartigenSubstanz aus Kalluszellen von Laubmoosporophyten, *Z. Pflanzenphysiol.,* 54, 241, 1966.
45. **Beutelmann, P.,** Studies on the biosynthesis of a cytokinin in callus cells of moss sporophytes, *Planta,* 112, 181, 1973.
46. **Beutelmann, P. and Bauer, L.,** Purification and identification of a cytokinin from moss callus cells, *Planta,* 113, 215, 1977.
47. **Hartmann, E.,** Über den Stoffwechsel und das Differenzierungshalten des Laubmooskallus in Abhängigkeit vom Licht, Ph.D. thesis, Johannes Guttenberg Universität, Mainz, Federal Republic of Germany, 1970.
48. **Menon, M. K. C. and Hartmann, E.,** Secretion of protein-bound hydroxyproline from moss callus cells, *J. Plant Physiol.,* 132, 569, 1988.
49. **Hartmann, E.,** Evidence of a neurohormone in moss callus and its regulation by the phytochrome, *Planta,* 101, 159, 1971.

50. **Hartmann, E. and Kilbinger, H.,** Gas-liquid-chromatographic determination of light-dependent acetylcholine concentrations in moss callus, *Biochem. J.*, 137, 249, 1974.
51. **Hartmann, E.,** Über die Wirkung des Phytochroms beim Laubmooskallus. II. Sauerstoffmetabolismus und Acetylcholinwirkung, *Z. Pflanzenphysiol.*, 71, 349, 1974.
52. **Hartmann, E.,** Über die Wirkung des Phytochrome beim Laubmooskallus. I. Photomorphogenese und Stoffwechsel, *Beitr. Biol. Pflanz.*, 49, 1, 1973.
53. **Rothe, G.,** Unterschiede im Enzymmuster von Protonema, Moospflänzchen, Sporogon, und Kallus der Laubmooskreuzung *Funaria hygrometrica* × *Physcomitrium piriforme*, *Beitr. Biol. Pflanz.*, 48, 433, 1972.
54. **Spiess, L. D.,** Antagonism of cytokinin-induced callus in *Pylaisiella selwynii* by nucleosides and cyclic nucleotides, *Bryologist*, 82, 47, 1979.

Chapter 11

GROWTH AND SECONDARY METABOLITES PRODUCTION IN CULTURED CELLS OF LIVERWORTS

Yoshimoto Ohta, Kenji Katoh, and Reiji Takeda

TABLE OF CONTENTS

I. INTRODUCTION

Secondary metabolites of plant origin have long been used as drugs, perfumes, spices, or pigments, and there is an increasing demand for these natural products. In contrast to the extensive utilization of substances from higher plant sources, bryophytes have rarely been considered as a source of substances useful for human beings. However, progress in the phytochemistry of bryophytes has revealed that they contain substances with various biological activities.[1] *Marchantia polymorpha* and *Conocephalum conicum* have been used in folk remedies as a diuretic drug and for treating gallstones, respectively. *Polytrichum commune* and *Marchantia* spp. are included in folk remedies for cancer.[2] Another moss, *P. juniperinum,* has also been shown to have some tumor-damaging activity.[3] Spjut et al.[4] screened bryophytes for antitumor activity and found that a considerable number of mosses and liverworts had it. Antibacterial, antifungal, and antiviral activities have been reported more frequently for a wide range of bryophytes.[5-7] For example, Banerjee and Sen[7] observed that out of 52 species of bryophytes examined 29 (56%) showed some antimicrobial activities. In spite of the presence of the various chemically and biologically interesting compounds in bryophytes, the development of their practical uses has been retarded, primarily for the following reasons: (1) the difficulty in collecting a large quantity of pure materials from a particular species and (2) the fact that bryophytes are generally regarded as bioaccumulators of exogenous substances.[8] Various microorganisms such as blue-green algae or fungi grow in close association with most bryophytes, and we cannot rule out the possibility that some constituents in bryophytes may be a result of a close association with microorganisms.

Since the pioneering work of White in the 1930s plant tissue and cell culture techniques have been applied to an enormous number of vascular plants, aiming at propagation of plants with desirable characteristics. Introduction of modern technologies like gene manipulation has accelerated the developments in this field. Plant cell culture has also been developed as a promising method of producing secondary metabolites, and in some cases (e.g., shikonins) production on an industrial scale has been realized. It can reasonably be expected that cell culture will also be advantageous for the production of secondary metabolites of bryophytes. Application of cell culture techniques to bryophytes started about 20 years after their application to vascular plants. In 1957 Allsopp[9] first reported the induction of calluses from spores of two species of liverworts, *Fossombronia pusilla* and *Reboulia hemisphaerica,* and induction from mosses, *P. commune* and *Atrichum undulatum,* was reported by Ward in 1960.[10] To date, callus formation has been observed in about 20 species of liverworts, in most cases during the course of studies examining the effects of various culture conditions such as nutrients, growth regulators, and light intensity on their growth, development, and regeneration. Induction of callus or suspension cells aiming at the production of secondary metabolites in bryophytes is very limited. Since liverworts seem to contain a larger variety of chemically and biologically interesting compounds than mosses or hornworts,[1,11] induction and growth physiology of cultured cells as well as production and biosynthesis of secondary metabolites in liverworts will be described.

II. INDUCTION OF CALLUS OF LIVERWORTS

Actively growing meristematic tissues in seedlings, roots, or stems are generally utilized as the starting material for the induction of callus in vascular plants. These tissues can be sterilized by successive or combined treatments with benzalkonium chloride, ethyl alcohol, chlorinated water, and sterilized water. However, the tissues or whole thalli of liverworts are too delicate to be sterilized with these reagents. Protonemata or gametophytes generated from the aseptic spores are generally used as the starting plant material. The capsules which contain mature spores are surface-sterilized with benzalkonium chloride and sodium hypochlorite solution. The spores are squeezed and spread over an agar medium supplemented with mineral salts to grow the protonema, which is eventually transferred to the callus-induction medium.[12,13] Callus of *M. polymorpha* was successfully obtained from multicellular gemmae after sterilizing with benzalkonium chloride and sodium hypochlorite solution.[14] However, mature spores or multicellular gemmae are not always available for most of the liverwort species. There is an urgent need to develop a moderate and reliable sterilization method to start callusing from the thallus of a liverwort which contains interesting and useful chemical compounds. A surface-sterilizing method for small plant parts developed by Basile[15] would be worthwhile to apply to liverworts. Dhingra-Babbar and Chopra[16] succeeded in using sterilized apical notches for the induction of callus in *M. palmata*.

Calluses from liverworts are generally induced on Knop[14,16,17] or modified Murashige-Skoog[12,18,19] medium supplemented with microelements, sugars, vitamins, and/or growth regulators. As noticed in the studies on regeneration of the Metzgeriales, *Noteroclada confluens,*[20] *Riccardia multifida,*[21] and *Riccardia pinguis,*[21] the types and concentrations of sugars included in the medium have a decisive influence on the growth and development of the liverworts. The increase in shoot length was enhanced with a low concentration (0.025 to 0.5%) of sugars, whereas at higher concentrations (2 to 4%) growth and normal development were suppressed considerably and callus formation was observed. In most of the liverworts examined so far, callus formation was observed in a medium containing 2 to 6% glucose or sucrose; an exception was *Riccia crystallina,* which formed callus in response to switching the nitrogen source in the medium from nitrate to ammonium.[22] *Riccardia multifida,*[21] *Riccardia pinguis,*[21] *N. confluens,*[20,23] *Calypogeia granulata,*[12] *Jungermannia subulata,*[19] *Lophocolea heterophylla,*[13] *Scapania parvidens,*[13] and *M. polymorpha*[24] formed calluses in response to an increase in sucrose or glucose content alone. Promotion of callus formation with various growth regulators such as auxins, cytokinins, and gibberellic acid has been observed in *M. nepalensis,*[25] *M. polymorpha,*[14] *Sphaerocarpus texanus,*[26] *Athalamia pusilla,*[27] *M. palmata,*[28,29] and *Asterella wallichiana*[30] by the inclusion of an appropriate amount of sugar in the medium.

The mode of action of comparatively high concentrations of sugars on the induction of callus formation in liverworts has not been fully understood. Formation of undifferentiated callus may be explained by the concept of polarity. Addition of a high concentration of sugar brings about a disturbance in the normal polarized flow of nutrients or hormones necessary for differentiated growth of the tissue while permitting a relatively unrestricted division of the thallus cells.[9,18,21,23] However, the observation that growth regulators like auxin and cytokinin strongly enhance cell multiplication and promote callus formation suggests that enhanced cell division in the thallus with sugar also may be responsible for the formation of callus.[31,32] A high osmotic potential caused by the addition of sugar to the medium also may be a trigger for the formation of callus. Enrichment of the sucrose media with mannitol enhanced the induction of stable calluses and their continued proliferation in *Asterella angusta*[17] and *M. palmata.*[16] Sugars are also important factors for induction of calluses in mosses. Callus formation from *P. commune,*[10] *Atrichum undulatum,*[10] *Bryum coronatum,*[31] *Barbula unguiculata,*[33] and *Sphagnum imbricatum*[34] was observed on media containing 2 to 4% sucrose or glucose. UV[31] or X-ray irradiation[10] was reported to be effective for induction and successive growth of stable calluses in some species of mosses. The mechanism of the promotion of callusing by irradiation was not understood, and this has not been applied for callus formation in liverworts.

For the study of the production of secondary metabolites suspension cultures are preferred to callus cultures on solid medium. This is advantageous because of the higher growth rate under homogeneous environmental conditions, availability of a larger amount of cell mass, accuracy in quantitative measurement of growth and productivity, and also ease of feeding the labeled precursors for biosynthetic studies. The cell suspension culture can be established by transferring the appropriate amount of callus to the liquid medium of the same nutritional composition and by culturing on a rotary or reciprocal shaker[19,35] or in an air-lift-type culture vessel.[12] Direct transfer of the protonema or gametophytes to the liquid medium can also afford cell suspension in some liverworts.

III. GROWTH PHYSIOLOGY OF CULTURED CELLS OF LIVERWORTS

The cultured cells of liverworts have several unique physiological attributes differing from those of vascular plants. First, the dedifferentiation of thallus and redifferentiation of callus into thallus are primarily regulated by the concentrations of sugars included in the medium. As described earlier, the high concentration of sugars is generally a prerequisite for callus formation. On the other hand, calluses of *Riccardia multifida,*[21] *Riccardia pinguis,*[21] *Noteroclada confluens,*[23] *Asterella angusta,*[17] and *M. polymorpha*[35] easily differentiate to form normal thalli when transferred to a medium devoid of or containing low concentrations of sugars. Callus of *Asterella wallichiana* also differentiated after a prolonged culturing period during which depletion of sugars in the medium might have occurred.[18] Second, the cultured liverwort cells, the callus, and cells in suspension culture usually contain well-developed chloroplasts and require light for growth even in a medium supplemented with appropriate sugars as carbon source. Third, some of the cultured liverwort cells of Jungermanniales prefer ammonium to nitrate as the nitrogen source. Fourth, although it has been reported in a very limited number of taxa, liverwort cells seem to retain the ability to biosynthesize secondary metabolites present in the mother plants.

Liverworts always form pale to dark green calli, and the chlorophyll content of the cultured cells is generally high, e.g., 13 to 25 mg per gram cell dry weight in *M. polymorpha,*[36] 18 mg in *J. subulata,*[13,19] and 7 to 12 mg in *Heteroscyphus bescherellei*[37] cells in suspension cultures. These cells can grow only in light, and growth is completely suppressed in darkness even in the presence of sugars in the medium. The only exception to these reported so far is the suspension cell line of *M. paleacea* var. *diptera.*[38] *M. paleacea* cells in glucose-supplemented media grow both in light and darkness at a comparable rate. Chlorophyll content and photosynthetic activity in the dark-grown cells are similar to those of the light-grown cells. Suspension cells of two species of mosses, *Barbula unguiculata*[33] and *Sphagnum imbricatum,*[34] are also rich in chlorophyll (ca. 12 and 15 mg per gram cell dry weight, respectively) and grow actively in light. However, these moss cells also can grow and synthesize chlorophyll in darkness when cultured in a medium containing 2% glucose.[31] Growth of *B. unguiculata* is accompanied by differentiation of the cells into protonemata.

The physiology of light-dependent growth of liverwort cells has been examined precisely in *M. polymorpha* cells in suspension culture.[36] The cells are unable to grow in darkness even when glucose is present in the medium, and the uptake of nutrients such as glucose, phosphate, nitrate, and ammonium is also suppressed completely.[35] The growth of cells in light is more sensitive to 3-(3,4-dichlorophenyl)-1,1-dimethylurea (DCMU), a specific inhibitor of photosystem II, than to antimycin A, a specific inhibitor of cyclic photophosphorylation. This implies that the energy required for growth and for uptake and metabolism of nutrients seems to be supplied through noncyclic photophosphorylation in the chloroplasts. The rate of glucose assimilation as revealed by the uptake of oxygen is nearly four times higher in light than in darkness, and ATP generated by the noncyclic photophosphorylation appears likely to prime glucose to enter the glycolytic pathway. *M. polymorpha* cells also can fix CO_2 photosynthetically. Even assuming that all glucose in the medium is assimilated by the cells, including reassimilation of CO_2 generated by glucose metabolism, nearly one third of the carbon atoms in the cellular constituents seem to be derived from atmospheric

CO_2. Thus, the liverwort cells cultured in the glucose-containing medium grow photomixotrophically, utilizing glucose by photoassimilation and CO_2 by photosynthesis.

The green cells of liverworts also can grow photoautotrophically utilizing CO_2 as the sole source of carbon under illumination.[39,40] The rate of photosynthetic carbon fixation of suspension-cultured *M. polymorpha* cells under optimal conditions was 1.30 μmol CO_2 per 10^6 cells per hour or 66 μmol CO_2 per milligram chlorophyll per hour, and the doubling time was approximately 1 day.[40] Although the establishment of several cell lines with high chlorophyll content has been reported for vascular plants,[41] the majority of the photoautotrophic cell lines exhibit low growth rates when compared to photomixotrophically or heterotrophically growing cells. The exceptions to this are asparagus cells growing in a turbidostat photoautotrophic culture[42] and soybean cells in a batch culture.[43] The chlorophyll content of the latter is approximately 13 mg per gram cell dry weight,[44] which is comparable to that of the cultured liverwort cells. The highest CO_2 fixation rate (curiously, unaccompanied by O_2 evolution under the conditions employed) in the soybean cells, 66 μmol CO_2 per milligram chlorophyll per hour, is also similar to that of *M. polymorpha* cells. In both cases, cell growth at the maximum rate does not last for a long time because the absorption of the incident light becomes larger in response to the increase in cell population density. Development of an effective illumination method is required to culture these chlorophyllous plant cells under photoautotrophic conditions.

Another interesting feature in the growth physiology of some liverwort cells is the preferential utilization of ammonium as a nitrogen source.[19,37] When the suspension cells of *J. subulata*[19] or *H. bescherellei*[37] were cultured in a modified Murashige and Skoog medium or in a medium containing ammonium as the sole nitrogen source, the cells grew actively only in the medium supplemented with organic acids of the tricarboxylic acid (TCA) cycle, like fumaric acid or citric acid, or with calcium carbonate. Ammonium in the medium was taken up preferentially by the cells, particularly at the earliest stage of the culture, while a negligible amount of nitrate was utilized as long as ammonium remained in the medium. This unbalanced utilization of the two nitrogen sources caused a drastic drop in the pH of the medium which eventually led to the death of the cells if the medium was not buffered with the organic acids or calcium carbonate. The suspension cells of *M. polymorpha* utilized ammonium together with nitrate, and this resulted in the maintenance of pH of the medium within the physiological range.[45] Nitrate is an unfavorable nitrogen source to some cultured liverwort cells.[19,37,44] For example, *J. subulata* cells cultured in nitrate medium are green, but the cell dry weight increases by only 18% during 15 days of culture.[19] This suggests that nitrogen metabolism in the cells may be somewhat different from that of other cells; for example, slow or negligible induction of nitrate reductase may occur. This enzyme has rarely been studied in liverworts. Examination of nitrate reductase from thalli[46] and suspension cells[47,48] of *M. polymorpha* and *M. paleacea* cells[48] revealed its unique nature in requiring reduced nicotinamide adenine dinucleotide phosphate (NADPH) as the specific electron donor. The amounts of nitrogen sources in Murashige and Skoog medium are 20 m*M* ammonium and 40 m*M* nitrate, and *M. polymorpha* cells consume all of the nitrogen to grow to about 13 mg cell dry weight per milliliter in 15 days. Although *J. subulata* cells utilized all of the ammonium, only about 10% of the nitrate was taken up by the cells in 15 days when the cell dry weight also reached the same value. This shows that *J. subulata* cells require only one third of the amount of nitrogen source to grow to the same extent as *M. polymorpha* cells. Similar growth behavior was also observed for the cell culture of other liverworts (*L. heterophylla*[13] and *Scapania parvidens*[75]). This unusual utilization of inorganic nitrogen sources during nitrogen metabolism in bryophyte cells seems to be an interesting research topic.

IV. PRODUCTION OF SECONDARY METABOLITES

Production of secondary metabolites in cultured cells of liverworts has been examined

TABLE 1
Sesquiterpenoids from *Calypogeia granulata*

	Intact plant	Cultured cells		Redifferentiated plant
		20 days	30 days	
Yield of essential oil (% of fresh weight)	0.15	0.30	0.20	0.23
Sesquiterpenoids (% of composition)				
Trinoranastreptene	0.1	1.5	1.3	0.7
Anastreptene	2.6	4.5	3.3	3.4
Ledene	1.0	1.5	1.2	1.6
Bicyclogermacrene	3.8	17.0	10.0	6.2
3,10-Dihydro-1,4-dimethylazulene	tr	0.6	tr	tr
Unknown compound	2.4	5.1	3.3	3.5
1,4-Dimethylazulene	51.0	43.0	48.0	51.0
3,7-Dimethylindene-5-carbaldehyde	2.8	2.8	3.9	6.7

Note: tr = trace.

in a very limited number of plant species. *J. subulata* cells contain a large amount of lipids, which comprise approximately 50% of the cell dry weight. About 40% of the lipid fraction is triglycerides composed of palmitic acid, linoleic acid, stearic acid, oleic acid, and arachidic acid as the organic acid components.[49] Kaurene, phytol, and stigmasterol were isolated in addition to the triglycerides.

In 1981 Takeda and Katoh[12,50] examined the production of sesquiterpenoids in suspension-cultured cells of *Calypogeia granulata*. The yield of the volatile essential oil, composed mainly of various sesquiterpenoids, was 0.3% of the fresh cells harvested after 20 days of inoculation, which was twice that of the intact liverwort (Table 1). The major component of the oil was 1,4-dimethylazulene **(1)**, an intensely blue-colored trinorsesquiterpene hydrocarbon. This compound had been isolated previously from *C. trichomanis*.[51] In addition to 1,4-dimethylazulene, five new sesquiterpenoids, 3,10-dihydro-1,4-dimethylazulene **(2),** 3,7-dimethylindene-5-carbaldehyde **(3),** trinoranastreptene **(4),** 2-acetoxy-3-hydroxybicyclogermacrene **(5),** and 3-acetoxy-2-hydroxybicyclogermacrene **(6),** were isolated together with a number of known sesquiterpenoids such as ledene and bicyclogermacrene from the cultured cells of *C. granulata*. 3,10-Dihydro-1,4-dimethylazulene, an immediate precursor of 1,4-dimethylazulene, was found only in the cultured cells. As can be seen in Table 1 and in gas chromatograms of the volatile oils (Figure 1), the sesquiterpenoid compositions of the cultured cells and of the intact plants are qualitatively and quantitatively very similar to each other. Furthermore, redifferentiated thallus obtained by transferring the callus onto agar medium devoid of glucose was proven to contain similar sesquiterpenoids. The accumulation of an equal or larger amount of terpenoids of the same composition in the intact plant, cultured cells, and the redifferentiated thallus of *C. granulata* is very unique when compared to vascular plants.

Accumulation of lower terpenoids in cultured cells of oil-producing vascular plants has not commonly been reported. In the known examples of terpenoid production, the characteristic components in the mother plants were very often not found in undifferentiated cells.[52-55] Failure of cultured cells to produce volatile terpenoids might be explained by one of the following reasons: (1) loss of the ability to biosynthesize this group of compounds or (2) loss of capacity to accumulate the products prevents their further metabolism or degradation. Banthorpe and Barrow[56] showed that cell-free extracts of the callus culture of *Rosa damascena,* which did not accumulate terpenoids characteristic of the mother plant, had several hundred times the amount of active enzymes for monoterpene biosynthesis than

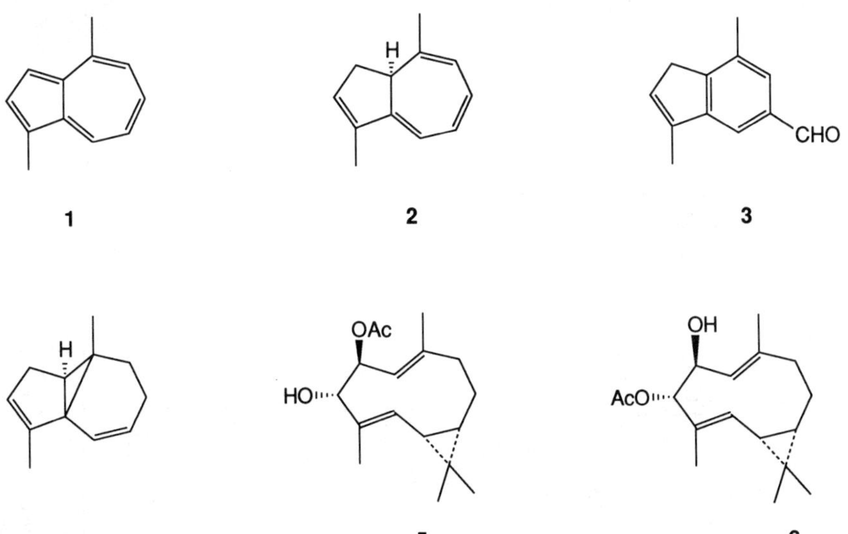

were found in the intact plant. They obtained similar results for several other species of oil-producing angiosperms.[55] Since the callus and suspension culture of *R. damascena* rapidly metabolized exogenously supplied monoterpenes via oxidative pathways,[57] nonaccumulation of lower terpenoids is considered to be due to further metabolism of nascent products formed endogenously. It is generally assumed that lower terpenoids are biosynthesized and accumulated in specially differentiated secretory structures such as oil glands, and the lack of development of these specialized organs in cultured cells is likely to be the reason for nonaccumulation of these compounds.

The volatile terpenoids biosynthesized by *Calypogeia granulata* cells are clearly seen to be deposited in oil bodies in the cells, as evidenced by the intense blue color of 1,4-dimethylazulene, the main constituent. The oil bodies are cellular organelles characteristic of liverworts of the order Jungermanniales and are known to contain mainly terpenoids and fatty acids. The fact that biosynthesized terpenoids are deposited in oil bodies and, thus, are protected from further degradation seems to be a plausible explanation for the production of the volatile oils in considerable quantities in cultured liverwort cells. Liverworts are considered to be at an early stage of evolution of terrestrial green plants, and their morphological differentiation is much simpler in comparison to vascular plants. Thus, it is reasonable to suppose that the morphologically dedifferentiated liverwort cells still retain the differentiation in function or metabolism as exemplified by the development of chloroplasts or oil bodies. The developed chloroplasts may be able to supply enough energy not only for the cell growth via photoassimilation of glucose or photosynthesis, but also for the production of secondary metabolites. The cultured cells of vascular plants generally consume the carbon source in the medium to supply substrates and energy primarily for cell proliferation. Production of secondary metabolites often occurs only when the flow of the substrates and energy is switched from primary to secondary metabolism by, for example, modification of the medium in a two-stage culture.

Incidentally, the cell culture of the moss *Leptobryum pyriforme* has been examined for the production of polyunsaturated fatty acids such as arachidonic and eicosapentaenoic acids.[58] The protonemal cells were cultured in a modified Bennecke liquid medium in light (16 h light/8 h darkness illumination regime). Of the total lipid fraction contained in the cells, 20% was arachidonic acid and 7% was eicosapentaenoic acid.

Most liverworts contain various types of aromatic compounds which are sometimes important as chemosystematic indicators.[59] Among them, lunularic acid **(7)**, a dihydrostilbene

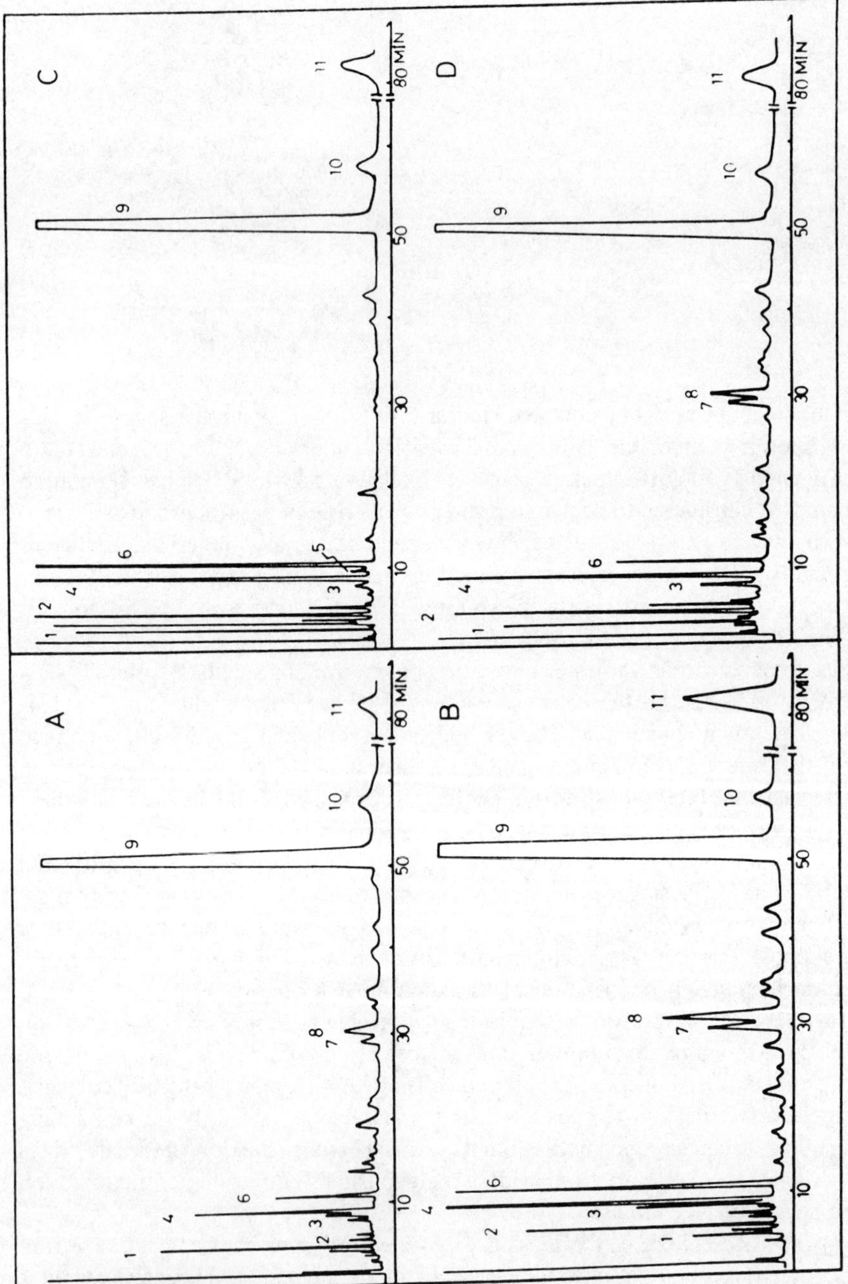

FIGURE 1. Gas chromatograms of the essential oil of *Calypogeia granulata*. (A) Intact plant; (B) redifferentiated plant; (C) suspension-cultured cells (20 days); (D) suspension-cultured cells (30 days). (1) Tetrahydro-1,4-dimethylazulene; (2) compound I; (3) ledene; (4) bicyclogermacrene; (5) dihydro-1,4-dimethylazulene; (6) compound II; (7) compound III; (8) compound IV; (9) 1,4-dimethylazulene; (10) compound V; (11) indene-type aldehyde. (From Takeda, R. and Katoh, K., *Planta*, 151, 525, 1981. With permission.)

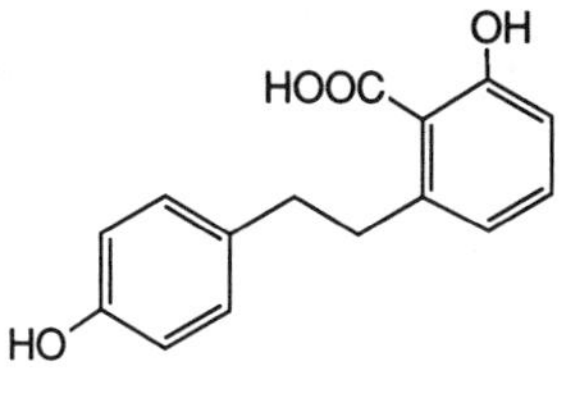

7

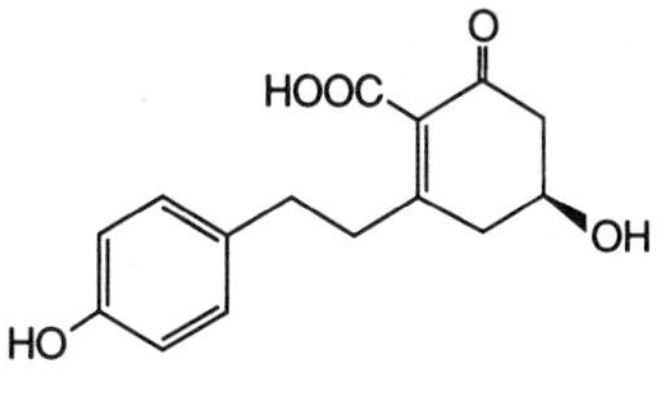

8

carboxylic acid, has drawn our attention because of its ubiquitous occurrence exclusively in liverworts (with the exception of one angiosperm species, *Hydrangea macrophylla*[60,61]) and because of its proposed role as an endogenous growth regulator (a dormancy factor) in liverworts.[60,62,63] Production of lunularic acid was examined in suspension-cultured cells of *M. polymorpha* and some other species of the Jungermanniales such as *J. subulata, Lophocolea heterophylla,* and *Calypogeia granulata.*[64] All of the cultured cell lines examined yielded a considerable amount of lunularic acid upon extraction with acetone and column purification of the strong acid fraction. This showed that the production of lunularic acid was a characteristic of cultured cells of liverworts as well as the intact plants. However, the amount of lunularic acid obtained from the cultured cells varied considerably depending on the extraction procedures employed. For example, extraction of *M. polymorpha* cells with boiling methanol containing 1% acetic acid gave 4.4 μg of lunularic acid per milligram dry weight of cells, whereas only 0.39 μg was obtained upon methanol extraction without acetic acid. This value further decreased to 0.12 μg by extraction with 60% acetonitrile in water under sonication at 4°C.[65] A similar effect of extraction methods on the efficiency of obtaining lunularic acid from intact liverworts has been described by Gorham,[61] who noticed that lunularic acid was extracted more efficiently by methanol containing hydrochloric acid instead of boiling ethanol. These results suggested the existence of a labile compound. This labile compound, prelunularic acid **(8),** was isolated from suspension-cultured cells of *M. polymorpha.*[66,67] Filter-harvested cells were extracted with 90% methanol, and successive column chromatography of the extract on Sephadex® LH-20, reverse-phase silica gel, and cellulose columns afforded prelunularic acid, which was directly converted into lunularic acid upon treatment with diluted sulfuric acid or sodium hydroxide solution, or even under neutral conditions (although at a slower rate). Prelunularic acid is a cyclization product of *p*-coumaryl β-triketo acid and is the first example of a prearomatic intermediate in the phenylpropanoid-polymalonate pathway. This compound is a plausible immediate precursor of lunularic acid.

Since then, prelunularic acid was proved to exist in several liverworts of Marchantiales and Jungermanniales.[68] These results clearly indicate that the formation and accumulation of prelunularic acid are not limited to the cells of *M. polymorpha* cultured under specific conditions, but are general to liverworts.

The question is whether lunularic acid is a real component in liverworts or is an artifact which originated from prelunularic acid during the isolation process. Sodium borohydride rapidly reduces prelunularic acid to give a mixture of diols which no longer convert into lunularic acid. The amount of lunularic acid extracted from cultured cells of *M. polymorpha* and the thallus of *Conocephalum conicum* in the presence of sodium borohydride was, as shown in Table 2, approximately 0.1% of the sum of prelunularic acid and lunularic acid,[65] and this value can be expected to be far smaller, possibly zero, since the conversion may still precede the reduction because of the basic nature of sodium borohydride. This leads to the conclusion that the endogenous lunularic acid is practically all prelunularic acid and is not a real constituent of liverworts as has been previously reported.[65]

The next question to be considered is the physiological role of prelunularic acid. Intra-

TABLE 2
Lunularic Acid (LNA) Content Determined in the Presence of $NaBH_4$

	LNA content				
	$NaBH_4$ treatment		Control		
	(a)	(b)	(a′)	(b′)	a/b′ × 100 (%)
Marchantia polymorpha cultured cells	0.004	0.005	0.127	3.97	0.1
Conocephalum conicum thallus	0.002	0.002	0.044	1.09	0.2

Note: Data are the means from two experiments. The cultured cells or thallus were ground with or without (control) $NaBH_4$. LNA contents in these homogenates were determined before (a and a′) and after (b and b′) alkaline treatment. Units are μg/mg dry weight.

cellular localization of the acid in cultured cells showed that this compound is not present in a bound form with cell walls or membranous components, but is almost equally distributed in vacuoles and cytoplasm.[69] This shows that prelunularic acid does not have any inhibitory effect comparable to lunularic acid on cellular processes. The existence of such active compounds as lunularic acid in high concentrations could reasonably be expected to exert a recognizable degree of inhibition on cellular growth, but growth of *M. polymorpha* cells is comparable to that of the cultured cells of other plants. Furthermore, the concentration of physiologically active compounds is generally regulated strictly as, for example, in *de novo* synthesis of abscisic acid in response to water stress,[70] while spontaneous and non-enzymic conversion of labile prelunularic acid into lunularic acid can be expected to occur even in living cells. Endogenously generated lunularic acid is probably metabolized into an unidentified methanol-insoluble substance, as was reported when ^{14}C-labeled lunularic acid was fed to *Lunularia cruciata*.[71]

V. BIOSYNTHESIS

Cell cultures of vascular plants have long been used for studies on the biosynthesis of steroids, terpenoids, phenolics, and alkaloids. Cultured cells have various advantages over intact plants as tools for biosynthetic studies, namely, homogeneity of the material, ease of feeding precursors, absence of microbial contamination, ease of obtaining a cell-free system, and the possibility of finding new labile intermediates which fill blanks in the biosynthetic pathway. These advantages are more desirable when examining the biosynthesis of metabolites of bryophytes because most bryophytes are small in size, causing difficulty in feeding experiments, and because they are generally regarded as bioaccumulators of various substances metabolized by closely associated microorganisms.[4,8] To date, only one example of the biosynthesis of secondary metabolites in cultured liverwort cells has been reported.

During the course of isolation of sesquiterpenoids from suspension cultured cells of *Calypogeia granulata* a labile hydrocarbon, 3,10-dihydro-1,4-dimethylazulene **(2)**, was isolated as the first example of dihydroazulene carrying a chiral center (C-10).[12] This dihydro derivative is easily converted into 1,4-dimethyl azulene **(1)** by exposure to air and is an immediate precursor of the latter. Only traces of dihydroazulene were detected in the intact plants or in the cells cultured for prolonged period. This suggests that the labile intermediate is metabolized to the final product with the aging of the plant materials. In addition to the dihydroazulene, another trinorsesquiterpenoid, 3,7-dimethylindene-5-carbaldehyde **(3)**, was isolated from *C. granulata* cells. A related compound, 3,7-dimethyl-5-methoxycarbonyl indene **(9)**, has been isolated together with 1,4-dimethylazulene and 4-methyl-1-methoxycarbonylazulene from *C. trichomanis*.[72] These rarely occurring indene-type natural com-

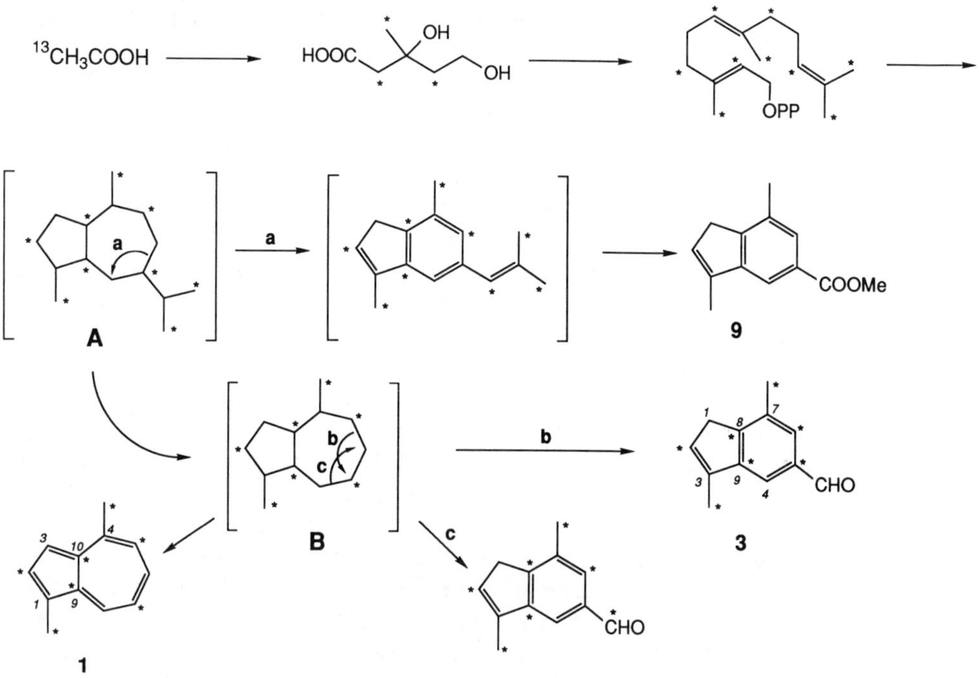

SCHEME 1.

pounds are interesting from the viewpoint of biosynthesis (Scheme 1). Meuche and Huneck[72] presented a hypothesis in which the indene carbon skeleton is originated from intermediate **A** by ring contraction via route **a** followed by dehydrogenation and oxidative cleavage of the side chain. In another possibility, the ring contraction via either route **b** or **c** in a hypothetical intermediate **B** formed from intermediate **A** by elimination of an isopropyl group also leads to the indene carbon skeleton. This was examined by feeding *C. granulata* cells with 2-^{13}C-acetate.[73] The cells were grown for 30 days in a medium supplemented with 5 m*M* 2-^{13}C-acetate instead of glucose. The locations of ^{13}C in 1,4-dimethylazulene **(1)** and 3,7-dimethylindene-5-carbaldehyde **(3),** as determined by ^{13}C-nuclear magnetic resonance (NMR), were carbons 2, 5, 7, 9, 10, 1-CH_3, and 4-CH_3 in the former and carbons 2, 5, 6, 8, 9, 3-CH_3, and 7-CH_3 in the latter, respectively. Thus, 3,10-dihydro-1,4-dimethylazulene and, subsequently, 1,4-dimethylazulene were clearly demonstrated to be formed from farnesyl pyrophosphate via a hypothetical intermediate **A** as shown in Scheme 1. The absence of label on the aldehyde carbon atom of 3,7-dimethylindene-5-carbaldehyde **(3)** indicates that this carbonyl carbon originates from C-6 by migration of the C–C bond in intermediate **B** via route **b.** The driving force for this ring contraction is the formation of a carbonium ion at C-7 by elimination of the isopropyl group. 3,7-Dimethyl-5-methoxycarbonylindene **(9)** in *C. trichomanis* should be formed in the same way.

VI. CONCLUSION

Cell culture is an advantageous tool for studies on primary metabolism as well as secondary metabolites of liverworts. Particularly, development of a number of chloroplasts in the cells makes them excellent experimental materials for studies on growth physiology under photoheterotrophic or photoautotrophic conditions. The fact that dedifferentiated liverwort cells retain the ability to synthesize secondary metabolites is very expedient for the production and biosynthetic studies of those metabolites. Liverworts are interesting sources of compounds with varying biological activity, compounds with novel structures, and compounds important for elucidation of biosynthesis. However, liverworts are generally tiny, and several species often grow together. This frequently makes it difficult to collect a quantity of one single species sufficient for the isolation of compounds required for structure elucidation and biological assay. Furthermore, the existence of varying degrees of symbiosis between liverworts and microbes and the tendency of liverworts to concentrate and accumulate compounds metabolized by closely associated microorganisms render the origin of the compounds isolated and the bioassay results ambiguous. Cell culture will solve most of these problems. Although a number of natural products have been detected in liverworts, most of them are the major components in their respective plants and are extractable in sufficient quantity for structure elucidation or are confirmed spectrometrically by gas chromatography-mass spectrometry. The cell culture will also serve in the search for minor constituents or nonvolatile compounds in liverworts.

Although the number of experiments is very limited, from the results of the formation of sesquiterpenoids by *Calypogeia granulata* cells or prelunularic acid (lunularic acid) by several cell lines it can be inferred that the amounts of the various metabolites in cultured cells do not remarkably exceed those in the intact plants. In some vascular plants the cultured cells can produce much larger quantities of metabolites than the mother plants. In *Lithospermum erythrorizon* the content of shikonins reaches several hundred fold that of the mother plant. The production of certain groups of secondary metabolites by cultured vascular plant cells has recently been reported to be greatly enhanced by treatment with various elicitors.[74] This implies that these compounds are stress metabolites similar in nature to phytoalexins which are synthesized in response to the invasion of plants by pathogenic microorganisms. At this moment we do not have any ideas as to the nature and role of secondary metabolites in liverworts. Once these points become clear we may be able to improve the production efficiency of metabolites in cultured cells of liverworts. Production of useful metabolites by plant cells growing under photoautotrophic conditions utilizing carbon dioxide as a sole carbon source and sunlight as an energy source could be our final goal. The green liverwort cells may make this dream come true in the near future.

REFERENCES

1. **Asakawa, Y.,** Chemical constituents of Hepaticae, in *Progress in the Chemistry of Organic Natural Products,* Vol. 42, Herz, W., Griesbach, H., and Kirby, G. W., Eds., Springer-Verlag, Vienna, 1982, 1.
2. **Hartwell, J. L.,** Plants used against cancer. A survey, *Lloydia,* 34, 386, 1971.
3. **Belkin, M., Fitzgerald, D. B., and Felix, M. D.,** Tumor-damaging capacity of plant materials. II. Plants used as diuretics, *J. Natl. Cancer Inst.,* 13, 741, 1952.
4. **Spjut, R. W., Suffness, M., Cragg, G. M., and Norris, D. H.,** Mosses, liverworts, and hornworts screened for antitumor agents, *Econ. Bot.,* 40, 310, 1986.
5. **McCleary, J. A., Sypherd, P. S., and Walkington, D. L.,** Mosses as possible sources of antibiotics, *Science,* 131, 108, 1960.
6. **Van Hoof, L., Vanden Berghe, D. A., Petit, E., and Vlietinck, A. J.,** Antimicrobial and antiviral screening of bryophyta, *Fitoterapia,* 52, 223, 1981.
7. **Banerjee, R. D. and Sen, S. P.,** Antibiotic activity of bryophytes, *Bryologist,* 82, 141, 1979.
8. **Glime, G. M. and Keen, R. E.,** The importance of bryophytes in a man-centered world, *J. Hattori Bot. Lab.,* 55, 133, 1984.
9. **Allsopp, A.,** Controlled differentiation in cultures of two liverworts, *Nature (London),* 179, 681, 1957.
10. **Ward, M.,** Callus tissues from the mosses *Polytrichum* and *Atrichum, Science,* 132, 1401, 1960.
11. **Huneck, S.,** Chemistry and biochemistry of bryophytes, in *New Manual of Bryology,* Schuster, R. M., Ed., Hattori Botanical Laboratory, Nichinan, Japan, 1983, 1.
12. **Takeda, R. and Katoh, K.,** Growth and sesquiterpenoid production by *Calypogeia granulata* Inoue cells in suspension culture, *Planta,* 151, 525, 1981.
13. **Ohta, Y. and Hirose, Y.,** Induction and characteristics of cultured cells from some liverworts of Jungermanniales, *J. Hattori Bot. Lab.,* 53, 239, 1982.

14. **Ono, K.,** Callus formation in liverwort, *Marchantia polymorpha, Jpn. J. Genet.*, 48, 69, 1973.
15. **Basile, D. V.,** A method for surface-sterilizing small plant parts, *Bull. Torrey Bot. Club,* 99, 313, 1972.
16. **Dhingra-Babbar, S. and Chopra, R. N.,** Induction of callus in the liverwort *Marchantia palmata* Nees, *J. Bryol.*, 13, 585, 1985.
17. **Rashid, A.,** Callusing and regeneration of cell-aggregates and free-cells of a Hepatic: *Asterella angusta* Aust., *Experientia,* 27, 1498, 1971.
18. **Kumra, S.,** Callus induction in the liverwort *Asterella wallichiana* (L. et L.) Grolle, *J. Plant Physiol.*, 115, 263, 1984.
19. **Ohta, Y., Ishikawa, M., Abe, S., Katoh, K., and Hirose, Y.,** Growth behavior of a liverwort, *Jungermannia subulata* Evans, in a cell suspension culture. The role of organic acids required for cell growth, *Plant Cell Physiol.*, 22, 1533, 1981.
20. **Allsopp, A. and Ilahi, I.,** Studies in the Metzgeriales (Hepaticae). III. The effects of sugars and auxins on *Noteroclada confluens, Phytomorphology,* 20, 9, 1970.
21. **Ilahi, I. and Allsopp, A.,** Studies in the Metzgeriales (Hepaticae). VI. Regeneration and callus formation in some thalloid species, *Phytomorphology,* 20, 126, 1970.
22. **Sood, S.,** In vitro studies in Marchantiales. II. Effect of mineral nutrients, chelates, and organic nitrogenous sources on the growth and sexuality in *Riccia crystallina, Phytomorphology,* 24, 186, 1974.
23. **Allsopp, A. and Ilahi, I.,** Studies in the Metzgeriales (Hepaticae). VII. Regeneration in *Noteroclada confluens* and *Blasia pusilla, Phytomorphology,* 20, 173, 1970.
24. **Katoh, K., Imoto, S., and Komemushi, S.,** Effect of sugars on the differentiation of *Marchantia polymorpha (Marchantiaceae),* in Abstr. XIV Int. Bot. Congr., West Berlin, July 24 to August 1, 1987, 2-110-6.
25. **Kaul, K. N., Mitra, G. C., and Tripathi, B. K.,** Responses of *Marchantia* in aseptic culture to well-known auxins and antiauxins, *Ann. Bot. (London),* 26, 447, 1962.
26. **Montague, M. J. and Taylor, J.,** The effects of glucose, indole-3-acetic acid and 2,4-dichlorophenoxy-acetic acid on the growth of *Sphaerocarpos texanus, Bryologist,* 74, 18, 1971.
27. **Mehra, P. N. and Pental, D.,** Induction of apospory, callus, and correlated morphogenetic studies in *Athalamia pusilla* Kash., *J. Hattori Bot. Lab.*, 40, 151, 1976.
28. **Dhingra-Babbar, S. and Chopra, R. N.,** Studies on callus induction, its growth and differentiation in *Marchantia palmata* Nees. II. Effect of some growth regulators, *Beitr. Biol. Pflanz.*, 61, 467, 1986.
29. **Chopra, R. N. and Dhingra-Babbar, S.,** Studies on callus induction, its growth and differentiation in *Marchantia palmata* Nees. I. Effect of some amino acids, complex organic substances, and activated charcoal, *J. Hattori Bot. Lab.*, 50, 193, 1986.
30. **Chopra, R. N. and Kumra, S.,** Studies on callus induction, its growth and differentiation in the liverwort *Asterella wallichiana* (Lehm. & Lindenb.) Grolle, *Cryptogamie Bryol. Lichenol.*, 7, 249, 1986.
31. **Kumra, P. K. and Chopra, R. N.,** Effect of some growth substances, vitamins, and ultraviolet radiation on callus induction in the moss *Bryum coronatum* Schwaegr., *Z. Pflanzenphysiol.*, 108, 143, 1982.
32. **Gautheret, R. J.,** Factors affecting differentiation of plant tissue grown *in vitro,* in *Cell Differentiation and Morphogenesis,* Beermann, W., Ed., Elsevier/North-Holland, Amsterdam, 1966, 55.
33. **Takio, S., Kajita, M., Takami, S., and Hino, S.,** Establishment and growth characterization of suspension cultures of cells from *Barbula unguiculata, J. Hattori Bot. Lab.*, 60, 407, 1986.
34. **Kajita, M., Takio, S., Takami, S., and Hino, S.,** Establishment and growth characterization of suspension culture of cells from the moss, *Sphagnum imbricatum, Physiol. Plant.*, 70, 21, 1987.
35. **Ohta, Y., Katoh, K., and Miyake, K.,** Establishment and growth characteristics of a cell suspension culture of *Marchantia polymorpha* L. with high chlorophyll content, *Planta,* 136, 229, 1977.
36. **Katoh, K.,** Photosynthesis and photoassimilation of glucose during photomixotrophic growth of *Marchantia polymorpha* cells in suspension culture, *Physiol. Plant.*, 57, 67, 1983.
37. **Takami, S. and Takio, S.,** Establishment of cell suspension culture of *Hedwigia ciliata* and *Heteroscyphus besherellei, J. Plant Physiol.*, 130, 267, 1987.
38. **Takio, S., Takami, S., and Hino, S.,** Photosynthetic ability of dark-grown *Marchantia paleacea* cells in suspension culture, *J. Plant Physiol.*, 132, 195, 1988.
39. **Katoh, K., Ohta, Y., Hirose, Y., and Iwamura, T.,** Photoautotrophic growth of *Marchantia polymorpha* L. cells in suspension culture, *Planta,* 144, 509, 1979.
40. **Katoh, K.,** Kinetics of photoautotrophic growth of *Marchantia polymorpha* cells in suspension culture, *Physiol. Plant.*, 59, 242, 1983.
41. **Barz, W. H. and Husemann, W.,** Aspects of photoautotrophic cell suspension cultures, in *Proc. Int. Congr. Plant Tissue Culture,* Fujiwara, A., Ed., Maruzen, Tokyo, 1982, 245.
42. **Peel, E.,** Photoautotrophic growth of suspension cultures of *Asparagus officinalis* L. cells in turbidostats, *Plant Sci. Lett.*, 24, 147, 1982.
43. **Horn, M. E., Sherrard, J. H., and Widholm, J. M.,** Photoautotrophic growth of soybean cells in suspension culture. I. Establishment of photoautotrophic cultures, *Plant Physiol.*, 72, 426, 1983.

44. **Rogers, S. M. D., Ogren, W. L., and Widholm, J. M.,** Photosynthetic characteristics of a photoautotrophic cell suspension culture of soybean, *Plant Physiol.*, 84, 1451, 1987.
45. **Katoh, K., Ishikawa, M., Miyake, K., Ohta, Y., Hirose, Y., and Iwamura, T.,** Nutrient utilization and requirement under photoheterotrophic growth of *Marchantia polymorpha:* improvement of the culture medium, *Physiol. Plant.*, 49, 241, 1980.
46. **Takio, S., Takami, S., and Hino, S.,** Nitrate reductase from thalli of *Marchantia polymorpha, J. Hattori Bot. Lab.*, 58, 131, 1985.
47. **Takio, S. and Hino, S.,** Nitrate reductase from suspension cultured cells of *Marchantia polymorpha* L., *J. Plant Physiol.*, 132, 470, 1988.
48. **Takio, S.,** Coenzyme requirements of nitrate reductase in extracts from suspension cultured cells of four bryophytes species, *J. Hattori Bot. Lab.*, 62, 269, 1987.
49. **Takeda, R. and Ohta, Y.,** Induction of and secondary metabolites production in cultured cells of *Jungermannia subulata,* in Abstr. 6th Symp. Plant Tissue Culture, Sapporo, Japan, July 18 and 19, 1978, 35.
50. **Takeda, R. and Katoh, K.,** Sesquiterpenoids in cultured cells of liverwort, *Calypogeia granulata* Inoue, *Bull. Chem. Soc. Jpn.*, 56, 1265, 1983.
51. **Meuch, D. and Huneck, S.,** Inhaltstoffe der Moose. II. Azulene aus *Calypogeia trichomanis* (L.) Corda, *Chem. Ber.*, 99, 2669, 1966.
52. **Tomita, Y., Uomori, A., and Minato, H.,** Sesquiterpenes and phytosterols in the tissue cultures of *Lindera strychnifolia, Phytochemistry,* 8, 2249, 1969.
53. **Witte, L., Berlin, J., Wray, V., Schubert, W., Kohl, W., Hofle, G., and Hammer, J.,** Mono- and diterpenes from cell cultures of *Thuja occidentalis, Planta Med.*, 49, 216, 1983.
54. **Banthorpe, D. V. and Wirz-Justice, A.,** Terpene biosynthesis. VI. Monoterpenes and carotenoids from tissue cultures of *Tanacetum vulgare* L., *J. Chem. Soc. Perkin Trans. 1,* p. 1769, 1972.
55. **Banthorpe, D. V., Branch, S. A., Njar, V. C. O., Osborne, M. G., and Watson, G.,** Ability of plant callus cultures to synthesize and accumulate lower terpenoids, *Phytochemistry,* 25, 629, 1986.
56. **Banthorpe, D. V. and Barrow, S. E.,** Monoterpene biosynthesis in extracts from cultures of *Rosa damascena, Phytochemistry,* 22, 2727, 1983.
57. **Banthorpe, D. V., Grey, T. J., Poots, I., and Fordham, W. D.,** Monoterpene metabolism in cultures of *Rosa* species, *Phytochemistry,* 25, 2321, 1986.
58. **Hartmann, E., Beutelmann, P., Vandekerkhove, O., Euler, R., and Kohn, G.,** . Moss cell cultures as sources of arachidonic and eicosapentaenoic acids, *FEBS Lett.*, 198, 51, 1986.
59. **Zinsmeister, H. D. and Mues, R.,** Moose als Resevoir bemerkenswerter sekundarer Inhaltsstoffe, *GIT Fachz. Lab.*, 31, 499, 1987.
60. **Pryce, R. J.,** The occurrence of lunularic acid and abscisic acid in plants, *Phytochemistry,* 11, 1759, 1972.
61. **Gorham, J.,** Lunularic acid and related compounds in liverworts, algae and *Hydrangea, Phytochemistry,* 16, 249, 1977.
62. **Valio, I. F. M., Burdon, R. S., and Schwabe, W. W.,** New natural growth inhibitor in the liverwort *Lunularia cruciata* (L.) Dum., *Nature (London),* 223, 1176, 1969.
63. **Valio, I. F. M. and Schwabe, W. W.,** Growth and dormancy in *Lunularia cruciata* (L.) Dum., *J. Exp. Bot.*, 21, 138, 1970.
64. **Abe, S. and Ohta, Y.,** Lunularic acid in cell suspension cultures of *Marchantia polymorpha, Phytochemistry,* 22, 1917, 1983.
65. **Abe, S. and Ohta, Y.,** The concentration of lunularic acid and prelunularic acid in liverworts, *Phytochemistry,* 23, 1379, 1984.
66. **Ohta, Y., Abe, S., Komura, H., and Kobayashi, M.,** Prelunularic acid, a probable immediate precursor of lunularic acid. First example of a "prearomatic" intermediate in the phenylpropanoid-polymalonate pathway, *J. Am. Chem. Soc.*, 105, 4480, 1983.
67. **Ohta, Y., Abe, S., Komura, H., and Kobayashi, M.,** Prelunularic acid, a probable immediate precursor of lunularic acid, in suspension-cultured cells of *Marchantia polymorpha, J. Hattori Bot. Lab.*, 56, 249, 1984.
68. **Ohta, Y., Abe, S., Komura, H., and Kobayashi, M.,** Prelunularic acid in liverworts, *Phytochemistry,* 23, 1607, 1984.
69. **Imoto, S. A. and Ohta, Y.,** Intracellular localization of lunularic acid and prelunularic acid in suspension cultured cells of *Marchantia polymorpha, Plant Physiol.*, 79, 751, 1985.
70. **Milborrow, B. V. and Noddle, R. C.,** Conversion of 5-(1,2-epoxy-2,6,6-trimethylcyclohexyl)-3-methylpenta-*cis*-2-*trans*-4-dienoic acid into abscisic acid in plants, *Biochem. J.*, 119, 727, 1970.
71. **Gorham, J.,** Metabolism of lunularic acid in liverworts, *Phytochemistry,* 16, 915, 1977.
72. **Meuche, D. and Huneck, S.,** Strukturaufklarung und Synthese des 3,7-Dimethyl-5-methoxycarbonylindens, eine neuen Inhalsstoffes aus *Calypogeia trichomanis* (L.) Corda, *Chem. Ber.*, 102, 2493, 1969.
73. **Takeda, R. and Katoh, K.,** 3,10-dihydro-1,4-dimethylazule, a labile biosynthetic intermediate isolated from cultured cells of liverwort *Calypogeia granulata* Inoue, *J. Am. Chem. Soc.*, 105, 4056, 1983.

74. **Watson, D. G., Brooks, C. J. W., and Freer, I. M.,** The elicitation of phytoalexins in cultures of *Capsicum annuum* and *Nicotiana tabacum,* in *Secondary Metabolism in Plant Cell Cultures,* Morris, P., Scragg, A., Stafford, A., and Fowler, M. W., Eds., Cambridge University Press, London, 1986, 128.
75. **Ohta, Y.,** unpublished data.

Chapter 12

MORPHOREGULATORY ROLE OF HYDROXYPROLINE-CONTAINING PROTEINS IN LIVERWORTS

Dominick V. Basile

TABLE OF CONTENTS

I. INTRODUCTION

Since the first report that hydroxyproline (Hyp) was present in hydrolysates of proteins from alfalfa leaves and potato tubers,[1] a steady increase in interest in the biosynthesis and physiological roles of proteins containing Hyp (Hyp-proteins) in plants has occurred. The earliest literature on these topics was reviewed by Lamport.[2] There have been a number of recent reviews by several others.[3-5] In all this research, that which implicates L-hydroxyproline (Hyp) and Hyp-protein in liverwort morphogenesis has been conducted by the author of this review and his collaborators. This review, therefore, is essentially a personal account of how hypotheses regarding the role of Hyp/Hyp-proteins in leafy liverwort morphogenesis were proposed and refined during the course of our investigations.

II. THE INFLUENCE OF EXOGENOUS ("FREE") L-HYDROXYPROLINE ON LEAFY LIVERWORT MORPHOGENESIS

The evidence that first prompted an investigation of a morphoregulatory role for Hyp in leafy liverworts actually came from a study on moss nutrition. Burkholder[6] reported that if Hyp was supplied to axenic cultures of protonemata of *Atrichum undulatum* (L.) Beauvois, diffuse growth, characteristic of the protonemal stage, was prolonged and the onset of localized, apical growth, characteristic of the leafy gametophore stage, was delayed. An experiment to find out if Hyp had the same influence on the morphogenesis of a leafy liverwort, *Scapania nemorosa* (L.) Dum., demonstrated that it did.[7] That is, it was found that by adding Hyp to "protonemal" cultures of *S. nemorosa* there resulted a prolonged period of diffuse growth and development and a delayed onset of localized, apical development. It should be pointed out that the protonemata of *S. nemorosa* are not filamentous, but globose-cylindrical cell masses (Figure 1). The finding that exogenous Hyp could influence this critical morphogenetic change during the course of ontogeny would suffice in itself to stimulate interest in its mode of action. However, the morphoregulatory influence of exogenous Hyp was not restricted to the "protonemal" stage. When Hyp was supplied to plants that already had entered the apical growth stage of ontogeny (Figure 1), extensive changes also occurred in the patterns of leaf and branch morphogenesis[7] (Figures 2 to 4).

That so many phenotypic changes could result solely from adding a single amino acid to the culture medium was difficult to believe or understand. Because the experiment was controlled and repeatable the conclusion that the altered morphogenetic patterns resulted from adding Hyp seemed unavoidable, however. Therefore, the question to be answered became "how" can exogenous Hyp bring about all the observed morphogenetic changes.

The original study with *S. nemorosa* showed that the changed patterns were temporary and only manifested while concentrations of 7 to 1 μg Hyp per milliliter were detectable in the culture medium.[7] When the detectable concentration of Hyp fell below 1 μg/ml the plants reverted to the morphogenetic patterns characteristic of *S. nemorosa*. This indicated that the exogenous Hyp was being taken up and metabolized. However, free Hyp was not (and still is not) known to be a common constituent/metabolite in plants. In fact, Steward et al.[8] had suggested earlier that free Hyp might act as a competitive inhibitor of proline synthesis and incorporation into protein. This suggestion was based on the capacity of L-proline (Pro) to reverse Hyp-induced growth inhibition. Since there were other reports in the literature that growth-inhibitory effects of exogenous Hyp were reversed by exogenous Pro,[9-11] experiments to find out if Pro could also reverse the morphoregulatory influences of Hyp in *S. nemorosa* were performed.[12] As suspected, exogenous Pro, provided in equimolar concentrations with Hyp, reversed its morphoregulatory effects — temporarily. In cultures provided with Pro at four to ten times that of Hyp the morphoregulatory influence

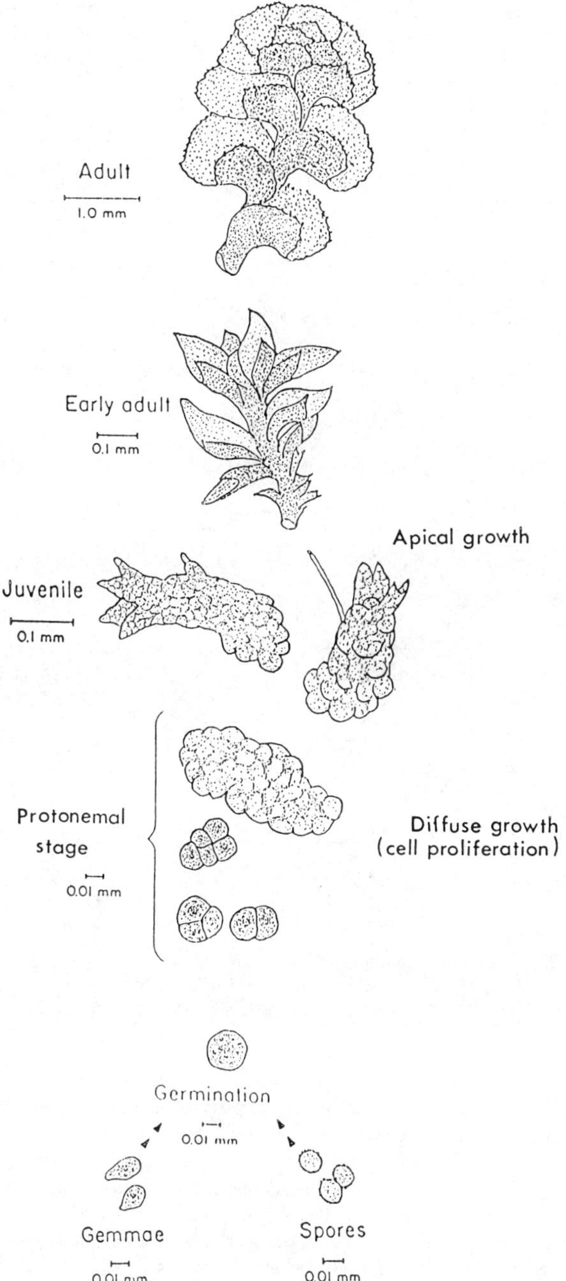

FIGURE 1. Ontogenetic stages characteristic of the vegetative development of leafy liverworts (Jungermanniales), exemplified by *Scapania nemorosa* (L.) Dum. The different stages are characterized by marked changes in the pattern of cell proliferation. The protonemal stage is distinguished by its diffuse pattern of cell proliferation throughout. In all other stages, axial growth and leaf initiation result from apically localized cell proliferation. Changes in leaf shape that occur from the juvenile to adult stages are due to changes in patterns of cell proliferation in localized populations of cells that comprise the developing leaf.

FIGURE 2. Scanning electron micrographs of *Scapania nemorosa* gametophytes exhibiting normally suppressed and experimentally "desuppressed" ventral leaf development. Bars on leaves correspond to ca. 0.2 mm. (A) Normal gametophyte, dorsal view showing two rows of complicate, bilobed lateral leaves. (B) Normal gametophyte, ventral view showing absence of ventral leaves. (C) Phenovariant gametophyte, lateroventral view showing ventral leaves (v) that were experimentally "desuppressed" by adding hydroxyproline to the culture medium after plants reached a juvenile, apically developing stage. (D) Phenovariant gametophyte, laterodorsal view showing experimentally induced branching. "Desuppressed" ventral leaf and branch development was experimentally induced by adding 2,2′-dipyridyl to the culture medium. (A, B, and C from Basile, D. V., *Bull. Torrey Bot. Club,* 107, 325, 1980. With permission.)

of Hyp was not observed at all. These results coupled with the fact that Hyp bears a close structural analogy to Pro prompted a further test of the hypothesis of Steward et al.[8] that exogenous Hyp is " . . . effective at a site where L-proline is synthesized and incorporated into protein . . . ".

Before proceeding, it would be helpful to put this phase of the research in proper historical perspective. Steward and collaborators[13-17] already had done much to bring attention first to the existence of Pro-Hyp-containing proteins in plants and then to their possible role in growth and morphogenesis. Added interest resulted from the discoveries by Dougall and Shimbayashi[18] and Lamport and Northcote[19] that some Hyp-proteins became covalently

FIGURE 3. Drawings to show the range of variation in leaf shape resulting from the addition of L-hydroxyproline or 2,2′-dipyridyl to *Scapania nemorosa* gametophytes in axenic culture. The lateral leaves labeled A are characteristic for *S. nemorosa.* Most of the leaves illustrated are atypical of the genus *(Scapania)* and family (Scapaniaceae), but phenocopy those of species belonging to other families in the Jungermanniales. (From Basile, D. V., *J. Hattori Bot. Lab.,* 38, 91, 1974. With permission.)

bound to plant cell walls. Lamport[20] speculated that wall-bound Hyp-protein helped regulate extension growth of plant cells. By 1965 extensive literature on Hyp-protein had accumulated and was reviewed.[2]

None of the pioneering research on the presence and possible morphoregulatory role of Hyp-proteins was conducted with bryophytes. None of this research, because it was conducted with excised cells and tissues of flowering plants, could be directed toward elucidating the possible morphoregulatory role of Hyp or Hyp-proteins with regard to leaf and branch initiation in any group of plants. Nevertheless, it was from the published reports of this research that we gained insight as to how Hyp and Hyp-proteins might be influencing leafy liverwort morphogenesis.

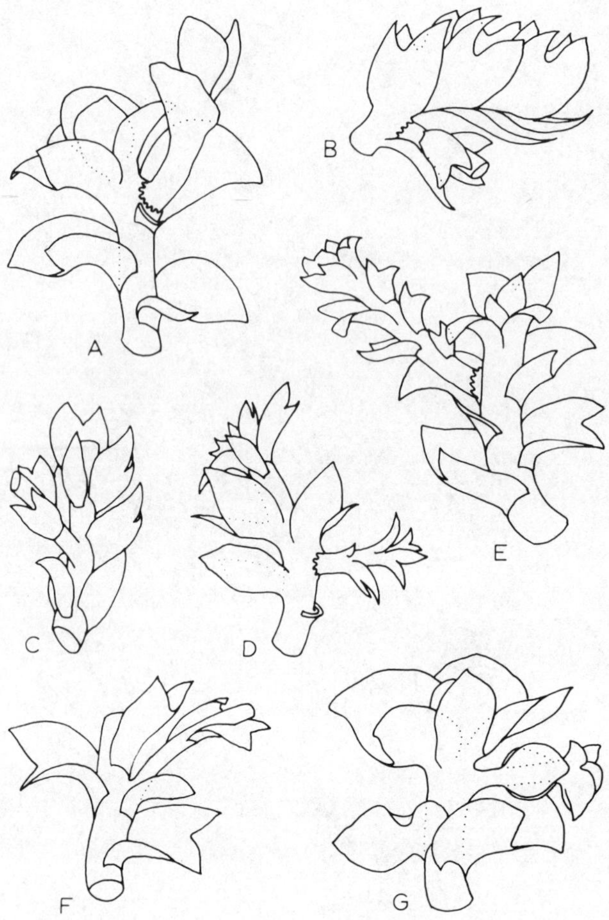

FIGURE 4. Semidiagrammatic representations of branching types exhibited by *Scapania nemorosa* (L.) Dum. gametophytes grown on hydroxyproline- or 2,2′-dipyridyl-containing culture medium. (A) *Anomoclada*-type; (B) *Bazzania*-type; (C) *Acromastigium*-type except the branch seems to replace the whole rather than one half of the ventral leaf; (D) *Bryopteris*-type; (E) *Plagiochila*-type; (F) *Frullania*-type; (G) *Radula*-type. Only the *Plagiochila*-type (E) is characteristic of *S. nemorosa*. (From Basile, D. V., *J. Hattori Bot. Lab.*, 38, 91, 1974. With permission.)

III. THE IMPLICATION OF HYDROXYPROLINE-CONTAINING PROTEINS IN REGULATING LEAFY LIVERWORT MORPHOGENESIS

Because protein-bound Hyp was only known to be synthesized from Pro after it was incorporated into peptides/proteins,[21] the suggestion by Steward et al.[8] and others[22-24] that Hyp interfered with one or more steps in the synthesis of protein containing Pro meant it also interfered with the synthesis of proteins containing both Pro and Hyp. To test a hypothesis that Hyp had its morphoregulatory effect by virtue of modifying the synthesis of proteins that contained *both* Hyp and Pro (i.e., not only containing Pro), an inhibitor of the post-translational conversion of Pro to Hyp was needed. Fortunately, one was available. Holleman[24] had recently reported that 2,2′-dipyridyl could completely block the conversion of peptide-bound Pro to peptide-bound Hyp without blocking Pro uptake and incorporation. The prolyl hydroxylase enzyme that oxidizes peptide-bound Pro to form peptide-bound Hyp requires ferrous iron as one of its essential cofactors. Dipyridyl binds ferrous iron, making it unavailable for the reaction.

When the morphoregulatory effects of Hyp and 2,2′-dipyridyl were compared, they were found to be the same.[25] That is, both antagonists induced the development of ventral leaves and a wide range of branching patterns (Figures 3 and 4). Because the only biosynthetic pathway that *both* Hyp and 2,2′-dipyridyl were shown capable of modifying was the synthesis of proteins containing both Pro and Hyp, the results of experiments with *S. nemorosa* were interpreted to mean that a protein containing both Pro *and* Hyp was probably involved in regulating morphogenesis in this plant.

A rigorous test of this hypothesis would have been to isolate and compare Hyp-proteins from control and experimentally modified plants of *S. nemorosa*. The difficulty being experienced at the time by workers who were attempting to isolate intact Hyp-proteins, even from relatively large starting quantities of tissue derived from flowering plants,[2,26] made an attempt with the limited material available from *S. nemorosa* seen impractical. Instead, a series of experiments were performed to find out if the two potential inhibitors of normal Hyp-protein synthesis had the same morphogenetic effects of other taxa.

The criteria for selecting taxa for these experiments were three. First, each species had to represent a different family of leafy liverworts. Second, each species had to manifest a one half phyllotaxy (i.e., fail to develop leaves from ventral merophytes). Third, each species had to be limited as to the type and frequency of branching. The taxa that met these criteria and with which experiments were conducted were: *Nowellia curvifolia* (Dicks.) Mitt. (Cephaloziaceae),[27] *Jungermannia leiantha* Grolle (= *J. lanceolata* L.) (Jungermanniaceae),[28] *Plagiochila arctica* Bryhn and Kaal. (Plagiochilaceae),[29] and *Gymnocolea inflata* (Huds.) Dum. (Lophoziaceae).[30] Each of these taxa responded to antagonists of normal Hyp-protein synthesis in the same three ways. Three rows of leaves developed instead of only two rows. Branching was more frequent and varied (Figures 5 to 8). The morphoregulatory influences of exogenous Hyp and 2,2′-dipyridyl were transitory, persisting only as long as their concentration in the culture medium remained above some minimal effective level. That level for Hyp was approximately 0.1 μg/ml. The minimal effective level for 2,2′-dipyridyl was not determined.

For reasons still not understood, two of these five species, *P. arctica* and *G. inflata,* could be maintained in an altered pattern of morphogenesis for more extended periods than the other three. Because of this it became possible to grow and accumulate enough material to perform chemical analysis. We therefore undertook to compare the relative amounts of Hyp in cell-wall-bound protein of plants with "suppressed" leaf and branch development (i.e., normal form) with those of plants with experimentally "desuppressed" leaf and branch development. The results of the comparative chemical analyses of both *G. inflata* and *P. arctica* are summarized in Figure 9. These data showed that experimental procedures that significantly alter the Hyp content of cell-wall-associated protein and the experimental procedures that significantly alter the pattern of cell division and organogenesis at the shoot apex are the same.

The results of the physiological experiments and chemical analyses were interpreted to indicate that Hyp-protein is important in regulating leaf and branch development in at least five families of leafy liverworts and probably all leafy liverworts. In the absence of any reason to reject this interpretation, it served as the basis of our revised working hypothesis: certain Hyp-proteins play a pivotal role in leafy liverwort morphogenesis.

FIGURE 5. Scanning electron micrographs of *Nowellia curvifolia* gametophytes exhibiting normally suppressed and experimentally ''desuppressed'' ventral leaf development. Bars on leaves correspond to ca. 0.1 mm. (A) Normal gametophyte, dorsal view, showing two rows of lateral leaves, i.e., one half phyllotaxy. (B) Normal gametophyte, ventral view, showing absence of ventral leaves. Rhizoids are developing from the only two files of superficial cells deriving from the ventral merophytes. The files of cells on either side derive from lateral merophytes. (C) Phenovariant gametophyte, ventral view, showing ventral leaves (v) that were experimentally ''desuppressed'' by adding hydroxyproline to axenic cultures in which plants had reached a juvenile, apically developing stage. (D) Phenovariant gametophyte, lateroventral view showing ventral leaves that were experimentally ''desuppressed'' by adding 2,2′-dipyridyl to cultures in which plants had reached a juvenile, apically developing stage. (A, B, and C from Basile, D. V., *Bull. Torrey Bot. Club,* 107, 325, 1980. With permission.)

IV. THE PROBABLE MODE OF ACTION OF HYDROXYPROLINE-CONTAINING PROTEINS IN LEAFY LIVERWORT MORPHOGENESIS

The next question addressed was what could explain the observation that by impairing the structure-function of this morphoregulatory Hyp-protein there resulted such apparently diverse developmental changes as a prolonged diffuse growth of protonemata,[7] a continued

FIGURE 6. Scanning electron micrographs of *Jungermannia leiantha* (= *J. lanceolata*) gametophytes exhibiting normally suppressed and experimentally "desuppressed" ventral leaf development. Bars on leaves correspond to ca. 0.25 mm. (A) Normal plant, dorsal view, showing two rows of lateral leaves, i.e., one half phyllotaxy. (B) Normal plant, ventral view. Slime papillae have developed from ventral merophytes. (C) Phenovariant gametophyte, lateral view, showing one third phyllotaxy and increased branching (br) due to experimentally induced "desuppressed" development. Ventral leaves (V) are isomorphic with lateral leaves. (D) Stem section showing a sequence of four leaves produced by an experimentally "desuppressed" plant. Note the relative size and shapes of lateral and ventral (v) leaves. Both hydroxyproline and 2,2′-dipyridyl had the same influence on ventral leaf development. (A and B from Basile, D. V., *Bull. Torrey Bot. Club,* 107, 325, 1980. With permission.)

growth of otherwise suppressed ventral leaf primordia,[7,27-30] and increased branching.[7,27-30] Our reasoning was, and remains, that if the impaired function of a Hyp-protein is to permit continued growth, then the unimpaired function is to suppress further growth. When viewed in this way, the morphoregulation of such diverse aspects of morphogenesis as protonemal, leaf, and branch development becomes comprehensible. The Hyp-proteins we are investigating do not specifically function to regulate organogenesis. Rather, they have a more general and basic function. They appear to be part of a correlative control *system* that acts to stop further growth of any part of a developing plant at times and places characteristic for a species.

Following this line of reasoning, the hypothesis regarding the role of Hyp-protein in leafy liverwort morphogenesis was refined. Instead of focusing on its influence on leaf and branch development (i.e., organogenesis) the hypothesis focuses on a more basic process,

FIGURE 7. Scanning electron micrographs of *Plagiochila arctica* gametophytes exhibiting normally suppressed and experimentally ''desuppressed'' ventral leaf development. Bars on leaves correspond to ca. 0.5 mm. (A) Leafy shoot, dorsal view, showing normal form, i.e., two rows of lateral leaves with rhizoids, the only ventral appendage. (B) Leafy shoot, ventral view, to show the absence of leaves produced from ventral merophytes. (C) Leafy shoot showing ''desuppressed'' leaf development resulting from exogenously suppled hydroxyproline. Leaves from ventral merophytes are indistinguishable from those developing from lateral merophytes. (D) Portion of leafy shoot showing both ''desuppressed'' leaf and branch development resulting from exogenously supplied hydroxyproline. The branch apices resemble those of normal plants (vide A and B) indicating that the supply of exogenous Hyp in the medium was almost depleted. (A, B, and D from Basile, D. V. and Basile, M. R., *J. Hattori Bot. Lab.*, 53, 221, 1982. With permission.)

cell proliferation.[31,32] The question that arises from refining the hypothesis so as to tentatively explain the more immediate regulatory function of the Hyp-protein of interest is, what factors regulate when and where the Hyp-protein functions to stop further cell proliferation? For example, if Hyp-protein functions to stop cell proliferation at the end of some predictable period of protonemal development, then what determines *when* this period should end? Likewise, if Hyp-protein functions to stop cell proliferation in a leaf after it is initiated, then what determines when this development should end? In seeking answers to these questions, a new phase of our research was begun.

FIGURE 8. Scanning electron micrographs of *Gymnocolea inflata* gametophytes exhibiting normally suppressed and experimentally "desuppressed" ventral leaf development. Bars on leaves correspond to ca. 0.1 mm. (A) Leafy shoot, dorsal view, showing typical form, i.e., two rows of lateral leaves and rhizoids, the only ventral appendage. (B) Portion of normal plant, ventral view, showing absence of ventral leaves. Occasional papillae (P) may develop from ventral merophytes. (C) Portion of leafy shoot showing "desuppressed" leaf development due to presence of ammonium ion in the culture medium. Ammonium ion apparently interferes with the posttranslational conversion of proline to hydroxyproline in this species (see Figure 9). The development of two ventral leaves (v) at the same level at the shoot apex suggests that this shoot may have branched dichotomously. Dichotomous branching, the most primitive type in all land plants, is not typical of *Gymnocolea* or any of the Jungermanniales. (D) A branch pair from an experimentally "desuppressed" gametophyte. The left branch in dorsal aspect shows changed size, shape, and spacing of leaves. The right branch in ventral aspect shows ventral leaves that are not isomorphic with the lateral leaves. This phenovariant resulted from adding azetidine-2-carboxylic acid to axenic cultures of *G. inflata*. Azetidine, like hydroxyproline, is a structural analogue of proline and is presumed to act in the same way as hydroxyproline. (A from Basile, D. V. and Basile, M. R., *J. Hattori Bot. Lab.*, 53, 221, 1982. With permission.)

V. THE INTERACTION OF AUXIN, ETHYLENE, AND HYDROXYPROLINE-CONTAINING PROTEINS IN LEAFY LIVERWORT MORPHOGENESIS

Again, it would be helpful to put this phase of our research into historical perspective. By the time the series of experiments showing that a variety of antagonists of Hyp-protein synthesis had the same morphogenetic effect on representatives of five distinct families of leafy liverworts were completed,[25,27-30] several reports that ethylene could trigger the synthesis and/or deposition of Hyp-protein into plant cell walls had been published.[33-36] These reports suggested the possibility that *when* Hyp-proteins functioned to suppress cell proliferation

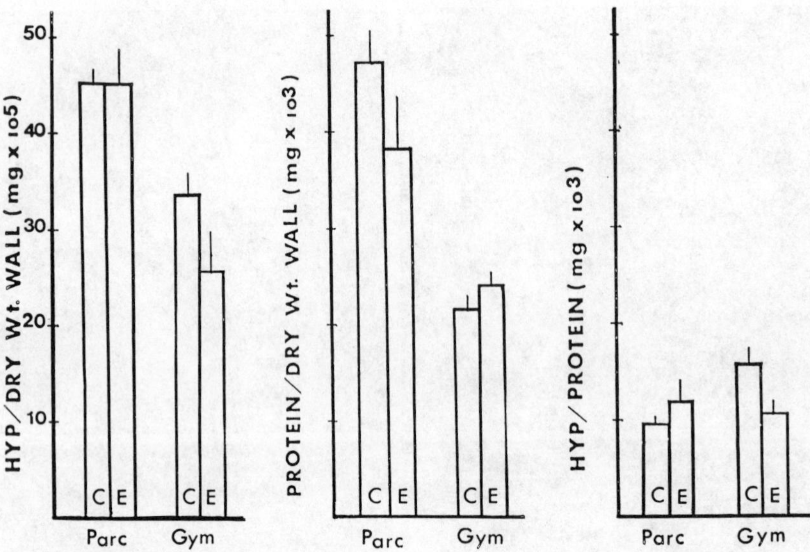

FIGURE 9. A comparison of peptide-bound hydroxyproline and of protein associated with the cell walls of *Plagiochila arctic* Bryhn & Kaal. (Parc) and *Gymnocolea inflata* (Huds.) Dum. (Gym). Mean values given for control (C) and experimentally "desuppressed" plants (E) are based on eight measurements. Lines extending from bars represent the standard deviation. The most biochemically important and statistically significant differences found were in the ratios of hydroxyproline to protein. In *P. arctica,* where direct incorporation of Hyp into Hyp-protein is correlated with "desuppressed" leaf development, there is significantly *more* ($p \geqslant 0.05$) Hyp/protein in the cell walls of experimentally "desuppressed" plants. In *G. inflata,* where ammonium ion is presumed to interfere with the conversion of Pro to Hyp by diverting the α-ketoglutarate required for the reaction, there is significantly *less* ($p \geqslant 0.01$) Hyp/protein. Apparently any significant alteration in the composition of the morphoregulatory Hyp-protein impairs its normal function.

might be related to *when* ethylene-mediated Hyp-protein synthesis and/or its deposition in walls of proliferating cells occurs.

To test this idea, antagonists of ethylene synthesis[37,38] and ethylene action[39] were used in experiments to find out if interfering with the synthesis or action of ethylene would concomitantly interfere with the morphoregulatory function of Hyp-protein (i.e., suppression). The antagonists of ethylene synthesis used were α-aminooxyacetic acid (AOA) and aminoethoxyvinylglycine (AVG). The antagonist of ethylene action used was silver nitrate. The results of these experiments supported our idea.[40] All three ethylene antagonists prevented the suppressed cell proliferation of specific leaf primordia (Figure 10). It appeared, then, that ethylene and Hyp-protein somehow act in an interrelated way to regulate cell proliferation. We assumed that we had identified molecules that functioned importantly to regulate when and how proliferation of cells is suppressed. Of our three questions, what regulatory factors determine how, when, and where cell proliferation is suppressed, one remained — what determines *where* suppressed cell proliferation occurs? Once ethylene was implicated, a possible candidate for regulating where ethylene synthesis might be induced quickly came to mind. Following the first report that native auxin, indole-3-acetic acid (IAA), " . . . accelerates the release of ethylene",[41] it had become well established that growth-suppressing levels of ethylene are often triggered by auxin.[42,43] Moreover, the transport of auxin is strongly directional and highly regulated in plants.[44] In short, it seemed to us that the phytohormone triggering ethylene-mediated suppression of leaf primordia in leafy liverworts might be the same one that triggers ethylene-mediated suppression of localized growth in dominance and tropic phenomena.[44,45]

FIGURE 10. Scanning electron micrographs showing examples of "desuppressed" ventral leaf development in *P. arctica* induced by three different ethylene antagonists. Examples of *P. arctica* exhibiting normal leaf development are given in Figure 7. (A) Apical portion of an $AgNO_3$-induced phenovariant showing a "three-sided" apical cell that has cut off three files of merophytes, each giving rise to isomorphic leaves. (B) $AgNO_3$-induced phenovariant exhibiting both "desuppressed" leaf and branch (br) development. Portions of leaves from the main stem and the branch were removed to show that the branch arose exogenously from a ventral merophyte. This mode of branching is not typical for *Plagiochila*. (C) AOA-induced phenovariant exhibiting one third phyllotaxy. (D) AVG-induced phenovariant that has developed three rows of isomorphic leaves. Phenovariations induced by ethylene antagonists are indistinguishable from those induced by antagonists of Hyp-protein synthesis. (A, B, and D from Basile, D. V. and Basile, M. R., *J. Hattori Bot. Lab.*, 55, 173, 1984. With permission.)

To test the hypothesis that the transport and action of auxin might be determining where growth inhibiting levels of ethylene were produced, experiments using antagonists of auxin transport, triiodobenzoic acid (TIBA) and *N*-1-naphthylphthalamic acid (NPA),[46,47] and of auxin action, *p*-chlorophenoxyisobutyric acid (PCIB),[48] were performed. The results of these experiments supported our hypothesis. Both inhibitors of auxin transport and the inhibitor of auxin action had the same effect — they prevented the suppression of cell proliferation in specific leaf primordia (Figure 11).[49]

FIGURE 11. Scanning electron micrographs showing examples of "desuppressed" ventral leaf development in *P. arctica* induced by three different auxin antagonists. Examples of *P. arctica* exhibiting normal leaf development are given in Figure 7. (A) PCIB-induced phenovariant, polar view, showing apical cell (A) and merophytes (M) at different stages of leaf development. Leaves develop equally from all three files, and there is no way to distinguish lateral from ventral in this view. (B) PCIB-induced phenovariant having leaves (v) in one file somewhat smaller than those in the other two. (C) TIBA-induced phenovariant with three rows of isomorphic leaves. (D) NPA-induced phenovariant with one row of leaves smaller and differently shaped. The phenovariation induced by auxin antagonists falls within the range induced by antagonists of Hyp-protein synthesis. (From Basile, D. V. and Basile, M. R., *J. Hattori Bot. Lab.*, 55, 173, 1984. With permission.)

Our experiments employing inhibitors yielded strong but indirect evidence that auxin, ethylene, and Hyp-protein are important components of a correlative control system that acts to suppress cell proliferation at specific sites. The most conspicuous morphogenetic consequences of this in leafy liverworts are that the development of one of every three leaf primordia initiated is suppressed and the plants manifest a one half instead of a one third phyllotaxy. A less conspicuous consequence of the action of this correlative control system is that the types and frequency of branching are also suppressed by it. Almost invariably, experimental procedures that prevented suppression of leaf development also prevented a suppression or promoted an increase in the types and frequency of branching (Figures 2, 4, 6 to 8, 10 and 12).[49] An investigation into the relationship between "desuppression" of cell proliferation in localized populations of cells at the stem apex and changed branching pattern has been postponed in favor of other lines of investigation.

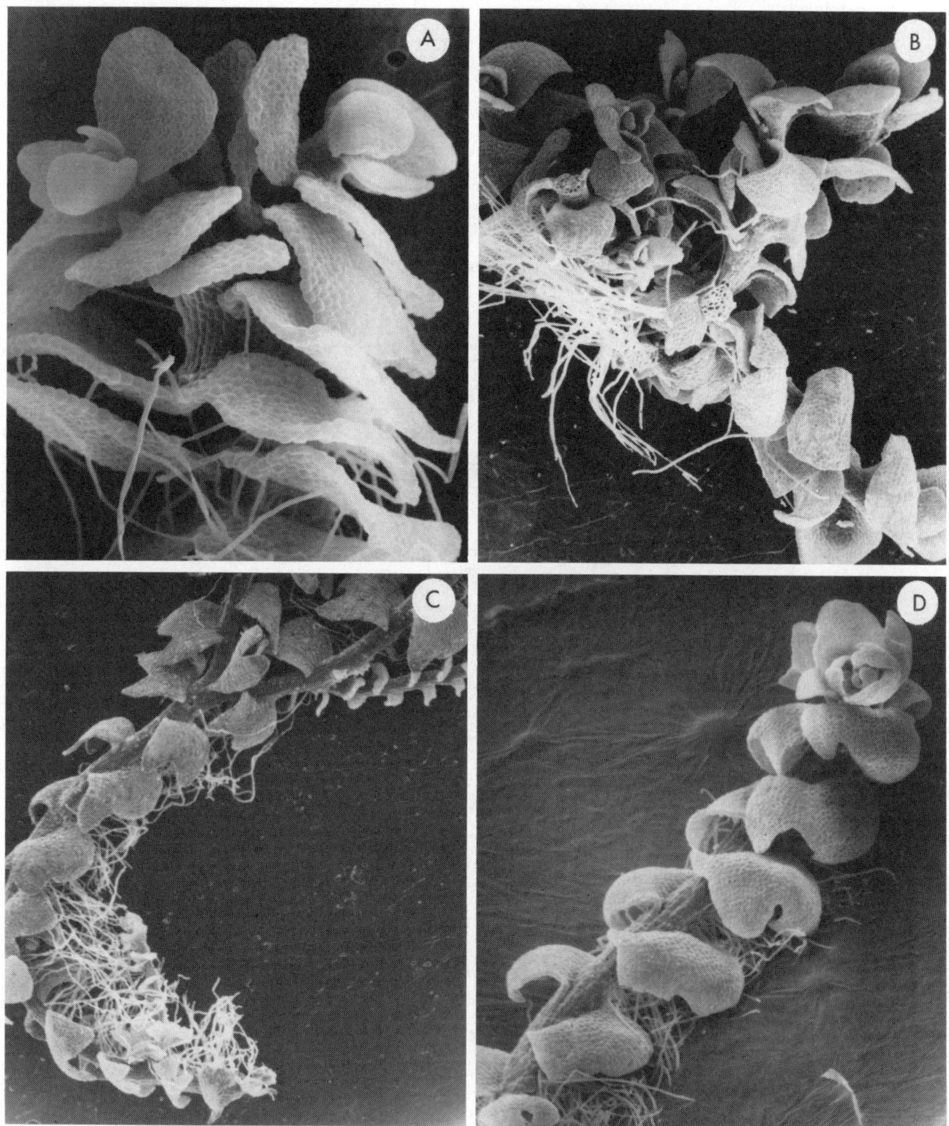

FIGURE 12. Scanning electron micrographs showing examples of experimentally induced "desuppressed" branching. All plants manifest branch types not characteristic for their species. (A) Hydroxyproline-induced phenovariant of *P. arctica* which has produced a terminal pair of branches that have the outward appearance of a dichotomy. Branching is rare in this species when cultured in the absence of antagonists. (B) Phenovariant of *Jungermannia leiantha* (= *J. lanceolata*) exhibiting profuse branching. (C) Azetidine-2-carboxylic acid-induced phenovariant of *Gymnocolea inflata* exhibiting profuse branching. (D) Ammonium ion-induced phenovariant of *G. inflata* showing branch pair having the outward appearance of "true" dichotomous branching. "True" dichotomous branching is not characteristic of *G. inflata*, but may be part of its evolutionary history. (From Basile, D. V. and Basile, M. R., *J. Hattori Bot. Lab.*, 55, 173, 1984. With permission.)

VI. THE DETERMINATIONS THAT INDOLE-3-ACETIC ACID AND ETHYLENE ARE NATURAL PRODUCTS OF LEAFY LIVERWORT GAMETOPHYTES

At the time the experiments with auxin and ethylene antagonists were being conducted there was no rigorously obtained evidence that either auxin or ethylene was produced naturally by leafy liverwort gametophytes. This evidence, which was essential for our hypothesis, was soon forthcoming. Evidence that ethylene is produced by a leafy liverwort, *P. arctica* Bryhn and Kaal., was determined by gas chromatography.[50,51] Evidence that IAA is an endogenously produced growth regulator of *P. arctica* was determined by high-performance liquid chromatography (HPLC) with electrochemical detection and confirmed by gas chromatography-mass spectrometry.[51] In addition, we were able to demonstrate that raised levels of IAA could trigger a statistically significant increase in ethylene production by our test plants.[51] We were also able to demonstrate[57] that the two antagonists of ethylene synthesis (AVG and AOA) which we found effective in preventing suppression of leaf development[40] acted to significantly depress ethylene production by *P. arctica*.

VII. THE SUPPRESSION HYPOTHESIS AND THE CORRELATIVE CONTROL OF PHYLLOTAXY AND BRANCHING PATTERNS

In order to sum up what we thought we had learned about the morphoregulatory role of Hyp-protein vis-à-vis other plant growth regulators, we again refined and expanded our hypothesis to include all components of the correlative control system. We refer to it as the ''suppression hypothesis'', which in its latest version may be expressed as follows:

1. Many of the differences in form by which different taxa of leafy liverworts are distinguished from one another are due to differences with respect to where and when cell proliferation and/or cell enlargement are suppressed in a developing organ or organ primordium.
2. The suppression of cell proliferation in a developing organ or an organ primordium is controlled correlatively in leafy hepatics by the interrelated actions of auxin, ethylene, and an Hyp-containing cell-wall-associated glycoprotein.

Auxin	▶	**Ethylene**	▶	**Hyp-protein**
(directional transport of auxin determines *where* suppression occurs)		(auxin-triggered ethylene synthesis determines *when* suppression occurs)		(ethylene-mediated deposition and binding of Hyp-protein to some unknown component(s) of the cell wall/cell surface determine *whether* suppression occurs)

The suppression hypothesis is, of course, a working hypothesis. Like all of those that preceded it, we have continued to test and, when appropriate, to refine it. Examples of our continued testing are our recent experiments with specific inhibitors of prolyl hydroxylase and its cofactors. The inhibitors of each component of the prolyl hydroxylase system all acted to prevent suppression of leaf development.[29,52,53] All these experiments strongly support the proposed role of Hyp-protein in the correlative control system. There are a number of critical points still to be determined if this hypothesis is to stand. It remains to be demonstrated that IAA is asymmetrically transported and accumulated at levels that trigger growth-inhibitory levels of ethylene at specific sites of suppressed cell proliferation. It also remains to be demonstrated that ethylene mediates increased and/or precocious Hyp-protein synthesis and its deposition in cell walls of leafy liverworts.

VIII. ARABINOGALACTANS AS POSSIBLE MORPHOREGULATORY HYDROXYPROLINE-CONTAINING PROTEINS

The most critical test of our suppression hypothesis, however, can be performed only after we have identified, purified, and characterized the species of Hyp-protein that is morphoregulatory. Fortunately, we have begun to make progress toward identifying some of the Hyp-proteins synthesized by liverworts. We have been able to detect one class of Hyp-proteins, arabinogalactan proteins (AGPs), in the walls of all mature (nonproliferating) cells of nine species of hepatics grown in axenic culture.[54] These nine species represent nine families, four orders, and both subclasses (i.e., Jungermanniidae and Marchantiidae) of the Hepaticae. This widespread occurrence suggests that AGPs may be ubiquitous in the Hepaticae and serve some basic biological function. Our attempts to separate and purify the AGPs extractable from *P. arctica* have enabled us to resolve three AGP fractions by agarose gel electrophoresis[55] and HPLC.[58] Whether any of these or still some other Hyp-protein is the long-sought-after morphoregulator of cell proliferation and organogenesis in liverworts remains to be determined. It appears now, however, that we will be able to test our hypothesized role for Hyp-proteins in liverwort morphogenesis rigorously in the near future.

REFERENCES

1. **Steward, F. C. and Thompson, J. F.,** The nitrogenous constituents of plants with special reference to chromatographic methods, *Annu. Rev. Plant Physiol.,* 1, 233, 1950.
2. **Lamport, D. T. A.,** The protein component of primary cell walls, in *Advances in Botanical Research,* Vol. 2, Preston, R. D., Ed., Academic Press, New York, 1965, 151.
3. **Fincher, G. B., Stone, B. A., and Clarke, A. E.,** Arabinogalactan-proteins: structure, biosynthesis and function, *Annu. Rev. Plant Physiol.,* 34, 47, 1983.
4. **Showalter, A. M. and Varner, J. E.,** Plant hydroxyproline-rich glycoproteins, in *The Biochemistry of Plants: A Comprehensive Treatise,* Vol. 15, Marcus, A., Ed., Academic Press, New York, 1990.
5. **Willson, L. G. and Fry, J. C.,** Extensin — a major cell wall glycoprotein, *Plant Cell Environ.,* 9, 239, 1986.
6. **Burkholder, P. R.,** Organic nutrition of some mosses growing in pure culture, *Bryologist,* 62, 6, 1959.
7. **Basile, D. V.,** The influence of hydroxy-L-proline on ontogeny and morphogenesis of the liverwort, *Scapania nemorosa, Am. J. Bot.,* 54, 977, 1967.
8. **Steward, F. C., Pollard, J. K., Patchett, A. A., and Witkop, B.,** The effects of selected nitrogen compounds on the growth of plant tissue cultures, *Biochim. Biophys. Acta,* 28, 308, 1958.
9. **Robbins, W. J. and McVeigh, I.,** Effects of hydroxyproline on *Trichophyton mentagrophytes* and other fungi, *Am. J. Bot.,* 33, 638, 1946.
10. **Cleland, R.,** Hydroxyproline as an inhibitor of auxin-induced growth, *Nature (London),* 200, 908, 1963.
11. **Norris, W. E., Jr.,** Reversal of hydroxyproline-induced inhibition of elongation of *Avena* coleoptiles, *Plant Physiol.,* 42, 481, 1967.
12. **Basile, D. V.,** The influence of L-proline and hydroxy-L-proline on ontogeny of *Scapania nemorosa, Bull. Torrey Bot. Club,* 95, 127, 1968.
13. **Steward, F. C., Wetmore, R. H., Thompson, J. F., and Nitsch, J. P.,** A quantitative chromatographic study of nitrogenous components of shoot apices, *Am. J. Bot.,* 41, 123, 1954.
14. **Steward, F. C., Bidwell, R. G. S., and Yem, E. W.,** Protein metabolism, respiration, and growth. A synthesis of results from the use of C^{14}-labeled substrates and tissue culture, *Nature (London),* 178, 734, 1956.
15. **Steward, F. C. and Pollard, J. K.,** ^{14}C-Proline and hydroxyproline in the protein metabolism of plants, *Nature (London),* 182, 828, 1958.
16. **Pollard, J. K. and Steward, F. C.,** The use of C^{14}-proline by growing cells: its conversion to protein and to hydroxyproline, *J. Exp. Bot.,* 10, 17, 1959.
17. **Lyndon, R. F. and Steward, F. C.,** The incorporation of ^{14}C-proline into the proteins of growing cells, *J. Exp. Bot.,* 14, 42, 1963.

18. **Dougall, D. K. and Shimbayashi, K.,** Factors affecting growth of tobacco callus tissue and its incorporation of tyrosine, *Plant Physiol.,* 35, 396, 1960.
19. **Lamport, D. T. A. and Northcote, D. H.,** Hydroxyproline in primary walls of higher plants, *Nature (London),* 188, 665, 1960.
20. **Lamport, D. T. A.,** Oxygen fixation into hydroxyproline of plant cell wall protein, *J. Biol. Chem.,* 238, 1438, 1963.
21. **Steward, F. C., Thompson, J. F., and Pollard, J. K.,** Contrasts in the nitrogenous constituents of rapidly growing and nongrowing plant tissues, *J. Exp. Bot.,* 9, 1, 1958.
22. **Cleland, R.,** Possible mechanisms of inhibition by hydroxyproline of auxin-induced growth, *Plant Physiol.,* Vol. 41 (Suppl.), xlvi, 1966.
23. **Norris, W. E., Jr.,** Reversal of hydroxyproline-induced inhibition of elongation of *Avena* coleoptiles, *Plant Physiol.,* 42, 481, 1967.
24. **Holleman, J.,** Direct incorporation of hydroxyproline into protein of sycamore cells incubated at growth inhibitory levels of hydroxyproline, *Proc. Natl. Acad. Sci. U.S.A.,* 57, 50, 1967.
25. **Basile, D. V.,** Inhibition of proline-hydroxyproline metabolism and phenovariation of *Scapania nemorosa, Bull. Torrey Bot. Club,* 96, 577, 1969.
26. **Lamport, D. T. A.,** Cell wall metabolism, *Annu. Rev. Plant Physiol.,* 21, 235, 1970.
27. **Basile, D. V.,** Hydroxy-L-proline and 2,2′-dipyridyl-induced phenovariations in the liverwort, *Nowellia curvifolia, Science,* 170, 1218, 1970.
28. **Basile, D. V.,** Hydroxy-L-proline and 2,2′-dipyridyl-induced phenovariation in the liverwort, *Jungermannia lanceolata, Bull. Torrey Bot. Club,* 100, 350, 1973.
29. **Basile, D. V.,** Hydroxyproline-induced changes in form, apical development, and cell wall protein in the liverwort, *Plagiochila arctica, Am. J. Bot.,* 66, 753, 1979.
30. **Basile, D. V. and Basile, M. R.,** Ammonium ion-induced changes in form and hydroxyproline content of wall protein in the liverwort, *Gymnocolea inflata, Am. J. Bot.,* 67, 500, 1980.
31. **Basile, D. V.,** A possible mode of action for morphoregulatory hydroxyproline-proteins, *Bull. Torrey Bot. Club,* 107, 325, 1980.
32. **Basile, D. V. and Basile, M. R.,** Evidence for a regulatory role of cell surface hydroxyproline-containing proteins in liverwort morphogenesis, *J. Hattori Bot. Lab.,* 53, 221, 1982.
33. **Ridge, I. and Osborne, D. J.,** Hydroxyproline and peroxidases in cell walls of *Pisum sativum:* regulation by ethylene, *J. Exp. Bot.,* 21, 843, 1970.
34. **Ridge, I. and Osborne, D. J.,** Role of peroxidase when hydroxyproline-rich protein in plant cell walls is increased by ethylene, *Nature (London),* 229, 205, 1971.
35. **Osborne, D. J., Ridge, I., and Sargent, J. A.,** Ethylene and the growth of plant cells: role of peroxidase and hydroxyproline-rich proteins, in *Plant Growth Substances 1970,* Carr, D. J., Ed., Springer-Verlag, New York, 1972, 534.
36. **Sadava, D. and Chrispeels, M. J.,** Hydroxyproline-rich cell wall protein (extensin): role in the cessation of elongation in excised pea epicotyls, *Dev. Biol.,* 30, 49, 1973.
37. **Amrhein, N. and Wenker, D.,** Novel inhibitors of ethylene production in higher plants, *Plant Cell Physiol.,* 20, 1635, 1979.
38. **Yu, Y.-B. and Yang, F. Y.,** Auxin-induced ethylene production and its inhibition by aminoethoxyvinylglycine and cobalt ion, *Plant Physiol.,* 64, 1074, 1979.
39. **Beyer, E. M.,** A potent inhibitor of ethylene action in plants, *Plant Physiol.,* 58, 268, 1976.
40. **Basile, D. V. and Basile, M. R.,** Desuppression of leaf primordia of *Plagiochila arctica* (Hepaticae) by ethylene antagonists, *Science,* 220, 1051, 1983.
41. **Morgan, P. W. and Hall, W. C.,** Accelerated release of ethylene by cotton following application of indolyl-3-acetic acid, *Nature (London),* 201, 99, 1964.
42. **Abeles, F. B.,** *Ethylene in Plant Biology,* Academic Press, New York, 1973.
43. **Burg, S. P.,** Ethylene in plant growth, *Proc. Natl. Acad. Sci. U.S.A.,* 70, 591, 1973.
44. **Thimann, K. V.,** *Hormone Action and the Life of the Whole Plant,* University of Massachusetts Press, Amherst, 1977.
45. **Burg, S. P. and Burg, E. A.,** Auxin stimulated ethylene formation; its relationship to auxin-inhibited growth, root geotropism, and other plant processes, in *Biochemistry and Physiology of Plant Growth Substances,* Wrighman, F. and Setterfield, G., Eds., Ruge Press, Ottawa, Ontario, Canada, 1968, 1275.
46. **Neidergang-Kamien, E. and Leopold, A. C.,** Inhibitors of polar transport, *Plant Physiol.,* 10, 29, 1957.
47. **Jacobs, M. and Gilbert, S. F.,** Basal localization of the presumptive auxin transport carrier in pea stem cells, *Science,* 220, 1297, 1983.
48. **Nakamura, T. and Takahashi, N.,** The role of auxin in growth of the plumular hook section of etiolated pea seedlings, *Plant Cell Physiol.,* 9, 681, 1968.
49. **Basile, D. V. and Basile, M. R.,** Probing the evolutionary history of bryophytes experimentally, *J. Hattori Bot. Lab.,* 55, 173, 1984.

50. **Basile, D. V., Basile, M. R., and Koning, R. E.,** Ethylene production and its possible role in regulating leaf and branch development in *Plagiochila arctica,* a leafy liverwort, *Am. J. Bot.*, 70, S3, 1983.
51. **Law, D. M., Basile, D. V., and Basile, M. R.,** Determination of endogenous indole-3-acetic acid in gametophytes of *Plagiochila arctica* (Hepaticae), *Plant Physiol.*, 77, 926, 1985.
52. **Basile, D. V., Basile, M. R., and Li, Q.-Y.,** Desuppression of cell division in leaf primordia in *Plagiochila arctica* (Hepaticae) by 3,4-dehydroproline, a selective inhibitor of prolyl-hydroxylase, *Bull. Torrey Bot. Club,* 112, 445, 1985.
53. **Basile, D. V., Basile, M. R., and Varner, J. E.,** Lycorene, ascorbate, prolyl-hydroxylase, and hydroxyproline proteins in relation to cellular suppression/desuppression, *Ann. N.Y. Acad. Sci.*, 494, 175, 1987.
54. **Basile, D. V. and Basile, M. R.,** The occurrence of cell wall-associated arabinogalactan proteins in the Hepaticae, *Bryologist,* 90, 401, 1987.
55. **Basile, D. V., Basile, M. R., and Varner, J. E.,** The occurrence of arabinogalactan protein in *Plagiochila arctica* Bryhn & Kaal. (Hepaticae), *Am. J. Bot.*, 73, 603, 1986.
56. **Basile, D. V.,** A possible role for hydroxyproline-proteins in leafy liverwort phylogeny, *J. Hattori Bot. Lab.*, 38, 91, 1974.
57. **Basile, D. V. and Basile, M. R.,** unpublished data.
58. **Basile, D. V., Cowan, J., Kushner, B. K., and Basile, M. R.,** unpublished data.

Chapter 13

LUNULARIC ACID IN GROWTH AND DORMANCY OF LIVERWORTS

W. W. Schwabe

TABLE OF CONTENTS

I. INTRODUCTION

Early work on the photoperiodic responses of liverworts had shown a very clear induction of dormancy by long days in *Lunularia cruciata,* especially in the Mediterranean strains, while in short-day conditions growth of this species continued rapidly without interruption.[1,2] These results indicated that there was a specific day-length-induced growth inhibition. Moreover, the observation was confirmed that the gemmae produced in the gemma cups of the mother thalli fail to ''germinate'' unless they are removed from the cups by rain splashes or some other mechanical means or after the death of the mother thallus under natural conditions.[3] Such gemmae, when placed on the soil or in a moist medium, started to grow almost at once. Again there was a strong indication of an inhibition, although at that time it seemed entirely feasible that two distinct factors were involved in gemma inhibition of growth and thallus dormancy caused by long days. Detailed studies on the photoperiodic induction followed, and it was established that with a high degree of probability a phytochrome effect was involved. Thus, Wilson and Schwabe[4] showed that the effect of long days could be simulated by light-break treatment and also that red light was particularly effective for long-day effects while subsequent irradiation with far-red light caused complete or partial reversal of the red effect. Furthermore, it was established that there is a strong interaction with temperature; higher temperatures bring about cessation of growth much more rapidly (Table 1). This latter effect was found to vary quantitatively with the strain of *Lunularia,* those collected from the British Isles being less sensitive at somewhat lower temperatures.

On the strength of this background information attempts were made to discover whether a specific growth-inhibitory substance or substances could be detected or isolated.

II. BIOASSAY

An obvious requirement for such an attempt is to devise a suitable bioassay method which will demonstrate the presence or absence of the suspected inhibitory or other substance. For this purpose the gemmae were chosen as the most suitable material. Gemmae possess two growing points in the case of *Lunularia* and have on an average about 8,000 to 10,000 cells, as determined by maceration and cell count studies; they are roughly 1 mm^2 in area. These gemmae are able to restart growth within a very short period (3 days), and the amount of growth can be assessed very readily by area measurement performed under a dissecting microscope with a suitably standardized eyepiece grid. Moreover, the gemmae are very suitable material for ''planting'' directly on paper chromatographs to detect bands of activity of growth inhibitors.

III. ISOLATION AND CHEMICAL CHARACTERIZATION

In the original extraction and purification procedure ca. 1 kg of fresh material was used, extracted with ethanol, acidified to pH 3, further extracted with ethyl acetate, washed with 5% $NaHCO_3$, chromatographed on paper with distilled water, eluted with wet methanol, and further chromatographed on silica gel with chloroform:ethyl acetate:acetic acid (60:40:5). The active band, detected by its blue fluorescence in ultraviolet (UV) light, was again eluted with wet ethyl acetate and methanol and rechromatographed on thin-layer chromatography (TLC) plates in benzene:ethyl acetate:acetic acid (50:5:2). After further elution with wet methanol the final extract was allowed to crystallize from methanol and water in a refrigerator. The crystalline material (about 2 mg), yellow needles with a melting point of 192°C, was shown by bioassay to be the pure inhibitor. Its structure was then identified by mass spectrometry as described by Valio et al.[5] and Valio and Schwabe[6] (Figure 1).

TABLE 1
Effect of Photoperiod and Temperature on the Number of Cycles Required to Induce Dormancy in *Lunularia cruciata*

Photoperiod (h)	Temperature (°C) 12	18	24
19	22	11	5—6
14	>30	>30	10
8	∞	∞	∞

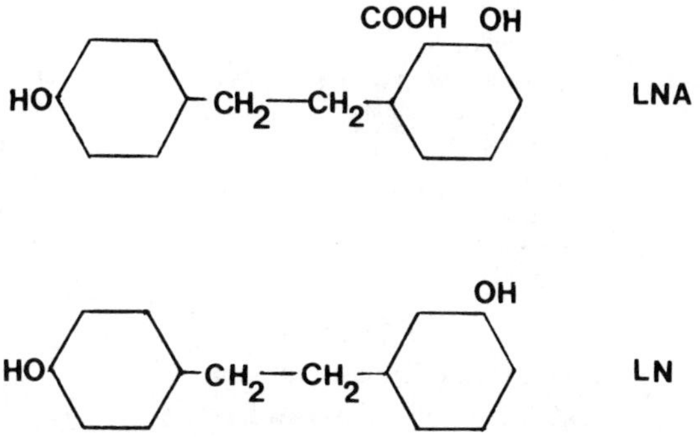

FIGURE 1. Structures of lunularic acid (LNA) and lunularin (LN).

Subsequently, a number of solvent extractions were tested and the following method was adopted for maximum yields. The fresh or dried plant material was refluxed for 1 h in 2 *N* methanolic HCl. The extract was dissolved in diethyl ether (aqueous solutions were acidified to pH 2 with concentrated HCl and then partitioned three times against ether). Partitioning the ether against 5% sodium carbonate extracted the strong acids, while weak acids and phenolic compounds were removed from the ether phase with 1 *N* NaOH. Weak and strong acids were then recovered into ether by acidification with HCl to pH 2. Where 1 *N* NaOH was used alone all the acidic compounds were contained in this fraction (combined acids). For routine assay the combined acids fraction was used directly in gas-liquid chromatography (GLC) analysis. Further purification, where required, was carried out by TLC using silica gel G/UV_{254} plates with the solvent systems (ethyl acetate/chloroform/acetic acid [15:5:1]; benzene/methanol/acetic acid [20:4:1]), although this led to losses of lunularic acid (LNA) (up to 50%). LNA and lunularin (LN) could be detected by their quenching effects in UV light at 254 nm or by spraying the plates with 4% ceric sulfate in 10% sulfuric acid.

IV. OCCURRENCE

A. WITHIN SPECIES

Of the 80 species of liverworts examined by Gorham[7] from a total of 47 genera, every one contained LNA and LN, i.e., yielded a positive qualitative test (Table 2). The quantities found in different orders differed markedly (see Table 3). However, the Anthocerotales gave consistently negative tests in all four species of *Anthoceros* investigated.

Tests were also carried out with several species of algae (i.e., *Chlorella pyrenoidosa, Monodus subterraneus, Anabaena inaequalis, Ulva lactuca, Sargassum muticum,* and *Fucus*

TABLE 2
Families of Liverworts in which Lunularic Acid (LNA) and Lunularin (LN) were Positively Identified[7]

Order	Family	Order	Family
Marchiantiales	Aytoniaceae	Calobryales	Haplomitriaceae
	Conocephalaceae	Jungermanniales	Adelanthaceae
	Corsiniaceae		Atheliaceae
	Exormothecaceae		Cephaloziaceae
	Lunulariaceae		Calypogeiaceae
	Marchantiaceae		Jungermanniaceae
	Monocleaceae		Gymnonmitriaceae
	Oxymitraceae		Jubulaceae
	Ricciaceae		Lepidoziaceae
	Targioniaceae		Lophocoleaceae
Sphaerocarpales	Sphaerocarpaceae		Plagiochilaceae
Metzgeriales	Aneuraceae		Pleuroziaceae
	Blasiaceae		Porellaceae
	Cadoniaceae		Radulaceae
	Pelliaceae		Scapaniaceae
	Metzgeriaceae		Trichocoleaceae

TABLE 3
Contents of Lunularic Acid (LNA) and Lunularin (LN) in Liverworts per Gram Fresh Weight

	LNA	LN
Jungermanniales	1—50 mg	ca. 1 mg
Marchantiales	>100 mg	ca. 30 mg

vesiculosus) using substantial quantities ranging from 5 g to 1 kg. These tests proved negative in all cases, and the claim by Pryce[8] of having detected LNA in algal species must be regarded as disproved. In higher plants some closely related stilbenes have been found in *Hydrangea* spp., and small amounts of LNA as well as its decarboxylation product (LN) were detected, although the former occurred in bound form, probably as the 3-β-D-glucopyranoside. Small amounts of LN have also been identified in the heartwood of several species of *Morus*.[9]

B. WITHIN THE INDIVIDUAL PLANT

Tests with *Conocephalum conicum* on the localization of LNA within the thallus have yielded quantitative differences in different parts of the thallus (Figure 2). However, it appears that all parts did contain some LNA.[7] No clear-cut conclusions could therefore be drawn as to whether LNA synthesis occurred primarily in the youngest part, nearest the growing tip of the thallus, although this seems likely due to loss by diffusion into the medium from the older parts.

C. WITHIN THE CELL

Attempts to localize LNA within different parts of the cell involved maceration and filtration techniques, fractionating the material into soluble portions, wall, and protein material. Although only preliminary work was carried out on this aspect, it seems clear that LNA probably occurs in the soluble fraction and is also associated with the wall material (Table 4). Unfortunately, the work was not taken sufficiently far to form any view as to the

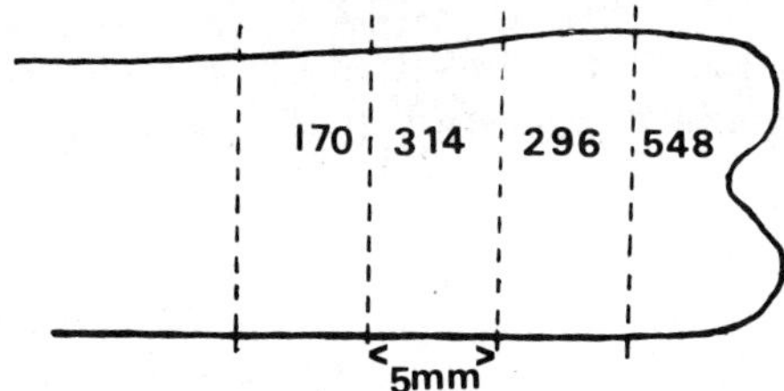

FIGURE 2. Concentration gradient of lunularic acid from the tip along the thallus of *Conocephalum conicum* (μg/g fresh weight).

TABLE 4
Extraction[a] of LNA by Various Techniques from Thalli of *Conocephalum conicum*

	% of total recovered
Mechanical[b]	
1000 × g	60 (wall material)
3000 × g	11
Supernatant	27 (soluble)
Wall enzymes	
Cellulase	0.8 μg/g fresh weight
Pectinase	0.8
Residual	1.0
Aqueous	48.2
Enzymes in comparison with 2 *N* HCl in methanol extraction	
Cell debris Pronase	7.4 μg/g fresh weight
Macerozyme	trace
Residue	8.4
2 *N* HCl in methanol	45.9
Aqueous	13.6

[a] In all cases a substantial amount of LNA was lost compared with reflux extraction.
[b] Maceration followed by centrifugation.

possible form of association of LNA with the wall material or perhaps wall-associated protein fractions.[10]

V. PHYSIOLOGICAL FUNCTION AND RELATION TO THE ENVIRONMENT

The basic physiological response of the thallose liverworts tested was an inhibition of growth due to natural, internal accumulation of LNA in the plant or as the result of LNA supplied exogenously. LNA is therefore clearly in the category of a growth inhibitor. The existence of such an inhibitor had been postulated for *Lunularia* by Nachmony-Bascomb and Schwabe[1,2] and also by Fries[11] for *Marchantia polymorpha*. The mode of action was studied in a series of experiments by Gorham,[7] but no firm conclusion could be drawn (see Section VI). In this respect, work on LNA unfortunately does not differ greatly from the situation concerning all other natural plant growth regulator studies. Other physiological effects are described and discussed below under specific headings.

TABLE 5
Effects of Day Length on Growth in *Lunularia cruciata* and Effects of LNA and Temperature on *Marchantia polymorpha*

Lunularia

	Day length		
	8 h SD	**16 h LD**	**24 h LD**
New branches on old thalli (after 35 days)			
Area	120 mm²	12.1 mm²	
Fresh weight	38.60 mg	5.20 mg	
Dry weight	2.48 mg	0.54 mg	
Mean cell length (thallus margin cells)	36.9 μm		28.2 μm
Mean cell length (second cell row)	47.4 μm		34.2 μm
Mean cell length (third cell row)	61.5 μm	±1.71 μm	38.4 μm
Rhizoids per gemmaling	9.7	±4.3	14.1
Cell no. per gemmaling	114,000		12,000

***Marchantia* (area, mm²)**

		LNA		
Temperature	**Water control**	**2 × 10⁻⁸ *M***	**8 × 10⁻⁸ *M***	**2 × 10⁻⁷ *M***
12°C	3.35	3.05	0.92	0.89
20°C	6.10	6.87	0.54	0.52
28°C	13.44	9.18	0.90	0.96

Note: Abbreviations are SD — short day, LD — long day, LNA — lunularic acid.

A. EFFECTS ON GROWTH

As a result of the accumulation of LNA the growth of thalli is arrested in long days and they enter into a state of dormancy. This is accompanied by a number of changes, which have been described in detail by Nachmony-Bascomb and Schwabe.[1,2] The most important is the cessation of cell proliferation by the apical cell, which interestingly in *Lunularia* does not resume meristematic activity again when growth is restarted in favorable photoperiods, but is replaced by the activity of a group of small cells posterior to it. Another major phenotypic change at the onset of dormancy is the death of the edge cells on the mature part of the thalli, which become brown and senescent, again an irreversible change. Internally, there is an accumulation of starch and free amino acids. It has been shown in nutritional experiments with nitrogen-starved plants that this accumulation is probably the consequence of protein hydrolysis rather than uptake of N from the medium in the absence of growth. Some quantitative assessments of the effects of long-day conditions on growth leading to high internal levels of LNA or applied LNA are shown in Table 5. These tests have also shown that very high concentrations of LNA may actually kill the thalli. However, it also has become apparent that LNA may accumulate in the thallus during active growth in short days. Under natural conditions it is clear that there is a steady loss of LNA (by diffusion?) from the thallus, and this appears to be the mechanism for reducing the internal concentrations at the growing tip to levels below that at which serious inhibition would occur. To demonstrate this effect, thalli were allowed to grow on slightly greased glass coverslips which prevented the normal loss of LNA into the moist growth medium from the tip region. This treatment caused an arrest of growth which could, however, be reversed almost immediately when the cover glasses were withdrawn. Similarly, the young gemmae, which have two growing tips at 180° to each other, grew faster in area when they were cut across so that the two halves were separated.

TABLE 6
Effect of Day Length, Spectrum, and Light Levels on Growth of *Lunularia cruciata* Gemmalings

Short-day control	5.11 mm^2
10 min red + 1 min far-red	4.13 mm^2
1 min far-red	3.90 mm^2 (LSD = 0.49 mm^2)
Continuous white light control	3.14 mm^2
10 min red	2.89 mm^2
1 min far-red + 10 min red	2.77 mm^2

Note: Final area of gemmalings after 24 days (cycles of treatment). Red light from narrow band interference filter 650 mμ (0.9 W m^{-2}); far-red light from narrow band interference filter 735 mμ (1.1 W m^{-2}).

B. GERMINATION OF GEMMAE

The gemmae of *Lunularia* are produced on stalks in the half-moon-shaped gemma cups characteristic of this genus. When they attain a size of about 8000 cells they tend to break away from the stalks and lie free in the cups. Under these conditions they fail to grow and remain inert for long periods. However, as soon as they are removed from the cup and placed on a moist growth medium (e.g., after being splashed by rain out of the gemma cup) they commence to grow. If replaced in the cup no growth occurs. Here, too, the clear assumption is that LNA from the mother thallus prevents germination; i.e., growth of gemmae over the surface of the mother thallus is prevented, and gemmae only succeed in germinating when removed from the source of LNA production. Inhibition of gemma germination in *Lunularia* also can be brought about by auxin (IAA), as recorded by LaRue and Narayanaswamy.[12]

C. PHOTOPERIODIC EFFECTS

As pointed out above, the earliest experiments with *Lunularia* (especially the Mediterranean strain from Israel) had revealed that the cessation of growth in long days and the onset of dormancy were paralleled by similar quantitative differences between the inhibitor contents of thalli grown in short days (8 h light per day) and thalli exposed to long days (16 or 24 h light per day). Those in long days contained substantially greater amounts. In a bioassay of 10 days duration the long-day thalli contained more than twice as much inhibitor.[5] The rate of response was shown to be dependent on both photoperiod and temperature. The rate of response depends on the actual length of the photoperiod as well as the temperature (Table 1). The photoperiodic response has also been shown to be clearly phytochrome mediated.[4] The specific effects of both red and far-red light as well as reversibility have been demonstrated (Table 6), thus meeting all the classical criteria for a phytochrome effect. Another effect also under strict phytochrome control in *Lunularia* is that of gemma germination, which is strictly light dependent except over a narrow temperature range when gemmae can also germinate in the dark.[13] Similar effects have been described in *Marchantia polymorpha* by Otto and Halbsguth.[14] Another interesting but probably unrelated phytochrome-mediated effect on the growth habit of *Marchantia* has been investigated and described by Fredericq and De Greef.[15] Growth becomes orthotropic rather than flat, and chlorophyll content is reduced after irradiation with far-red light compared with red irradiation.

D. DROUGHT RESISTANCE

Thalli which have been arrested in their growth by long-day treatment were found to exhibit resistance to drought. *Lunularia* thalli growing actively on a moist medium (e.g.,

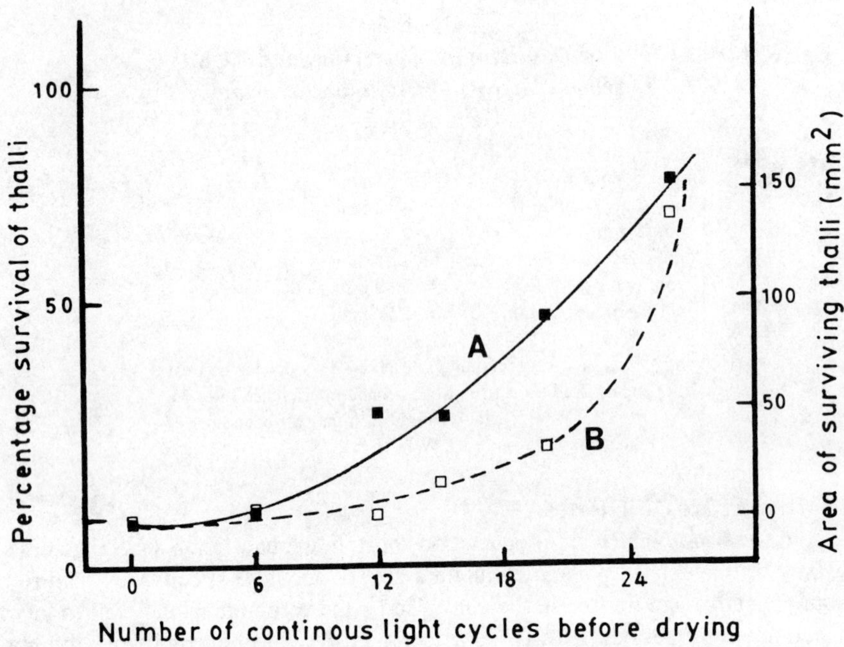

FIGURE 3. Effect of duration of continuous light treatment in inducing drought resistance in *Lunularia cruciata.* (A) Percentage survival of thalli; (B) area (in mm^2) of surviving thalli.

TABLE 7
Effect of Long-Day Treatment on Composition of Thalli of *Lunularia cruciata*

	SD controls (Only in old center of gemma)	16—21 days LD
Starch		+ + +
Free amino acids	+	+ + + (esp. glutamine and arginine)
Water content (% dry matter)	1450	950
Osmotic value (conc. of sucrose solution plasmolyzing 50% of cells)	0.45 *M*	0.55 *M*

Note: Abbreviations — SD = short day; LD = long day.

soil) will desiccate and die if removed from the moist substrate and exposed to atmospheres of relative humidity less than 90%; i.e., their drought resistance is negligible. By contrast thalli exposed to long-day conditions can be air dried and nevertheless will survive when stored in the dry state for years, able to resume growth almost immediately (i.e., within 3 to 4 days) when planted on a suitable moist medium. The degree of drought resistance is positively correlated with the previous period of long-day exposure (Figure 3). Again there is a positive correlation of drought resistance with the content of LNA, but it is not easy to determine experimentally whether the drought resistance is a direct consequence of the arrest of growth or whether LNA can induce intracellular biochemical changes which confer drought resistance. The changes in levels of soluble carbohydrates, amino acids, and osmotic values (Table 7) may then be secondary consequences which themselves affect the resistance.

E. ALLELOPATHIC EFFECTS

The continuous diffusion of LNA from the thallus suggested that this might in itself

TABLE 8
Effect of Lunularic Acid on Fungal Spore Germination and Colony Growth

Species	Spore germination (% of control)				Colony diameter (% of control)
	500 ppm LNA	250 ppm LNA	100 ppm LNA	50 ppm LNA	2×10^{-4} *M* (ca. 50 ppm) LNA
Fusarium culmorum	—	—	—	—	71
Colletotrichum lindemuthianum	—	—	—	—	93
Ascochyta fabae	—	—	—	—	97
Glomerella cingulata	0	0	23	100	85
Aspergillus niger	—	—	—	—	89
Alternaria brassicicola	0	0	71	100	80
Botrytis cinerea	0	65	100	100	—
Sclerotinia fructigena	0	0	35	100	—

Note: Dashes indicate data not determined.

have effects on other competing organisms. The initial experiments extracting soil on which *Lunularia* had been grown for some time did not suggest that large amounts of LNA could be accumulated stably in the soil, although a minimal amount might be active at most times which could not be extracted easily from the soil. However, with synthetic LNA being available, it was possible to test the effects on several types of organisms. Tests with bacterial suspensions (from a selection of pure cultures) indicated that there was no noticeable inhibition. In fact, from the limited amount of testing possible, most of the cultures tested were able to metabolize LNA as a carbon source, i.e., destroying it.

Liverworts as a rule appear to suffer remarkably little from fungal attack or diseases, in spite of the fact that the humid environmental conditions in which most species thrive are obviously particularly well suited to fungal attack. The tests carried out with several species of fungi (using the spore germination and colony growth tests) indicated that LNA is indeed inhibitory to the germination of spores, but there were large differences between several fungi tested over a wide LNA concentration range. Spore germination was totally prevented and colony growth reduced in some, whereas in others there was a limited effect. Furthermore, the amounts needed for such inhibition were relatively high (Table 8). Nevertheless, it would still seem entirely probable that the internal LNA content of thalli may be high enough to prevent fungal attack in the actively growing liverwort, and perhaps even more so in the case of dormant thalli which might be particularly vulnerable but which have higher internal LNA contents.

Algae, which may be one of the major competitors of liverworts in their natural habitats, were also tested for their response to LNA. In this case axenic cultures of nine species of unicellular and filamentous algae (data for some are shown in Table 9) were grown in the presence of different concentrations of LNA.

Very drastic effects were found in some species at concentrations as low as 1 ppm (see Table 9 and Figures 4 and 5). Thus, algae are likely to be strongly inhibited in their growth by LNA, and many algal colonies would not be able to survive when growing over the surface of liverwort thalli. Clearly, LNA has an allelopathic action on some microorganisms, and this must be ecologically advantageous to the liverwort. Another highly competitive group is, of course, all those phanerogams which can exist in the natural habitat of liverworts. It seemed of interest, therefore, to test whether seed germination might also be interfered with by LNA. Here again the effects were rather variable, but generally the presence of LNA in the medium had little effect over a concentration range from 2×10^{-5} to 10^{-3} *M*

TABLE 9
Effect of Different Concentrations of Lunularic Acid on Some Unicellular and Filamentous Algae

Chorella vulgaris (cells/ml)	
2×10^{-5} *M*	65.8%
4×10^{-5} *M*	57.3%
Scenedesmus sp.	
10^{-4} *M*	93.10%
2×10^{-4} *M*	80.12%
4×10^{-4} *M*	24.4%[a]
Chlamydomonas sp. (cell number increase from inoculation)	
10 ppm	39.7%
1 ppm	71.6%
Anabaena sp. (cell number per filament)	
10 ppm	49.1%
1 ppm	47.9%
Tribonema sp. (increase in filament length)	
10 ppm	13.9%
1 ppm	29.2%

Note: All values are shown as % of controls.

[a] Significant at 0.1% level.

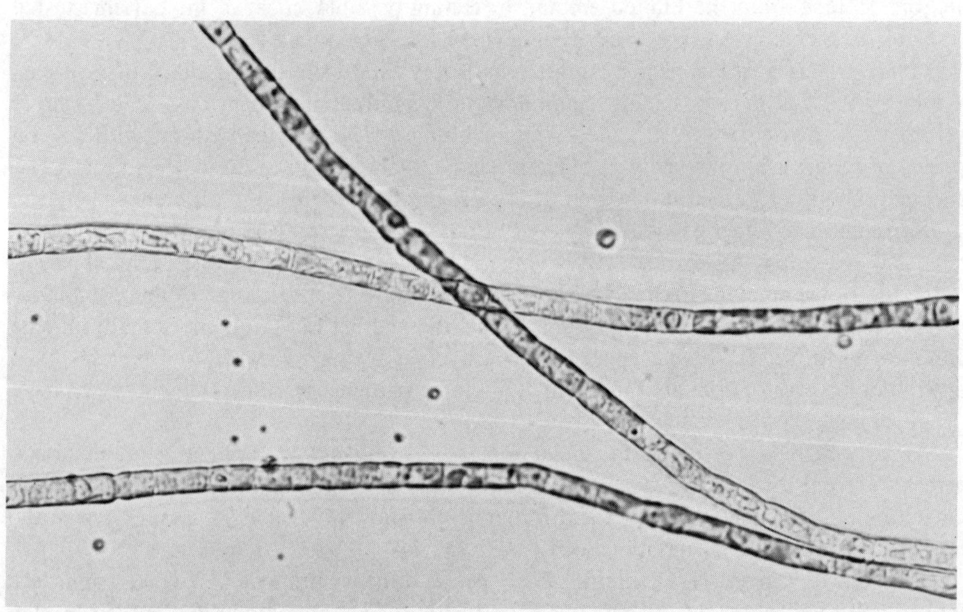

FIGURE 4. Filaments of *Tribonema* sp. grown for 2 weeks in axenic nutrient solution.

LNA in preventing seed germination, although in cress *(Lepidium)* seeds and lettuce seeds the reduction in the germination percentage was from 25 to 30% compared with the controls. Root growth in cress seedlings was reduced by about 35%. However, the concentrations needed for any effect seemed rather too high to suggest that this effect would be of ecological significance, although it may not be entirely negligible.

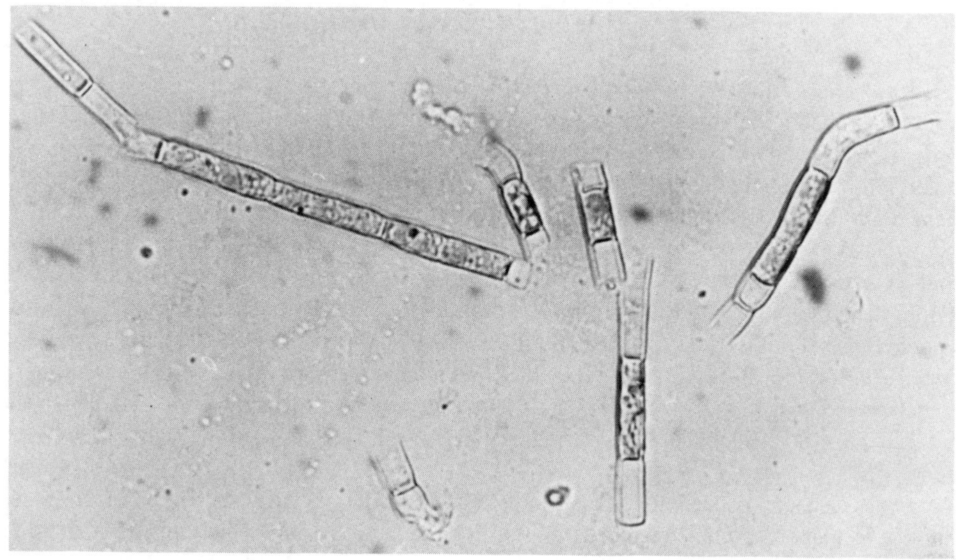

FIGURE 5. Filament pieces of *Tribonema* sp. grown for 2 weeks in axenic nutrient solution in the presence of LNA at a concentration of 100 ppm.

VI. METABOLIC AND BIOCHEMICAL ASPECTS

A number of tests were carried out with LNA to assess its possible role in controlling growth in relation to other naturally occurring plant hormones. These tests also involved the search for the occurrence of other hormones in *Lunularia* and some other liverwort species. From the literature it was not entirely certain that these hormones were natural components in the Hepaticae. The identification of auxin by Schneider et al.[16] and Maravolo and Voth[17] has been questioned by Sheldrake.[18] In *Marchantia polymorpha* Melstrom et al.[19] have firmly established the occurrence of gibberellins. For *Conocephalum* it was established that both indole-3-acetic acid (IAA) and gibberellin were natural constituents. However, the presence of abscisic acid (ABA) could not be detected in any species examined, confirming Weiler's[20] report of the absence of ABA in *Marchantia polymorpha.*[7] On the other hand, exogenously applied ABA had a similar effect on *Lunularia* as in higher plants; i.e., it inhibited growth quantitatively. It would seem, then, that functionally LNA represents the abscisin of the liverworts with the exception of the Anthocerotales.

Tests on the possible mechanism of LNA action revealed that the activity of IAA oxidase could be inhibited to a significant extent and also that glucose-6-phosphate dehydrogenase was partly inhibited by LNA concentrations of 2×10^{-4} *M*. However, there is no clear evidence that these effects are the key reactions involved in growth inhibition *in vivo* (see also the article by Mato and Calvo,[21] who investigated the effect of LNA on auxin oxidase).

The natural occurrence of an LNA decarboxylase was demonstrated by Pryce and Linton,[22] and this has been confirmed by Gorham[23] in several analyses. An extensive study of the inhibitory effects of LNA analogues was also made by Gorham,[24] who examined some 80 different compounds. Several of these had considerably greater inhibitory effects than the naturally occurring LNA and LN, but the underlying mechanism of this activity has not been elucidated as yet. The compounds tested included a number of substitution patterns: OH/OH, OH,COOH/OH, OH, *m*-diOH, COOH, COOH/OH, polyphenols, and methylated compounds.[24] In summary, none of these studies have as yet resulted in a clear-cut idea of how LNA and LN may function *in vivo* and, thus, these two natural inhibitors have joined the ranks of virtually all natural plant growth regulators whose basic function still needs to be established.

VII. SYNTHESIS

It is believed that the biosynthesis of LNA is from phenylalanine, acetate, and hydrangenol,[25] probably via cinnamic acid (both phenylalanine ammonia lyase and cinnamic acid-4-hydroxylase activity having been detected in *Conocephalum* material by Gorham[7,26]). The phenylpropanoid-polymalonate route was first suggested by Birch and Donovan.[27] The evidence for this is based on incorporation experiments with universally labeled U-^{14}C-L-phenylalanine, 1-^{14}C-sodium acetate, and ^{14}C-hydrangenol into LNA by *Lunularia cruciata* thallus material. This has been further confirmed by Gorham.[26] *In vitro,* lunularin has been synthesized by two methods: (1) the Oglialoro route, with intermediates sodium (*p*-hydroxyphenyl) acetate, 3,4′-diacetoxy-*trans*-stilbene-α-carboxylic acid, 3,4′-diacetoxy-*trans*-stilbene, and 3,4′-dihydroxy-*trans*-stilbene; or (2) a modified Wittig synthesis, with intermediates 4-methoxybenzylalcohol, 4-methoxybenzylbromide, diethyl (4-methoxybenzyl) phosphate + 3-methoxybenzaldehyde, 3,4′-dimethoxy-*trans*-stilbene, and 3,4′-dihydroxy-*trans*-stilbene.[7,28]

It was also shown that exogenously supplied LNA (^{14}C labeled) was rapidly incorporated into the methanol-insoluble residue from which it could not be solubilized by acid or alkaline hydrolysis or by lignin solvents. Thomas[10] also failed to release LNA from cell wall material by pectinase, cellulase, or macerozyme and concluded that it may be bound to proteinaceous material itself associated with the cell wall.

VIII. SUMMARY AND ECOLOGICAL SIGNIFICANCE OF LUNULARIC ACID IN LIVERWORTS

While there are clearly quantitative and probably also qualitative differences between species which so far have not been detected, the generalizations below must be regarded as almost certainly too sweeping. However, it seems that LNA has several quite distinct functions:

1. It serves to regulate growth of the thallus quantitatively.
2. Its production being under phytochrome control (at least to some extent), it serves the photoperiodic sensitivity to be expressed in terms of dormancy induction.
3. Affecting growth as such, it also appears to control drought resistance of the thallus.
4. As growth depends on the diffusion of LNA from the plant, it prevents at least to some extent the ''overgrowth'' of one thallus over another, and perhaps more importantly it prevents the ''germination'' of gemmae until these have been removed from the cup of the mother thallus.
5. By being effective as an algicide it has a clear allelopathic effect on a competing class of organisms.
6. A somewhat similar, but weaker, effect appears to be exerted on fungal spore germination, although the virtual absence of fungal diseases among thallose liverworts suggests that internal concentrations may suffice to prevent fungal attack almost completely.
7. Effects on seed germination, albeit only at relatively high concentrations, also indicate a weak allelopathic effect.

Thus, while LNA appears to fill the role of abscisin in liverworts, it has a multiplicity of effects. Unfortunately, the resemblance to higher plant hormones does not end here. In the case of LNA we are still equally ignorant of the direct metabolic reaction(s) mediated by it.

REFERENCES

1. **Nachmony-Bascomb, S. and Schwabe, W. W.,** Growth and dormancy in *Lunularia cruciata* (L.) Dum. I. Resumption of growth and its continuation, *J. Exp. Bot.,* 14, 1153, 1963.
2. **Schwabe, W. W. and Nachmony-Bascomb, S.,** Growth and dormancy in *Lunularia cruciata* (L.) Dum. II. The response to day length and temperature, *J. Exp. Bot.,* 14, 353, 1963.
3. **Oppenheimer, H.,** Das Unterbleiben der Keimung in den Behältern der Mutterpflanze, *Sitzungsber. Akad. Wiss. Wien Math. Naturwiss. Kl. Abt. 1,* 131, 59, 1922.
4. **Wilson, J. R. and Schwabe, W. W.,** Growth and dormancy in *Lunularia cruciata* (L.) Dum. III. The wavelengths of light effective in photoperiodic control, *J. Exp. Bot.,* 15, 368, 1964.
5. **Valio, I. F. M., Burden, R. S., and Schwabe, W. W.,** New natural growth inhibitor in the liverwort *Lunularia cruciata* (L.) Dum., *Nature (London),* 223, 1176, 1969.
6. **Valio, I. F. M. and Schwabe, W. W.,** Growth and dormancy in *Lunularia cruciata* (L.) Dum. VII. The isolation and bioassay of lunularic acid, *J. Exp. Bot.,* 21, 138, 1970.
7. **Gorham, J.,** Some Aspects of the Distribution, Metabolism and Physiological Role of Lunularic Acid in Liverworts, Ph.D. thesis, University of London, London, England, 1975, 206.
8. **Pryce, R. J.,** The occurrence of lunularic and abscisic acid in plants, *Phytochemistry,* 13, 2497, 1972.
9. **Deshpande, V. H., Srinivasan, R., and Rao, A. V. R.,** Wood phenolics of *Morus* species. IV. Phenolics of the heartwood of five *Morus* species, *Indian J. Chem.,* 13, 453, 1975.
10. **Thomas, B. D.,** Studies of Physiological Effects of Lunularic Acid in Cryptogams and Higher Plants, Ph.D. thesis, University of London, London, England, 1977, 215.
11. **Fries, K.,** Über einen genuinen keimungs- und streckungswachstumsaktiven Hemmstoff bei *Marchantia polymorpha* L., *Beitr. Biol. Pflanz.,* 40, 177, 1964.
12. **LaRue, C. D. and Narayanaswami, S.,** Auxin inhibition in the liverwort *Lunularia, New Phytol.,* 56, 61, 1957.
13. **Valio, I. M. F. and Schwabe, W. W.,** Growth and dormancy in *Lunularia cruciata* (L.) Dum. IV. Light and temperature control of rhizoid formation in gemmae, *J. Exp. Bot.,* 20, 615, 1969.
14. **Otto, K. and Halbsguth, W.,** Die Förderung der Bildung von Primärrhizoiden an Brutkörpern von *Marchantia polymorpha* L. durch Licht und IES, *Z. Pflanzenphysiol.,* 80, 197, 1976.
15. **Fredericq, H. and De Greef, J.,** Photomorphogenic and chlorophyll studies in the bryophyte *Marchantia polymorpha.* I. Effect of red, far-red irradiations in short and long-term experiments, *Physiol. Plant.,* 21, 346, 1968.
16. **Schneider, M. J., Troxler, R. F., and Voth, P. D.,** Occurrence of indoleacetic acid in the bryophytes, *Bot. Gaz. (Chicago),* 128, 174, 1967.
17. **Maravolo, N. C. and Voth, P. D.,** Morphogenic effects of three growth substances on *Marchantia* gemmalings, *Bot. Gaz. (Chicago),* 127, 79, 1966.
18. **Sheldrake, A. R.,** The occurrence and significance of auxin in the substrata of bryophytes, *New Phytol.,* 70, 519, 1971.
19. **Melstrom, C. E., Maravolo, N. C., and Stroemer, J. R.,** Endogenous gibberellins in *Marchantia polymorpha* and their possible physiological role in thallus elongation and orthogeotropic growth, *Bryologist,* 77, 33, 1974.
20. **Weiler, E. W.,** Radioimmunoassay for the detection of free and conjugated abscisic acid, *Planta,* 144, 255, 1979.
21. **Mato, M. C. and Calvo, R.,** Effect of lunularic acid on auxin oxidase activity, *Biol. Plant.,* 19, 394, 1977.
22. **Pryce, R. J. and Linton, L.,** Lunularic acid decarboxylase from the liverwort *Conocephalum conicum, Phytochemistry,* 13, 2497, 1974.
23. **Gorham, J.,** Recent research on lunularic acid and its influence, *Bull. Br. Biol. Soc.,* 31, 11, 1978.
24. **Gorham, J.,** Effect of lunularic acid analogues on liverwort growth and IAA oxidation, *Phytochemistry,* 17, 99, 1978.
25. **Pryce, R. J.,** Biosynthesis of lunularic acid — a dihydro-stilbene endogenous growth inhibitor of liverworts, *Phytochemistry,* 10, 2679, 1971.
26. **Gorham, J.,** Metabolism of lunularic acid in liverworts, *Phytochemistry,* 16, 915, 1977.
27. **Birch, A. J. and Donovan, F. W.,** Studies in relation to biosynthesis. I. Some possible routes to derivatives of orcinol and phloroglucinol, *Aust. J. Chem.,* 6, 360, 1953.
28. **Gorham, J.,** Lunularic acid and related compounds in liverworts, algae, and *Hydrangea, Phytochemistry,* 16, 249, 1977.

Chapter 14

BIOLOGICALLY ACTIVE SUBSTANCES FROM BRYOPHYTES

Yoshinori Asakawa

TABLE OF CONTENTS

I. INTRODUCTION

Until the last 15 years the phytochemistry of bryophytes had been completely neglected. There are no descriptions concerning the chemical constituents of bryophytes in Japanese textbooks of pharmacognosy and medicinal plants, although a number of biologically active compounds found in algae, mushrooms, lichens, and ferns have been recorded. There are some reasons why the chemical constituents of bryophytes have remained essentially unknown. First, the bryophytes have been considered to be a small plant group almost useless for the human diet. Second, collection of a large quantity of pure material of certain species is quite difficult. Third, identification of each species is also difficult. Fourth, very frequently other species, soil and decayed leaves, bark, stems, and roots of higher plants are intermingled with a mat of the desired species, and separation of such mixtures is a time-consuming and troublesome process. However, a number of bryophytes have been used as medicinal plants in China,[1] Europe,[2] and North America,[3] as shown in Table 1. In China, more than 400 years ago, Ri[4] reported that some *Fissidens* and *Polytrichum* species resulted in diuretic activity and promoted growth of hair on the human head. Most of the medicinal bryophytes have been used as decoctions to cure disease. Some bryophytes are crushed and the resulting powder is mixed with oil to make a paste which is applied to burns, cuts, and external wounds. North American Indians have used *Bryum, Mnium, Philonotis* spp., and *Polytrichum juniperinum* as medicinal bryophytes to heal burns, bruises, and wounds. *Marchantia polymorpha* was used as a diuretic drug in France.[2] In this case, the fresh specimens were soaked in white liquor and patients drank the liquor containing the extracts. In spite of the presence of such medicinally and pharmacologically interesting substances in bryophytes, their isolation and structure determination had not been carried out fully until the present decade. Müller[5] reported the presence of sesquiterpenoids in Hepaticae. Over 50 years later Fujita et al.[6] found that the essential oil of *Bazzania pompeana* was composed of sesquiterpene hydrocarbon. In 1967, Huneck and Klein[7] isolated two sesquiterpenoids, (−)-longifolene (**1**) and (−)-longiborneol (**2**), from *Scapania undulata*. This is the first record of the isolation of terpenoids from oil bodies of the Hepaticae. The recent development of analytical instruments, gas chromatography, high-pressure liquid chromatography, gas chromatography-mass spectrometry, ^{1}H- and ^{13}C-nuclear magnetic resonance (NMR) spectrometry using super magnets (400 and 500 MHz), and high-resolution mass spectrometry makes it easy to isolate pure substances and to establish the absolute stereochemistry of even milligram samples. A number of new compounds have been isolated from bryophytes and their structures elucidated.[8-10] In addition to the physiological activity of bryophytes shown in Table 1, some bryophytes produce a characteristic odor or have an intensely hot and bitter or saccharine-like taste. Generally, bryophytes are not infected by bacteria and fungi and are not damaged by insects, snails, and other small animals. Furthermore, some liverworts cause intense allergenic contact dermatitis and allelopathy. We have been interested in these bioactive substances present in bryophytes and have studied about 700 species from the viewpoint of chemistry and pharmacology as well as their application as a source of cosmetics and medicinal or agricultural drugs.[9,11-18] The isolation and chemical structures of some biologically active substances found in bryophytes and their activity have been summarized in this chapter.

II. CHARACTERISTIC ODORS OF BRYOPHYTES

Some bryophytes emit volatile terpenoids or simple aromatic compounds which are responsible for intense turpentine, mushroomy, sweet woody, sweet mossy, fungal, or carrot-like odor. In Table 2 some bryophytes possessing characteristic odors are listed.[3,9,13,14,16,18-20]

TABLE 1
Medicinal Bryophytes and their Physiological Activity and Effects

Medicinal bryophytes	Physiological activity and effects
Hepaticae	
Conocephalum conicum	Antimicrobial, antifungal, antipyretic, antidotal; for cuts, burns, scalds, fractures, swelling, poisonous snake bite, gallstone
Frullania tamarisci	Antiseptic
Marchantia polymorpha	Antipyretic, antidotal, antihepatic, diuretic; for cuts, fractures, poisonous snake bite, boils, burns, scalds, open wounds
Reboulia hemisphaerica	For blotches, hemostasis, external wounds, bruises
Musci	
Bryum argenteum	Antipyretic, antidotal, antirhinitic; for bacteriosis
Cratoneuron filicinum	For malum cardis
Ditrichum pallidum	For convulsions
Fissidens japonicus	Diuretic; for growth of hair, burns, choloplania
Funaria hygrometrica	For hemostasis, pulmonary tuberculosis, hematemesis, bruises, athlete's foot dermatophytosis
Haplocladium calillatum	Antipyretic, antidotal; for adenopharyngitis, uropathy, mastitis, erysipelas, pneumonia, urocystitis, tympanitis
Leptodictyum riparium	Antipyretic; for choloplania, uropathy
Mnium cuspidatum	For hemostasis, nose bleeding
Oreas martiana	For anodyne, hemostasis, external wound, epilepsy, menorrhalgia, neurasthenia
Philonotis fontana	Antipyretic, antidotal; for adenopharyngitis, boils
Plagiopus oederi	Sedative; for epilepsy, apoplexy, cardiopathy
Pogonatum sp. or *Polytrichum* sp.	Diuretic; for growth of hair
Polytrichum commune	Antipyretic, antidotal; for hemostasis, cuts, bleeding from gingivae, hematemesis, pulmonary tuberculosis
Rhodobryum giganteum	Antipyretic, diuretic, antihypertension, sedative; for neurasthenia, psychosis, cuts, cardiopathy, expansion of blood vessels of heart
Rhodobryum roseum	Sedative; for neurasthenia, cardiopathy
Taxiphyllum taxirameum	Antiphlogistic; for hemostasis, external wound
Weisia viridula	Antipyretic, antidotal; for rhinitis
Sphagnum cymbifolium, *S. magellanicum*, *S. palustre*, *S. squarrosum*	Substitute for sanitary cotton

So that these substances could be used in cosmetics we studied the chemical structures of fragrant substances of some liverworts. Ando and Matsuo[3] confirmed that *p*-ethylanisol **(3)** was the characteristic odorous substance of *Leptolejeunea elliptica. Conocephalum conicum,* particularly the male thallus, and *Wiesnerella denudata* emit a sweet mushroomy odor. (+)-Bornyl acetate **(4)** and (−)-sabinene **(5)** were isolated from the ether extract of *C. conicum* as the major components, together with methyl cinnamate **(6)**, 1-octen-3-ol **(7)**, and its acetate **(8)**, which are the important fragrant substances of the most expensive Japanese mushroom, Matsutake (= *Trichloma matsutake*).[20] The fragrance of this liverwort is due to the mixture of these simple monoterpenoids and the special mushroom components. The intense sweet mushroomy odor of *W. denudata* is due to (+)-bornyl acetate **(4)** and the mixtures of monoterpene hydrocarbons α-pinene **(9)**, β-pinene **(10)**, camphene **(11)**, α-terpinene **(12)**, β-phellandrene **(13)**, and terpinolene **(14)**.[21,22] A small French thalloid liverwort, *Targionia hypophylla,* emits a very intense fragrant odor when crushed. *Cis-* **(15)** and *trans*-pinocarveyl acetates **(16)**, which comprise the characteristic fragrance of *T. hypophylla,* were isolated from the methanol extract (Figure 1).[23]

Herout[24] reported that the special mossy odor of *Lophocolea heterophylla* was $C_{12}H_{20}O$,

TABLE 2
Characteristic Odor of Bryophytes

Species	Odor
Asterella sp.	Higher plant *Houttuynia cordata*-like
Conocephalum conicum	Higher plant *Houttuynia cordata*-like; strong mushroomy
Conocephalum japonicum	Higher plant *Houttuynia cordata*-like
Frullania sp.	Turpentine-like
Funaria hygrometrica	Sea-star-like
Geocalyx graveolens	Turpentine-like
Jungermannia sp.	Turpentine-like
Lejeunea sp.	Turpentine-like
Leptolejeunea elliptica	Mixed smell of naphthaline and dried bonito
Lophocolea heterophylla	Strong and distinct mossy
Lophocolea minor	Strong mossy
Lophozia bicrenata	Pleasant odor like cedar oil
Moerkia sp.	Intense unpleasant smell
Pellia endiviifolia	Dried sea weed-like
Plagiochila acanthophylla subsp. *japonica*	Sweet mossy and woody
Plagiochila sp.	Turpentine-like
Porella cordaeana	Dried sea weed-like
Porella sp.	Turpentine-like
Riella sp.	Anise-like
Solenostoma obovata	Carrot-like
Takakia lepidozioides	Mixed smell of cinnamon and burnt wheat powder
Trichocolea tomentella	Sulfur-like
Wiesnerella denudata	Strong sweet mushroomy

which was suggested by its mass spectrum. However, the structure remained to be elucidated. Silica gel chromatography of the ether extract of *L. heterophylla* collected in West Germany resulted in the isolation of (−)-2-methyl-isoborneol **(17)**, which is responsible for the characteristic mossy fragrance of *L. heterophylla.*[25] The compound isolated by Herout[24] may be identical to the above homomonoterpenoid. Some *Frullania, Jungermannia, Lejeunea, Plagiochila,* and *Porella* spp. emit a strong turpentine-like odor which is very similar to that of conifer leaves. This smell is due to the mixture of the simple monoterpene hydrocarbons. The fresh *Trichocolea tomentella* emits a sulfur-like smell when crushed. Surprisingly, this liverwort accumulates sulfur.[26] Two oxygenated sesquiterpenoids, tamariscol **(18)** and bicyclohumulenone **(20)**, were isolated from the European *Frullania tamarisci*[27] and *Plagiochila acanthophylla* subsp. *japonica,*[28] respectively. Both of them emit mysterious sweet mossy and woody odors, and they have been highly esteemed in some European, American, and Japanese perfumary companies as excellent sources of perfume. Epoxy-tamariscol **(19)** derived from **18** emits the same fragrant odor as that of **18**. These components are applied as perfumes themselves or used as perfumes in various cosmetics.[29-33] Tamariscol **(18)** is present not only in European *F. tamarisci,* but also in Japanese *F. tamarisci* subsp. *obscura,* Taiwanese *F. nepalensis,* and American *F. asagrayana.*[34]

III. HOT AND BITTER SUBSTANCES

Some genera of bryophytes produce intensely hot or bitter substances which show interesting biological activity (to be described in Sections IX and X). It is known that most bryophytes growing in North America contain substances with unpleasant taste, and some of them taste like immature green peas or pepper.[35] Mizutani[36] reported that *Porella vernicosa* contained very hot substances, and *Jamesoniella autumnalis* possessed an intense bitter substance which tasted like the leaf of the lilac, *Swertia japonica,* or the root of *Gentiana*

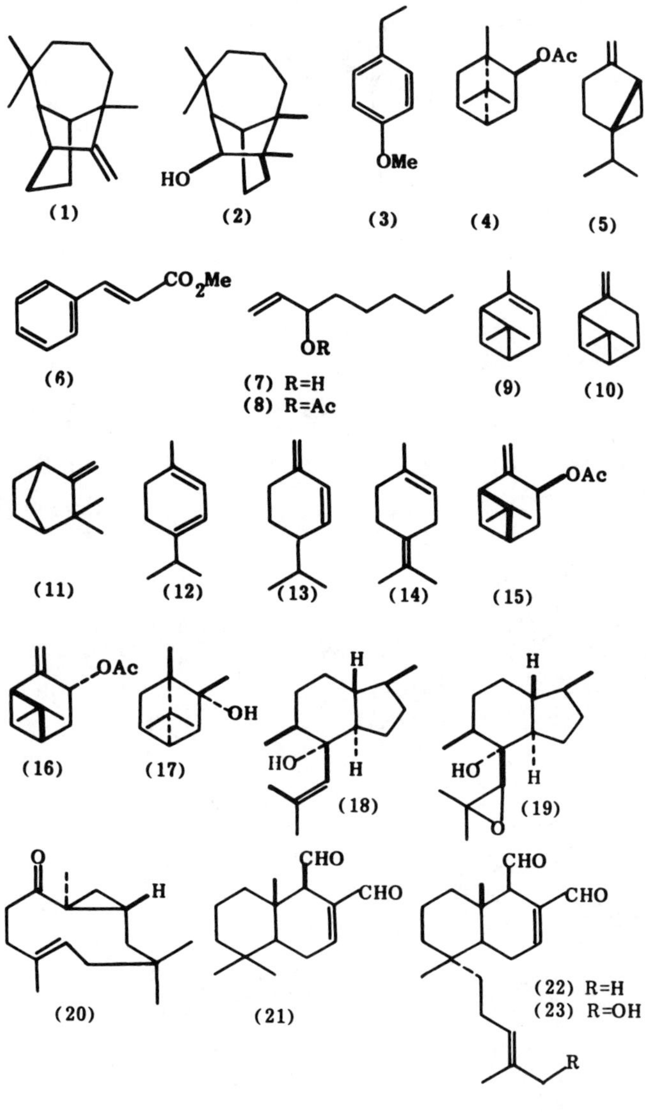

FIGURE 1.

scabra var. *orientalis* (= Gentianae Scabrae), and *Rhodobryum giganteum* showed saccharine-like sweetness. Polygodial **(21)**, a sesquiterpene dialdehyde, is the true pungent substance of *P. vernicosa* complex, *P. arboris-vitae, P. fauriei, P. gracillima, P. obtusata* subsp. *macroloba, P. roerii,* and *P. vernicosa.*[9,37] The crude extracts of these species contain ca. 10 to 30% polygodial. This compound has already been isolated from the higher plant *Polygonum hydropiper* (Polygonaceae), which is a useful spice for rough and grilled fishes in Japan. Three new sacculatane-type diterpene dialdehydes, sacculatal **(22)**, 19-hydroxysacculatal **(23)**, and 18-hydroxysacculatal **(24)**, were isolated from *Trichocoleopsis sacculata,* together with isosacculatal **(25)**, the C-9 epimer of **22.**[9] The compounds **(22 to 24)** have a persistent pungent taste. *Pellia endiviifolia* has a persistent pungent taste when one chews fresh material. This pungency is also due to sacculatal **(22)**.[9] The dried *P. endiviifolia* lacks hot taste. Isosacculatal **(25)** is tasteless. *Chiloscyphus polyanthus* collected in southern France produces intensely pungent substances. This pungency is due to the mixtures of *ent*-diplophyllolide **(26)** and *ent*-7α-hydroxydiplophyllolide **(27)**.[9] *Diplophyllum albicans* con-

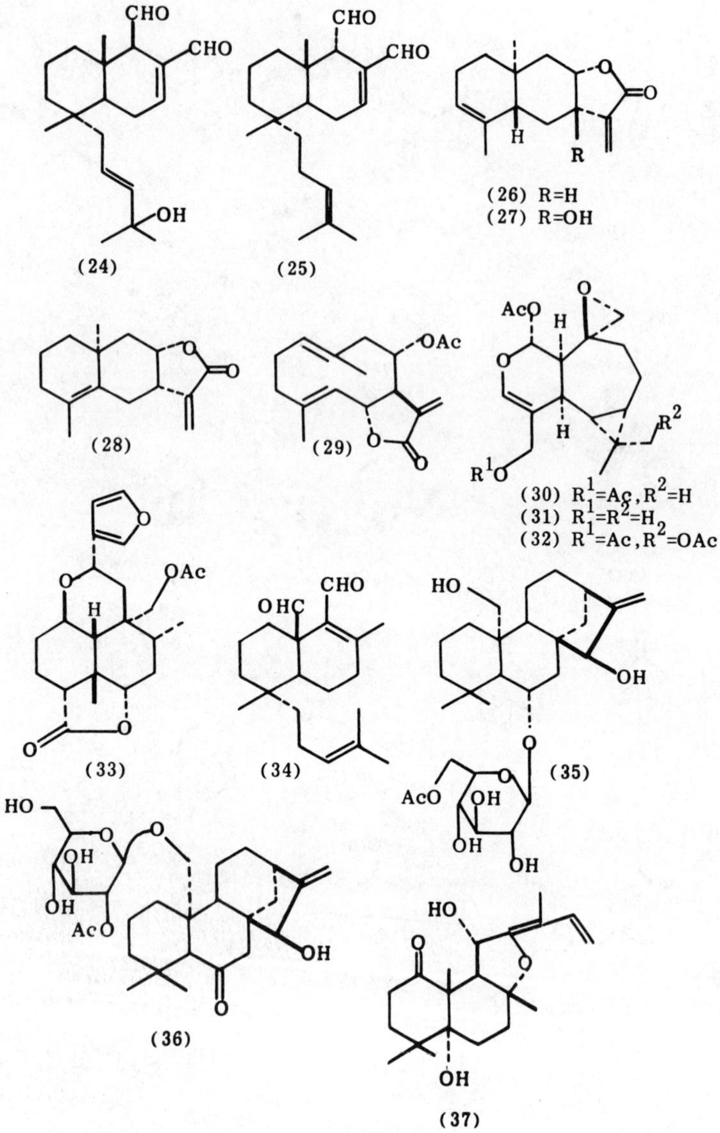

FIGURE 2.

tains slightly pungent substances. Diplophyllolide **(26)** and diplophyllin **(28)** were isolated from this liverwort, but the content of the former compound is very low.[9] The double-bond isomer **(28)** of **26** is almost tasteless. *Wiesnerella denudata* produces the hot sesquiterpene lactone tulipinolide **(29)**,[9] which has been isolated from the higher plant *Liriodendron tulipifera*.[38] At present, more than 500 sesquiterpene lactones have been isolated from the plant kingdom, but there are no records of a powerful pungent taste for these natural products.

Plagiochila species are widely distributed in the world, and there are more than 3000 species. These liverworts are divided into two chemotypes: pungent species and nonpungent species. The strong pungency of the former group is mainly due to plagiochiline A **(30)**, a 2,3-secoaromadendrane-type sesquiterpene hemiacetal (Figure 2).[9] *Plagiochila yokogurensis* produces **30** as well as the other pungent component, plagiochiline I **(31)**.[9] *P. hattoriana* elaborates plagiochiline A and a bitter hemiacetal, plagiochiline B **(32)**.[9] Some Lophoziaceae species produce much more intensely bitter terpenoids. *Gymnocolea inflata* has a tremen-

TABLE 3
Allergy-Inducing Bryophytes

Hepaticae	Musci
Frullania asagrayana	*Eurhynchium oreganum*
F. bolanderi	*Isothecium stoloniferum*
F. dilatata	*Leucoleptis menziesii*
F. eboracensis	*Mnium cuspidatum*
F. franciscana	*Mnium undulatum*
F. inflata	*Polytrichum juniperinum*
F. kunzei	*Rhytidiadelphus loreus*
F. nisquallensis	*Sphagnum palustre*
F. riparia	
F. tamarisci	
Marchantia polymorpha	
Metzgeria furcata	
Radula complanata	

dously persistent bitter taste and induces vomiting when it is chewed for some time. The active substance was extracted with ether, and chromatography of the crude extract on silica gel resulted in the isolation of the bitter substance gymnocolin **(33)**, a *cis*-clerodane-type diterpene lactone having a furane ring whose structure was established by X-ray crystallographic analysis.[9,39] The large-stem leafy liverwort *Porella perrottetiana* contains bitter substances. From the ether extract a diterpene dialdehyde, perrottetianal **(34)**, which is the isomer of sacculatal **(22)**, was isolated as the bitter substance.[9] Perrottetianal **(34)** was totally synthesized by Hagiwara and Uda.[40] *Jungermannia infusca* shows an amazingly intense bitter taste which is due to a mixture of kaurene-type diterpene glucosides. Two strong bitter substances, infuscaside A **(35)** and infuscaside B **(36)**, were isolated from the methanol extract.[41] This is the first record of the isolation of the terpene glycosides from the Hepaticae, although a number of flavonoid glycosides have been found in the bryophytes.[9] Huneck and Overton[42] reported the isolation of several bitter compounds from a few liverworts, i.e., barbilycopodin from *Barbilophozia lycopodioides,* anastreptin ($C_{20}H_{24}O_5$) from *Anastrepta orcadensis,* floerkein A and B ($C_{20}H_{34}O_3$) from *Barbilophozia floerkei,* and scapanin ($C_{20}H_{30}O_4$) from *Scapania undulata*. The structure of barbilycopodin was established to be the dolabellane diterpenoid **(38)** which was also present in the other *Barbilophozia* species, *B. floerkei* and *B. attenuata*.[43,44] The bitter diterpenoids scapanin **(37)**, floerkein B **(39)**, and anastreptin **(40)** have been isolated and their structures elucidated by Connolly.[43]

IV. ALLERGENIC CONTACT DERMATITIS

Table 3 shows the species causing allergenic contact dermatitis.[12] The *Frullania* species are the special liverworts causing allergy. In Canada, the U.S., and Europe, it has been known that some occupational allergy is associated with the handling of wood on which some epiphytic liverworts grow. The hapten of this allergy is the mixture of eudesmane- and eremophilane-type sesquiterpene lactones found in *Frullania tamarisci,*[45,46] *F. dilatata,*[47] and *F. nisquallensis*.[48] The patch test of the isolated sesquiterpene lactones, (+)-frullanolide **(41)**, (−)-frullanolide **(42)**, (+)-oxyfrullanolide **(43)**, *cis*-β-cyclocostunolide **(44)**, and (+)-eremofrullanolide **(45)**, indicated that patients sensitive to *Frullania* are all sensitive to at least one of the sesquiterpene lactones **(41 to 45)**, as shown in Table 4 and Figure 3.[47] This is the first record of the isolation from bryophytes of biologically active substances for the human body.

(−)-Frullanolide **(42)** has been synthesized from (−)-santonin.[49,50] Kido et al.[51] reported the total synthesis of (±)-frullanolide. All allergy-inducing sesquiterpenoids possess an α-

TABLE 4
Patch Test of Selected Sesquiterpene Lactones[a] Isolated from European *Frullania dilatata*

Compounds tested	Patient no. 1—5	6—14	15	16	17	18,19	20,21	22	23
(+)-Frullanolide (**41**)	+	+	+	+	+	0	0	0	0
(−)-Frullanolide (**42**)	+	+	+	+	+	0	0	0	0
(+)-Oxyfrullanolide (**43**)	+	+	+	+	+	+	+	0	0
(+)-*cis*-β-Cyclocostunolide (**44**)	+	+	0	0	0	+	0	0	+
Eremofrullanolide (**45**)	+	x	+	x	0	+	+	+	x

Note: Abbreviations — +: active; 0: inactive; x: not tested. Numbers in parentheses = structures found in text.

[a] EtOH solution (1%)

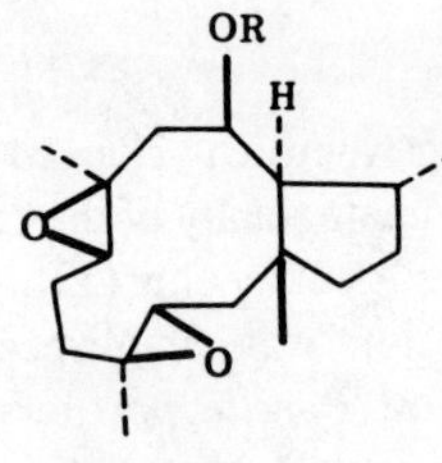

(38) R=Ac
(39) R=H
(40) 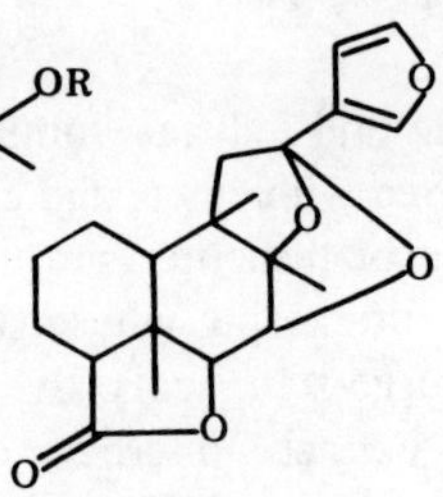
(41) 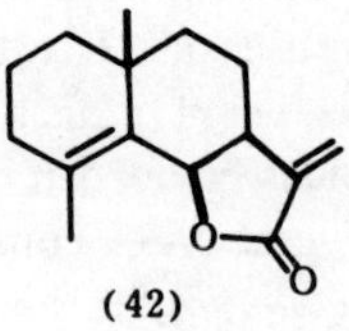 (42) 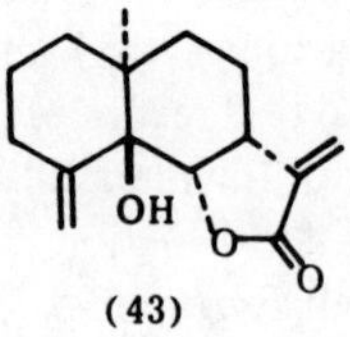 (43) 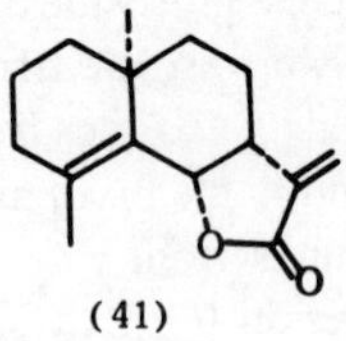
(44) 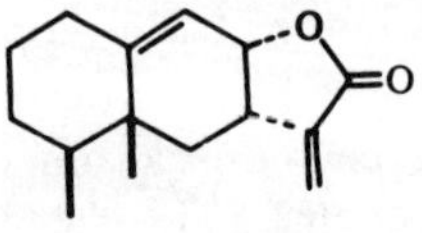 (45) 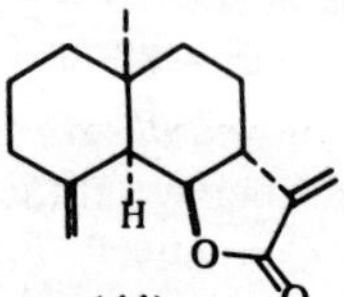
(46) R=H 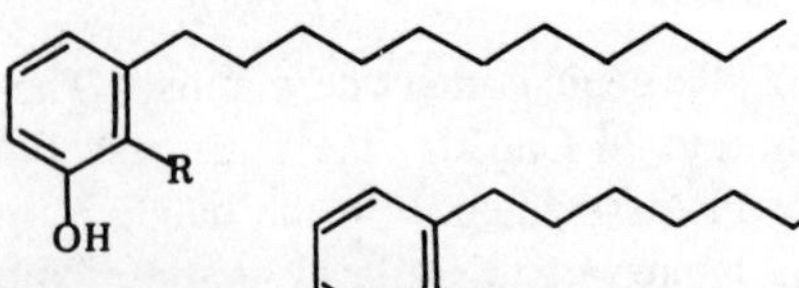
(47) R=COOH
(48) R=COOK
(49) 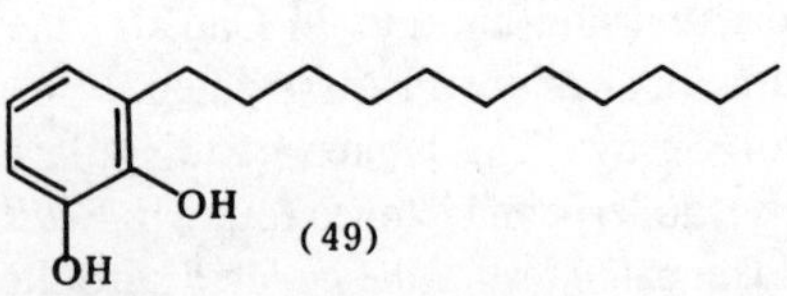
(50) R=H 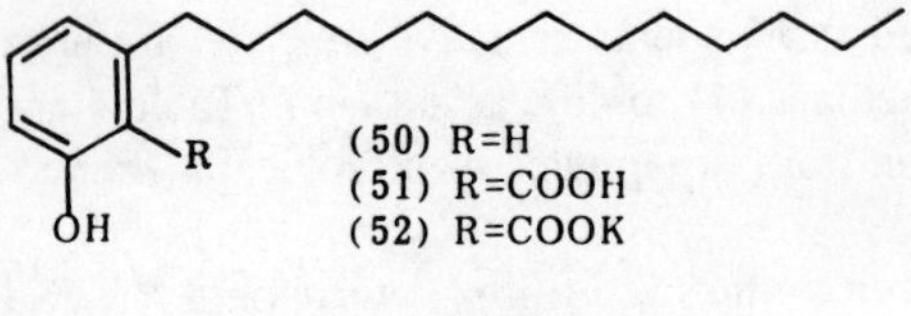
(51) R=COOH
(52) R=COOK

FIGURE 3.

methylene-γ-lactone group. This group is unambiguously the active site causing allergenic contact dermatitis, since the sesquiterpenoids possessing the α-methyl-γ-butyrolactone group cause no allergy. Allergenic intensities of (+)-frullanolide **(41)** and (−)-frullanolide **(42)** and of (+)-epoxyfrullanolide **(126)** and its enantiomer are almost equal, meaning there is no chiral specificity for the allergenic activity.[9,12] There are about 500 species of *Frullania*. There are two chemotypes of *Frullania* spp.; type A contains sesquiterpenoids having an α-methylene-γ-butyrolactone group, and type B does not contain any sesquiterpene lactones.[9,52] When patients sensitive to *Frullania* spp. come in contact with a type A an amazingly intense allergenic contact dermatitis is induced. It has been known that some *Radula* spp. cause allergy.[11] *R. complanata* is often intermingled with *F. dilatata;* in fact, (+)-frullanolide **(41)** has been isolated from the crude extract of the mat of *R. complanata.*[53] We reexamined the remaining mat of *R. complanata* and found a small quantity of *F. dilatata* intermingled in the *R. complanata* sample. Thus, it is considered that the hapten of *R. complanata* is not due to the chemical components of this liverwort, but to (+)-frullanolide from *F. dilatata.*[53] Potentially allergenic eudesmanolides, germacranolides, guaianolides, and drimanolides have been isolated not only from leafy liverworts, but also from thalloid liverworts.[12] The pungent polygodial **(21)** caused allergy on the skin of the guinea pig.[54] A large, beautiful liverwort, *Schistochila appendiculata,* causes contact dermatitis. This liverwort elaborates the long-chain alkyl phenols 3-undecyl phenol **(46)**, 6-undecyl salicylic acid **(47)**, and potassium 6-undecyl salicylate **(48)** as the major components.[55] 6-Undecyl catechol **(49)**, 3-tridecyl **(50)** and 3-pentadecyl phenol **(53)**, 6-tridecyl **(51)** and 6-pentadecyl salicylic acid **(54)**, and potassium 6-tridecyl **(52)** and potassium 6-pentadecyl salicylate **(55)** were also detected in *S. appendiculata* as the minor components.[55] The allergenic reaction brought about by *S. appendiculata* might be due to the presence of one of these long-chain phenolic compounds, such as catechol **(49)**.

V. CYTOTOXIC AND ANTITUMOR ACTIVITY AND 5-LIPOXYGENASE, CALMODULIN, AND THROMBOXANE SYNTHETASE INHIBITORY ACTIVITY

Some liverworts produce marchantin-, riccardin-, plagiochin-, isomarchantin-, isoriccardin-, and perrottetin-type bis(bibenzyls) possessing cytotoxicity.[9,13-18] The female, male, and sterile thalli of the Japanese, Indian, and German *Marchantia polymorpha,* the Japanese *M. paleaceae* var. *diptera,* and *M. tosana* produce marchantin-type cyclic bis(bibenzyls) **(56** to **67)** of which marchantin A **(56)** is the major component (Figure 4). About 100 g of compound **56** are obtained from 2 kg of the dried *M. paleaceae* var. *diptera* and from 8 kg of the dried *M. polymorpha.* The structure of the new natural product **(56)** was established by chemical reaction and by extensive 400-MHz ^{1}H-NMR and 100-MHz ^{13}C-NMR spectra (spin decoupling, nuclear Overhauser effect [NOE], ^{1}H-^{1}H and ^{13}C-^{1}H 2D correlation spectroscopy [COSY], ^{13}C-^{1}H long-range 2D COSY NMR). The stereochemistry of **56** was established by the single crystallographic X-ray analysis of a trimethyl ether of **56**. The total synthesis of **56** has been accomplished in 12 steps by Kodama et al.[61] The structures of the other marchantins were also elucidated by the same methods described above. The similar cyclic bis(bibenzyls), riccardins **(68** to **73)** having a biphenyl ether and a biphenyl linkage, have been found in *Marchantia,*[59] *Riccardia,*[62] *Reboulia,*[63] and *Monoclea* (Figure 5).[64] The structure of riccardin A **(68)** was determined by the X-ray crystallographic analysis of a triacetate of **68**.[62] Shiobara et al.[65] synthesized riccardin B **(69)** possessing ether linkages by an intramolecular Wittig-type reaction. Iyoda et al.[66] synthesized the same compound by a nickel-catalyzed intramolecular cyclization reaction in high yield. The constitution of riccardin B **(69)** was further confirmed by the synthesis of its di-*ortho* methyl ether using Ullmann coupling, Wittig reaction, and intermolecular Wurz reaction.[67] *Marchantia poly-*

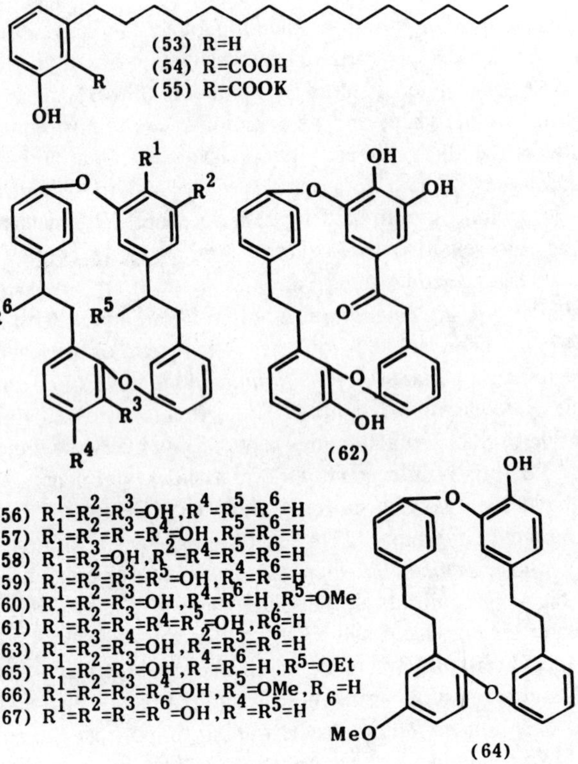

(53) R=H
(54) R=COOH
(55) R=COOK

FIGURE 4.

morpha and *M. palmata* contain isomarchantin C **(74)** and isoriccardin C **(75)**, whose structures have a different ether or biphenyl linkage from those of marchantin C **(58)** and riccardin C **(70)**.[59] *Radula* spp. contain linear bis(bibenzyl) ethers which may be precursors of the cyclic bis(bibenzyls) found in the Hepaticae. Perrottetin E **(76)**, perrottetin F **(77)**, and perrottetin G **(78)** were obtained from *Radula perrottetii*.[15,56,68] The former compound **(76)** has been synthesized in seven steps from 3,4-dihydroxybenzaldehyde.[68]

Plagiochila acanthophylla subsp. *japonica* elaborates plagiochin-type cyclic bis(bibenzyls) **(79** to **82)** (Figure 6).[69] Marchantin A **(56)** indicated cytotoxicity against KB cells (Table 5) and intense 5-lipoxygenase and calmodulin-inhibitory activity (Tables 6 and 7).[13-18,70] The other bis(bibenzyls) also showed cytotoxic activity against KB cells and 5-lipoxygenase and calmodulin inhibitory activity, as shown in Tables 5 and 7. In addition to the linear bis(bibenzyls), *Radula complanata*, *R. perrottetii*, and *R. kojana* elaborate prenyl bibenzyls whose structures have been established by spectroscopic methods and syntheses.[71] The compounds **(83, 84, 86, 87,** and **90)** possess 5-lipoxygenase and calmodulin inhibitory activity (Table 7).[18,70] The bibenzyls **(89** and **91)** isolated from *R. complanata*, the simple bibenzyls **(92** to **95)** from some *Radula* and *Frullania* species, and a labdane-type diterpene diol **(96)** show weak calmodulin inhibitory activity (Table 7; Figure 7).[18,70] Plagiochiline A **(30)**, guaianolides **(98** to **100)** isolated from *Wiesnerella denudata*, pinguisenol **(101)** from *Trocholejeunea sandvicensis*, and 4-epi-arbusculin A **(102)** from *Frullania tamarisci* subsp. *obscura* possess cytotoxic activity against KB cells (Table 5).[13-18] Belkin et al.[72] reported that an alcoholic or acidic extract of *Polytrichum juniperinum* showed antitumor activity against carcinoma caused by sarcoma 37 transported into the muscle of CAF_1 mice. However, no active substance has been isolated from *P. juniperinum*. Yamada et al.[73] found that the ether extract of *Isothecium subdiversiforme* showed intense antitumor activity (IC_{50} = 3

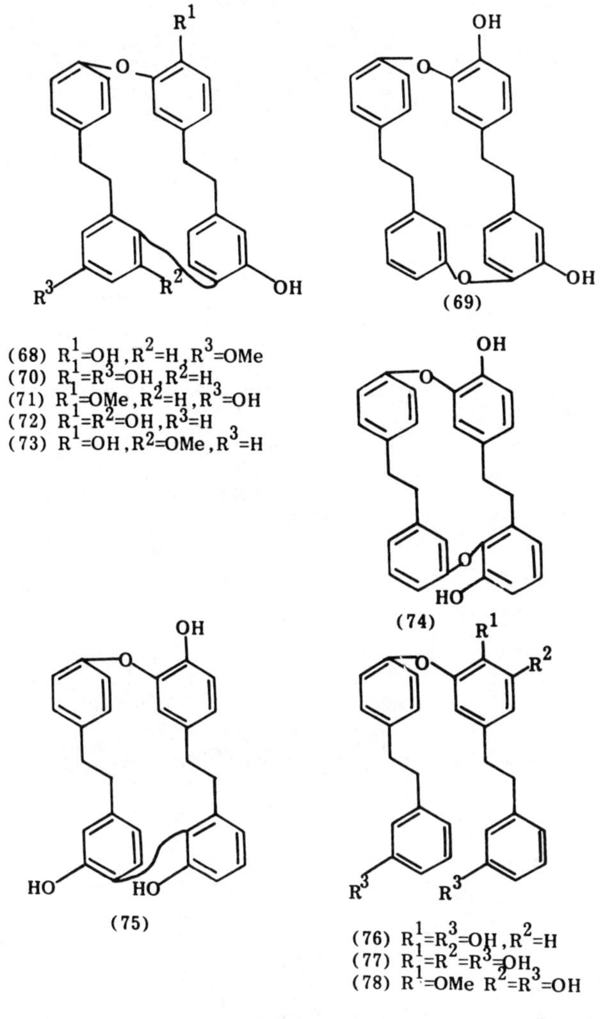

(68) R^1=OH, R^2=H, R^3=OMe
(70) R^1=R^3=OH, R^2=H
(71) R^1=OMe, R^2=H, R^3=OH
(72) R^1=R^2=OH, R^3=H
(73) R^1=OH, R^2=OMe, R^3=H

(69)

(74)

(75)

(76) R^1=R^3=OH, R^2=H
(77) R^1=R^2=R^3=OH
(78) R^1=OMe R^2=R^3=OH

FIGURE 5.

μg/ml) against mouse lymphocytic leukemia (P388). The active fraction concentrated ten times gave T/C values of 149% in a dosage of 40 mg/kg for 1 to 5 days against P388. From the fresh mosses (20.3 kg) four maytansinoids **(103** to **106)** have been obtained, with yields between 2.0 × 10^{-5} and 8.0 × 10^{-6} g. The compounds **(104** to **106)** have been isolated from the fermentation broth of a *Nocardia* species,[74] the African plant *Maytenus buchananii,*[75] and the Indian plant *Trewia nudiflora,*[76] respectively (Figure 8). 15-Methoxyansamitocine P-3 **(103)** showed strong antitumor activity (IC_{50} = 2.0 × 10^{-5} μg/ml) against P388 *in vitro.*[73] This activity is 100 times stronger than that of the known antitumor drug adriamycin (IC_{50} = 1.7 × 10^{-3} μg/ml). However, it cannot be neglected that the above compounds might originate from the metabolites of microorganisms intermingled in the moss, since their content is low. Lunularic acid **(107)**, which is widely distributed in the Hepaticae, shows thromboxane synthetase inhibitory activity (ID_{50} = 5.6 × 10^{-3} mol).[77] The content of **107**, however, is very poor in many liverworts. A large amount of **107** was obtained from hydrangenol **(108)** which was easily prepared from a hydrangenol glucoside **(109)** isolated from the higher plant *Hydrangea macrophylla.* Compound **108** was treated with sodium borohydride and palladium chloride to give lunularic acid **(107)** in good yield (84%) (Figure 9).[78]

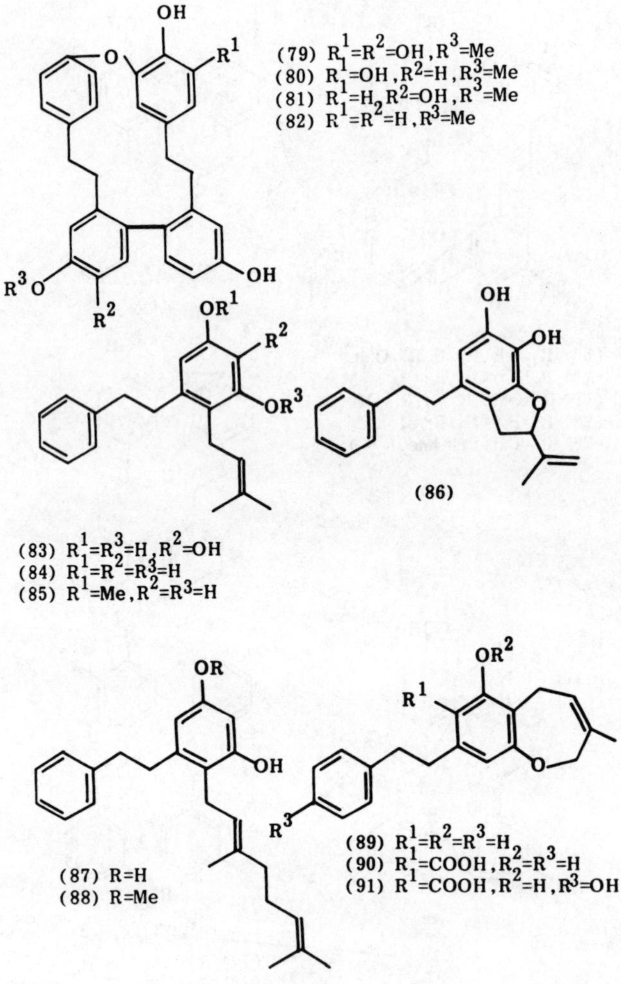

(79) $R^1=R^2=OH, R^3=Me$
(80) $R^1=OH, R^2=H, R^3=Me$
(81) $R^1=H, R^2=OH, R^3=Me$
(82) $R^1=R^2=H, R^3=Me$

(83) $R^1=R^3=H, R^2=OH$
(84) $R^1=R^2=R^3=H$
(85) $R^1=Me, R^2=R^3=H$

(86)

(87) R=H
(88) R=Me

(89) $R^1=R^2=R^3=H$
(90) $R^1=COOH, R^2=R^3=H$
(91) $R^1=COOH, R^2=H, R^3=OH$

FIGURE 6.

VI. VASOPRESSIN ANTAGONIST AND CARDIOTONIC ACTIVITY

Prenyl bibenzyl **(110)** isolated from *Radula complanata* collected in France exhibited vasopressin (VP) antagonist activity, as shown in Table 8. However, its isomer **(85)** obtained from *R. kojana* did not show the same activity.[79] Two synthetic prenyl bibenzyls **(88** and **111)**, which are the monomethylated derivatives of the natural bibenzyls **(87** and **112)** found in *Radula* spp., also showed the same activity. Again, marchantin A **(56)** increased coronary blood flow (2.5 ml/min at 0.1 mg);[79] hence, this compound may be useful as a coronary vasodilator.

VII. IRRITANCY AND TUMOR PROMOTING ACTIVITY

12-*O*-Tetradecanoylphorbol-13-acetate (TPA) **(113)**, teleocidin A **(114)**, and aplysiatoxin, which are potent tumor promoters, cause irritation on mouse ear, induction of ornithine decarboxylase (ODC), and adhesion of cultured human promyelocytic leukemia cells (HL-60).[80] Two hot sesquiterpenoids, polygodial **(21)** and plagiochiline A **(30)**, caused the same irritation of mouse ear as those of the above complex compounds (Table 9); however, neither

TABLE 5
Cytotoxic Compounds Isolated from Bryophytes

Compounds	Species	ED_{50} (μg/ml) in KB cell	T/C(%) P388
Diplophyllin **(28)**	*Chiloscyphus polyanthus*	2.1	
	Diplophyllum albicams		
Tulipinolide **(29)**	*Wiesnerella denudata*	0.45	
Plagiochiline A **(30)**	*Plagiochila fruticosa*	2.95	115
Oxyfrullanolide **(43)**	*Frullania dilatata*	0.80	
Eremofrullanolide **(45)**	*F. dilatata*	1.70	
Marchantin A **(56)**	*Marchantia polymorpha*	8.39	117
Marchantin B **(57)**	*M. polymorpha*	10.0	
Marchantin C **(58)**	*M. polymorpha*	10.0	
Riccardin A **(68)**	*Riccardia multifida*	10.0	
Riccardin B **(69)**	*R. multifida*	10.0	
Perrottetin E **(76)**	*Radula perrottetii*	12.5	
Zaluzanin D **(98)**	*W. denudata*	1.50	
8α-Acetoxyzaluzanin C **(99)**	*W. denudata*	1.60	
8α-Acetoxyzaluzanin D **(100)**	*W. denudata*	2.50	
Pinguisenol **(101)**	*Trocholejeunea sandvicensis*	12.5	
4-Epiarbusculin A **(102)**	*F. tamarisci* subsp. *obscura*	0.50	
15-Methoxyansamitocine **(103)**[73]	*Isothecium subdiversiforme*		145
(+)-Epoxyfrullanolide **(126)**[a]		2.65	

Note: Numbers in parentheses = structures found in text.

[a] Derived from (+)-frullanolide **(41)**.

TABLE 6
Screening of Marchantin A (56) Regulating *In Vitro* Arachidonate Metabolism

Concentration (*M*)	Inhibition (%)			
	LTB_4	5-HETE	HHT	12-HETE
10^{-4}			12	36
10^{-5}	89	99	3	19
10^{-6}	94	97		
10^{-7}	45	70		
10^{-8}	16	40		
10^{-9}				

Note: LTB_4 = 5(*S*),12(*R*)-dihydroxy-6,8,10,14-eicosatetraenoic acid; 5-HETE = 5-hydroxy-6,8,11-14-eicosatetraenoic acid; HHT = 12-hydroxy-5,8,10-heptadecatrienoic acid; 12-HETE = 12-hydroxy-5,8,10,14-eicosatetraenoic acid. Number in parentheses = structure found in text.

of them has ODC induction activity. On the other hand, sacculatal **(22)** and plagiochiline C **(115)**, which are slightly irritating substances, caused ODC induction.[9,13-18,70] Thus, these terpenoids are potential tumor promoters. This ODC induction is completely inhibited by 13-*cis*-retinoic acid.[18,70] (+)-Frullanolide **(41)**, obtained from *Frullania dilatata* in high yield, inhibited tumor promotion.[81]

VIII. ANTIMICROBIAL AND ANTIFUNGAL ACTIVITY

As shown in Tables 10 and 11, bryophytes display antibiotic activity.[3,11,82-85] The li-

TABLE 7
5-Lipoxygenase and Calmodulin Inhibitory Activity of Bibenzyls, Bis(bibenzyls), Prenylbibenzyls, and Diterpenoid Isolated from Liverworts

Compounds	Inhibition	
	5-Lipoxygenase (10^{-6} *M*)	Calmodulin (ID_{50}, μg/ml)
Marchantin A (**56**)	See Table 5	1.85
Marchantin D (**59**)	40	6.0
Marchantin E (**60**)	36	7.0
Riccardin A (**68**)	4	20.0
Perrottetin E (**83**)	76	3.5
Prenyl bibenzyl (**84**)	50	4.9
Perrottetin D (**86**)	40	2.0
Prenyl bibenzyl (**87**)	11	4.0
Prenyl bibenzyl (**89**)	x	95.0
Radulanin H (**90**)	15	17.0
Prenyl bibenzyl (**91**)	x	18.5
Bibenzyl (**92**)	x	100
Bibenzyl (**93**)	x	100
Bibenzyl (**94**)	x	100
Bibenzyl (**95**)	x	100
Labdane-type diterpenoid (**96**)	x	82.0

Note: x = not tested; numbers in parentheses = structures found in text.

pophilic extracts of several liverworts (*Bazzania, Frullania, Marchantia, Plagiochila, Porella,* and *Radula* spp.) show antibacterial and antifungal activity. Marchantin A **(56)** inhibited the growth of both bacteria and fungi (Tables 12 and 13).[14-18] Matsuo et al.[86,87] isolated a new dolabellane diterpenoid, acetoxyodontoschimenol **(118)**, from *Odontoschisma denudatum.* Compound **118** and its derivatives **(119** to **121)** inhibited the growth of the pathogenic fungus *Botrytis cinerea* (Figure 10). For example, **118** inhibited the growth of *B. cinerea* by 20 to 40% at a concentration of 100 μg/ml.[3] Ichikawa et al.[88,89] isolated cyclopentanoyl fatty acids **(145** to **150)** and their precursors, 9*Z*,12*Z*,15*Z*-octadecatriene-6-ynoic acid **(151)** and 1-monoglycerides **(152)** of 13-hydroxy-9*Z*,11*E*,15*Z*-octadecatriene-6-ynoic acid **(153)**, from *Dicranum scoparium* and *D. japonicum.* The compounds **(145** and **148)** were found to have antimicrobial and antifungal activity. The same activity has been found in compounds **151** and **152**.[89] The antibiotically active substances of *Atrichum, Dicranum, Mnium, Polytrichum,* and *Sphagnum* spp. were considered to be polyphenolic compounds.[85] Several antimicrobial and antifungal compounds isolated from the bryophytes are shown in Table 14.[14-18,90,91]

IX. INSECT ANTIFEEDANT ACTIVITY

Generally, liverworts and mosses are not damaged by larval and adult insects. The lipophilic extracts of pungent and bitter liverworts show potent antifeedant activity against the African army worm *(Spodoptera exempta).* Nakanishi and Kubo[92] reported that the potent pungent sesquiterpene dialdehyde warburganal **(122)**, isolated from the African tree *Warburgia ugandensis,* showed intense antifeedant activity against the above insect larvae (100 ng/cm², leaf disk method). Plagiochiline A **(30)**, possessing a powerful hot taste, showed 10 to 100 times stronger antifeedant activity (1 to 10 ng/cm², leaf disk method) than **122** against the same insect larvae described above.[9,12-18,93] The pungent sacculatal **(22)**, eudes-

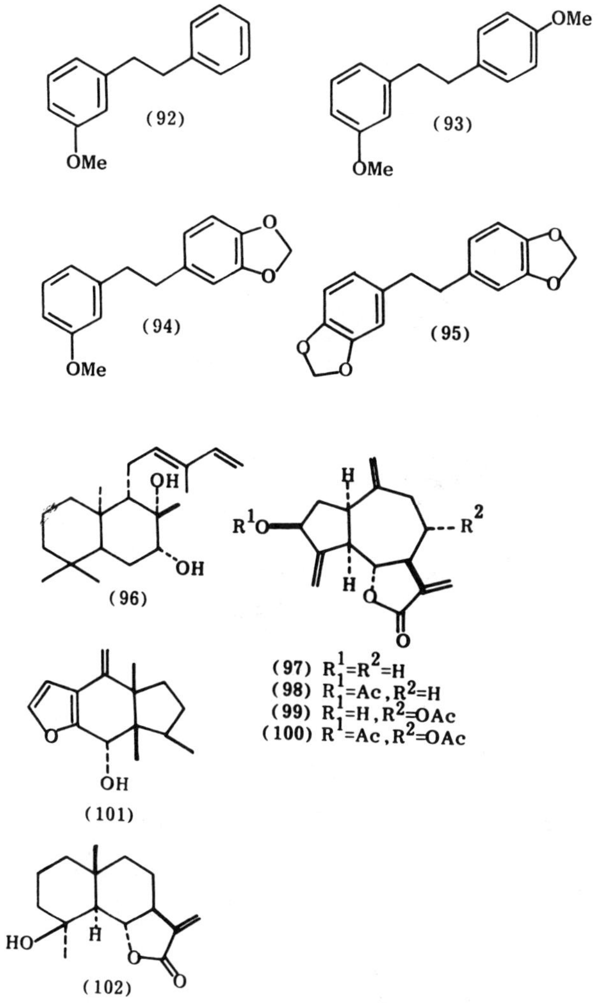

(92) (93) (94) (95) (96)

(97) $R^1=R^2=H$
(98) $R^1=Ac, R^2=H$
(99) $R^1=H, R^2=OAc$
(100) $R^1=Ac, R^2=OAc$

(101) (102)

FIGURE 7.

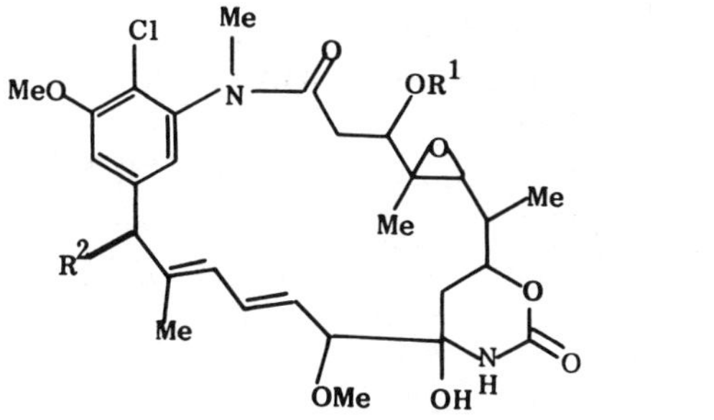

(103) $R^1=COCH(Me)_2, R^2=OMe$
(104) $R^1=COCH(Me)_2, R^2=H$
(105) $R^1=COCHMe\cdot NMe\cdot COCH(Me)_2, R^2=H$
(106) $R^1=COCHMe\cdot NMe\cdot COCH(Me)_2, R^2=OMe$

FIGURE 8.

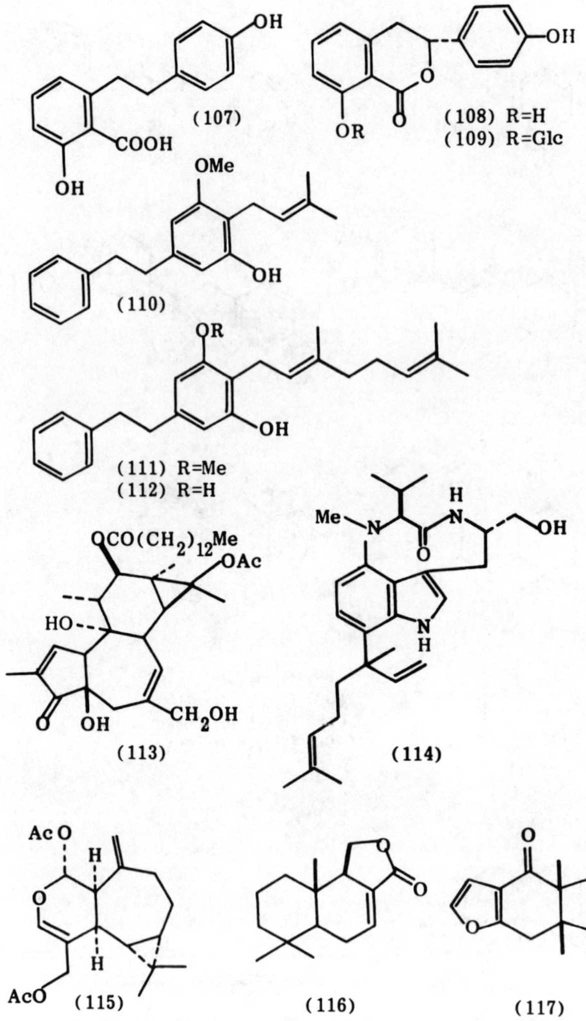

FIGURE 9.

TABLE 8
Vasopressin (VP) Antagonist Activity of Prenyl Bibenzyls

Compounds	ID_{50} (μg/ml)	Sources
(85)	—[a]	*Radula kojana*
(87)	—[a]	*R. complanata, R. kojana, R. japonica, R. javanica*
(88)	57	Synthetic
(110)	27	*R. complanata*
(111)	17	Synthetic
(112)	—[a]	*R. kojana*

Note: Numbers in parentheses = structures found in text.

[a] No activity.

TABLE 9
Irritancy and Ornithine Decarboxylase (ODC) Activity of Terpenoids Isolated from Liverworts

Compounds	Species	Irritancy (100 μg/ear of mouse)	ODC activity (nmol CO_2/mg protein)
Polygodial (**21**)	*Porella vernicosa* complex	+ + +	0
Sacculatal (**22**)	*Pellia endiviifolia*	+	1.7
	Trichocoleopsis sacculata		
Plagiochiline A (**30**)	*Plagiochila fruticosa*	+ + + +	0
	P. ovalifolia		
	P. semidecurrens		
TPA (**113**)	*Sapium sebiferum*[a]	+ + + +	7.2
Teleocidin A (**114**)	*Streptomyces mediocidicus*[b]	+ + + +	3.6
Plagiochiline C (**115**)	*P. fruticosa*	+	2.5
	P. ovalifolia		
	P. yokogurensis		

Note: Numbers in parentheses = structures found in text.

[a] Higher plant (Euphorbiaceae).
[b] Bacterium.

TABLE 10
Antimicrobial Bryophytes

Hepaticae	Musci
Conocephalum conicum	*Atrichum angustatum*
Dumortiera hirsuta	*A. undulatum*
Marchantia domingensis	*Dicranum scoparium*
Marchantia polymorpha	*Isothecium stolonigerum*
Metzgeria furcata	*Mnium punctatum*
Porella platyphylla	*Orthotrichum rupestre*
	Polytrichum commune
	Polytrichum juniperinum
	Thuidium recognitum var. *delicatulum*
	Tortula ruralis
	Sphagnum fimbriatum
	S. palustre
	S. strictum

manolide (**26** and **27**), germacranolides (**29**), the other sesquiterpene lactones (**41, 42,** and **116**), and the bitter gymnocolin (**33**) also have antifeedant activity against the larvae of the Japanese *Pieris* species; however, their activity is significantly less than that of plagiochiline A.[13,18] Pinguisone (**123**) isolated from *Aneura pinguis*[9] showed weak antifeedant activity against larvae of Japanese *Spodoptera littoralis*.[94] Compound **123** has been synthesized totally by Bernasconi et al.[95]

X. PISCICIDAL ACTIVITY

Table 15 shows the piscicidal activity of terpenoids isolated from some liverworts against the killifish *(Oryzia latipes)*.[16] The strongest piscicidal compounds are the pungent polygodial (**21**) and sacculatal (**22**).[13-18,96] The fish is killed within 2 h by a 0.4 ppm solution and within

TABLE 11
Antifungal Bryophytes

Hepaticae	Musci
Conocephalum conicum	*Atrichum undulatum*
Diplophyllum albicans	*Bryum pollens*
Lunularia cruciata	*Dicranella heteromalla*
Marchantia polymorpha	*Dicranella scoparium*
Porella vernicosa complex	*Mniobryum delicatulum*
	Mnium hornum
	Oligotrichum hercynicum
	Plagiothecium denticulatum
	Pogonatum aloides
	Pogonatum urnigerum
	Polytrichum commune
	Pseudoscleropodium purum
	Sphagnum nemoreum
	S. portoricense
	S. strictum
	S. subsecundum

TABLE 12
Antimicrobial Activity of Marchantin A (56)

Microorganisms	MIC (μg/ml)
Acinetobacter calcoaceticus	6.25
Alcaligenes faecalis	100
Bacillus cereus	12.5
B. megaterium	25
B. subtilis	25
Cryptococcus neoformans	12.5
Enterobacter cloacae	100
Escherichia coli	100
Proteus mirabilis	100
Pseudomonas aeruginosa	100
Salmonella typhimurium	100
Staphylococcus aureus	3.13—25

Note: Number in parentheses = structure found in text.

20 min by a 7 ppm solution of **21** and **22.** The killifish is also killed within 2 h by a 0.4 ppm solution of non-natural (+)-polygodial **(124)** possessing a potent hot taste.[18] Thus, it is confirmed that piscicidal activity against the killifish is not affected by chiral specificity of **21** and **124.** Polygodial **(21)** is very toxic against bitterling fish, which are killed within 3 min by a 0.4 ppm solution.[18] Some mosquito larvae are killed within 4 h by a 40 ppm solution of **21.**[16] On the other hand, isosacculatal **(25)** and isopolygodial **(125)** (which has been obtained from *Polygonum hydropiper*) lack fish-killing activity even in a 10,000 ppm solution.[13-18] From the above results it is obvious that the occurrence of piscicidal activity of polygodial and sacculatal is significantly related to hot taste, that is, the absolute configuration of a formyl group at C-9. Costunolide **(136)** isolated from *Marchantia polymorpha* also possesses weak piscicidal activity (TLm = 2 ppm) against the killifish.[97]

XI. PLANT GROWTH REGULATORY ACTIVITY

It is known that higher plants sometimes do not grow in places where some bryophytes

TABLE 13
Antifungal Activity of Marchantin A (56)

Fungi	MIC (μg/ml)
Alternaria kikuchiana	100
Aspergillus fumigatus	100
Aspergillus niger	25—100
Candida albicans	100
Microsprorum gypseum	100
Penicillium chrysogenum	100
Piricularia oryzae	12.5
Rhizoctonia solani	50
Saccharomyces cerevisiae	100
Sporothrix schenckii	100
Trichophyton mentagrophytes	3.13
T. rubrum	100

Note: Number in parentheses = structure found in text.

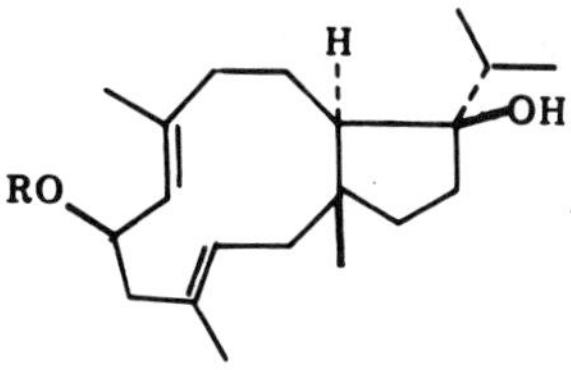

(118) R=Ac
(119) R=H
(120) 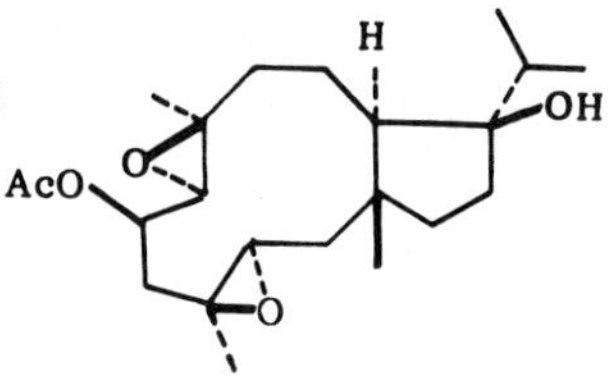
(121) 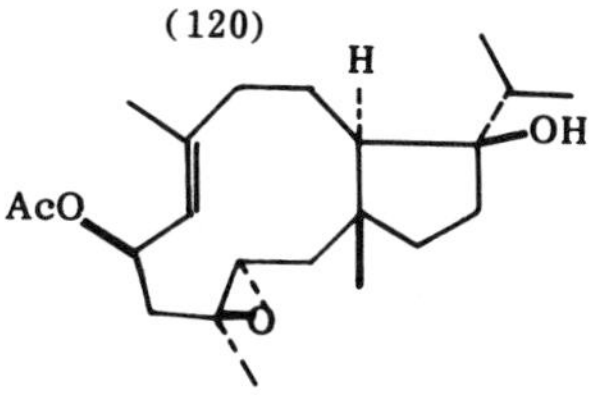
(122) 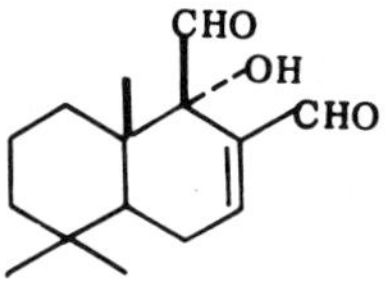
(123) 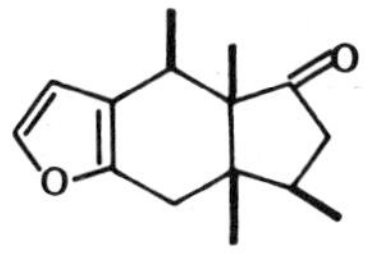
(124) 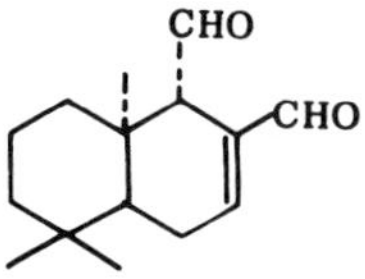
(125) 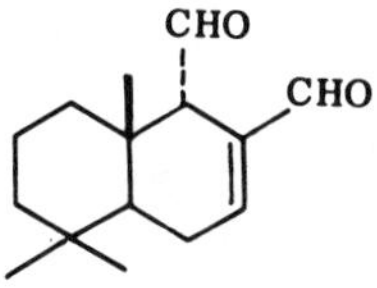
(126) 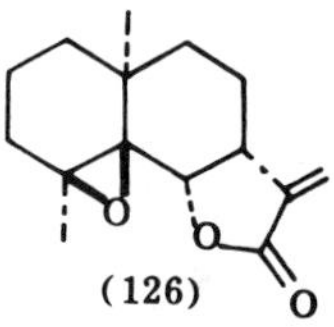
(127) 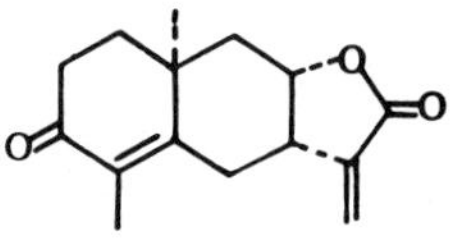
(128) 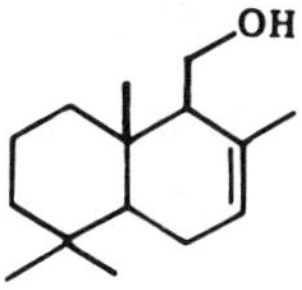
(129) 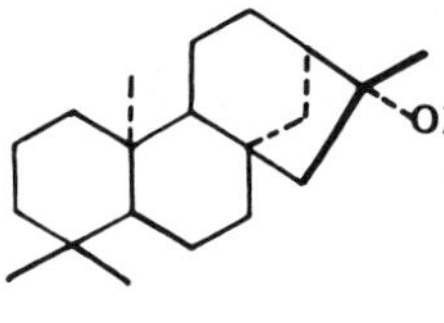
(130) 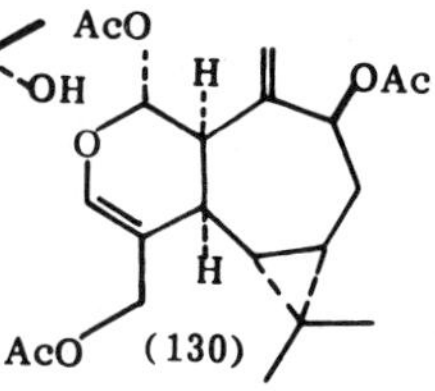

FIGURE 10.

TABLE 14
Antimicrobial and Antifungal Activity of Terpenoids and Bibenzyl Isolated from Bryophytes

Microorganisms	Inhibition (MIC µg/ml)									
	(21)	(22)	(87)	(107)[90]	(110)	(116)[91]	(117)	(118)[3]	(145)[88]	(148)[88]
Artenaria brassicicola		100		50						
Aspergillus fumigatus	100	100								
Aspergillus niger	25						100			
Bacillus cereus										
Botrytis cinerea				100				100	400	400
Candida albicans	100	25								
Cryptococcus neoformans	100	50								
Microsporum gypseum						20				
Piricularia oryzae										
Saccharomyces cerevisiae		25							60	60
Septoria nodorum				50						
Staphylococcus aureus	50—100		20		3					
Trichophyton mentagrophytes	50					10				
T. rubrum						20				
Uromyces fabae				25						

Note: Numbers in parentheses = structures found in text.

TABLE 15
Piscicidal Activity of Terpenoids Isolated from Liverworts

Compounds	Species	Fishes	ppm/min[a]
(−)-Polygodial (**21**)	*Porella vernicosa* complex	*Oryzia latipes*	0.4/120
		Bitterling fish	0.4/3
(+)-Polygodial (**124**)	Synthetic	*Oryzia latipes*	0.4/120
		Bitterling fish	0.4/3
Isopolygodial (**125**)	*Polygonum hydropiper*	*Oryzia latipes*	—[b]
Sacculatal (**23**)	*Pellia endiviifolia*	*Oryzia latipes*	0.4/120
	Trichocoleopsis sacculata		
Isosacculatal (**25**)	*Pellia endiviifolia*	*Oryzia latipes*	—[b]
	T. sacculata		
Diplophyllin (**28**)	*Diplopyllum albicans*	*Oryzia latipes*	6.7/240
Plagiochiline A (**30**)	*Plagiochila fruticosa*	*Oryzia latipes*	0.4/240
	Plagiochila ovalifolia		
(+)-Frullanolide (**41**)	*Frullania dilatata*	*Oryzia latipes*	0.4/240

Note: Numbers in parentheses = structures found in text.

[a] Concentration at which all fishes were killed.
[b] Inactive.

occur. We suggested that some liverworts elaborate allelopathic compounds.[98] In fact, most crude extracts of bryophytes show inhibitory activity against germination, root elongation, and second coleoptile growth of rice in husk, wheat, lettuce, and radish. The pungent or nonpungent sesquiterpene lactones, pungent sesqui- and diterpenoids, and 2,3-secoaromadendrane-type sesquiterpene hemiacetals exhibited plant growth inhibitory activity (25 to 500 ppm) against rice in husk, as shown in Table 16.[9,12] Polygodial (**21**) completely inhibited the germination of rice in husk at 100 ppm. At a concentration of <25 ppm it dramatically promoted root elongation of rice.[99] Epoxyfrullanolide (**126**), derived from (+)-frullanolide (**41**) isolated from *Frullania dilatata,* inhibited the germination of rice in husk ten times better than the original lactone.[12] Huneck and Schreiber[100] reported several plant growth regulatory active substances of the liverworts, as shown in Table 17. Longiborneol (**2**), gymnocolin (**33**), lunularic acid (**107**), drimenol (**128**), and 16-α-kaurenol (**129**) inhibited root elongation of cress. In contrast, longiborneol (**2**), gymnocolin (**33**), scapanin (**37**), and drimenol (**128**) promoted germination of wheat seeds at lower concentrations. Longifolene (**1**) promoted the growth of the coleoptile of wheat. Gorham[101] reported that lunularic acid (**107**) inhibited *Marchantia* gemmalings at 5×10^{-9} to 10^{-7} mol per gemmaling and 20% of the cress root growth at a concentration of 10^{-3} mol. Benesova and Herout[102] found that an antifeedant sesquiterpene ketone (**123**) showed intense inhibitory activity against growth of the wheat coleoptile. Arai et al.[103] reported that the coleoptile of wheat elongated in response to 0.3 to 0.03 ppm auxin was inhibited by lunularic acid (**107**) at a concentration of 10 to 30 ppm. Ando and Matsuo[3] focused on the plant growth inhibitory active substances of *Plagiochila* and *Lepidozia* species and found that five 2,3-secoaromadendrane-type sesquiterpenoids (**30, 115, 130** to **132**) and three new sesquiterpenoids [vitrenal (**133**), lepidozenal (**134**), and isobicyclogermacrenal (**135**)] inhibited the growth of rice seedlings (Table 18; Figure 11).

XII. MISCELLANEOUS

Conocephalum japonicum (= *C. supradecompositum*) contains (+)-costunolide (**136**) as the major component. This germacranolide and related guaianolides (**137**) inhibit pene-

TABLE 16
Inhibitory Activity Against Germination and Growth of Rice in Husk

Terpenoids	Species	Complete germination inhibition (ppm)	Complete inhibition of root (ppm)
Polygodial (**21**)	*Porella vernicosa* complex	100	100
Diplophyllolide (**26**)	*Chiloscyphus polyanthus* *Diplophyllum albicans*	200	100
ent-7α-Hydroxydiplophyllolide (**27**)	*C. polyanthus*	200	100
Diplophyllin (**28**)	*C. polyanthus* *D. albicans*	200	100
Plagiochiline A (**30**)	*Plagiochila asplenioides* *Plagiochila fruticosa* *Plagiochila ovalifolia* *Plagiochila semidecurrens*	100	100
Perrottetianal (**34**)	*Porella perrottetiana*	500	—[b]
(+)-Frullanolide (**41**)	*Frullania dilatata*	200	100
Zaluzanin C (**97**)	*Conocephalum conicum* *Wiesnerella denudata*	100	50
Zaluzanin D (**98**)	*Conocephalum conicum* *W. denudata*	100	50
8α-Acetoxyzaluzanin C (**99**)	*Conocephalum conicum* *W. denudata*	200	50
8α-Acetoxyzaluzanin D (**100**)	*Conocephalum conicum* *W. denudata*	100	50
(+)-Epoxyfrullanolide (**126**)	—[a]	20	10
3-Oxodiplophyllin (**127**)	*Chiloscyphus polyanthus*	200	100

Note: Numbers in parentheses = structures found in text.

[a] Derived from (+)-frullanolide (**41**).
[b] Not tested.

TABLE 17
Inhibitory and Promotory Effects of Terpenoids and Bibenzyl on Growth of Cress Root, Wheat Coleoptile, and Germination of Wheat Seed[100]

Compounds	Species	Inhibition, cress root (*M*)	Promotion, wheat seed (*M*)	Promotion, wheat coleoptile (*M*)
Longifolene (**1**)	*Scapania undulata*			10^{-4}—10^{-7}
Longiborneol (**2**)	*S. undulata*	10^{-3}—10^{-4}	10^{-6}—10^{-7}	
Gymnocolin (**33**)	*Gymnocolea inflata*	10^{-3}—10^{-4}	10^{-6}—10^{-7}	
Scapanin (**37**)	*S. undulata*		10^{-6}—10^{-7}	
Lunularic acid (**107**)	*Lunularia cruciata*	10^{-3}—10^{-4}	10^{-6}—10^{-7}	
Drimenol (**128**)	*Bazzania trilobata*	10^{-3}—10^{-4}	10^{-6}—10^{-7}	
16α-Kaurenol (**129**)	*Anthelia julacea* *A. juratzkana*	10^{-3}—10^{-4}		

Note: Numbers in parentheses = structures found in text.

TABLE 18
Inhibitory Activity of Terpenoids against Growth of Leaf and Root of Rice Seedlings[3]

Compounds	Species	Complete inhibition (ppm)
Plagiochiline A **(30)**	*Plagiochila ovalifolia*	50
Plagiochiline C **(115)** (= ovalifoliene)	*P. ovalifolia*	25
9α-Acetoxyovalifoliene **(130)**	*P. semidecurrens*	25
Ovalifolienalone **(131)**	*P. ovalifolia*	500
Ovalifolienal **(132)**	*P. ovalifolia*	50
Vitrenal **(133)**	*Lepidozia vitrea*	25
Lepidozenal **(134)**	*L. vitrea*	250
Isobicyclogermacrenal **(135)**	*L. vitrea*	50

Note: Numbers in parentheses = structures found in text.

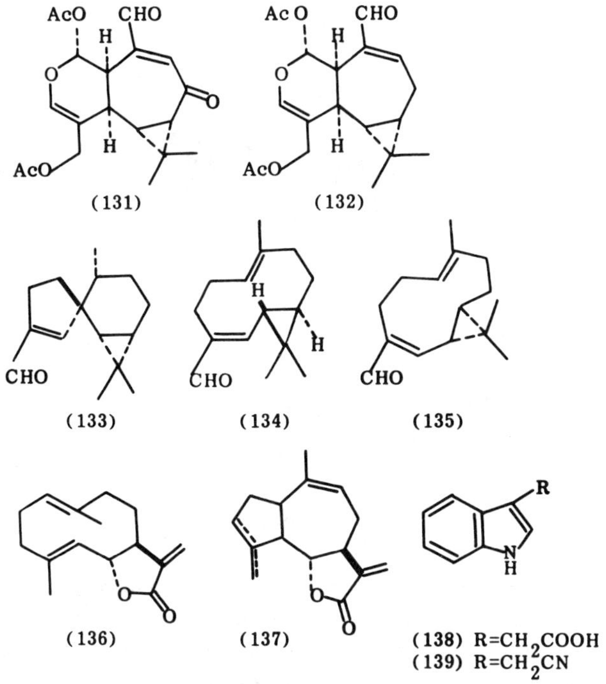

FIGURE 11.

tration of cercariae of the trematode *Schistosoma mansoni,* which causes the miserable disease schistosomiasis.[9,12,14,16,104] Wilson and Schwabe[105] isolated a hormone from *Lunularia cruciata* which inhibited growth and promoted dormancy. Independently, Fries[106] isolated the same compound from *Marchantia polymorpha.* Valio et al.[107] established the structure of the active compound to be lunularic acid **(107)**. This aromatic acid is widespread in the Hepaticae, but does not occur in the Musci, Anthocerotae, pteridophytes, and algae.[108,109] Pryce[108] reported the structural relationship between **107** and abscisic acid, which is a common hormone found in higher plants. The endogenous plant hormone indole-3-acetic acid (an auxin; **138**) was isolated from *Marchantia polymorpha.*[110] Compound **138** has also been found in *Asterella angusta* and *Pallavicinia canarus.*[11] Indole-3-acetonitrile **(139)** was detected in *A. angusta* and *P. canarus* and might be the precursor of **138.**[11] A cytokinin-like compound, N^6-(Δ^2-isopentenyl)adenine, was isolated from a culture medium of callus

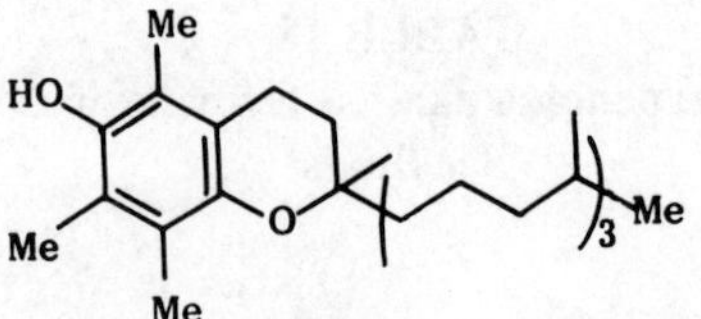

(140)

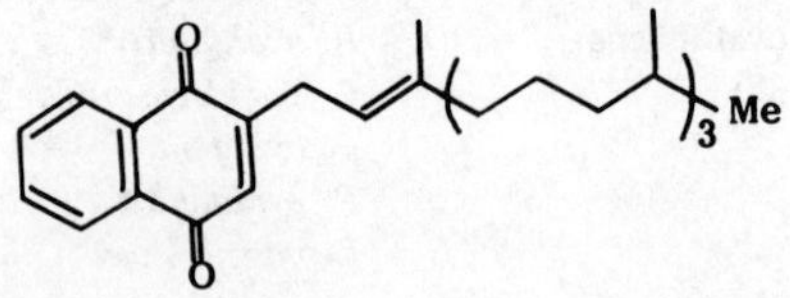

(141)

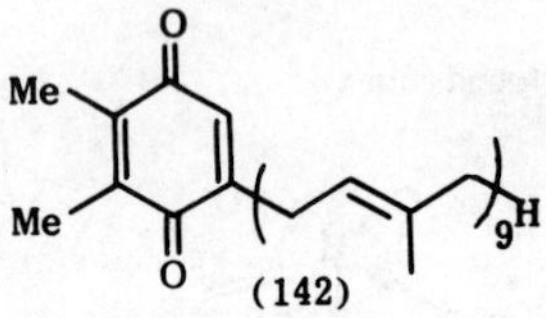

(142)

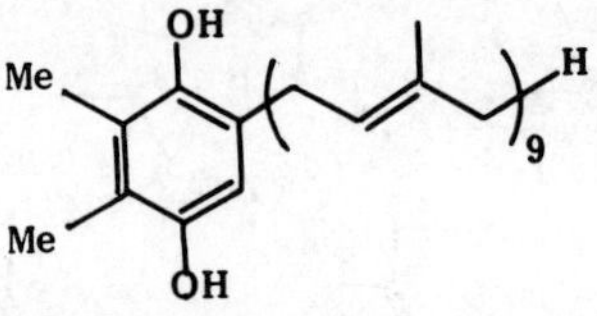

(143)

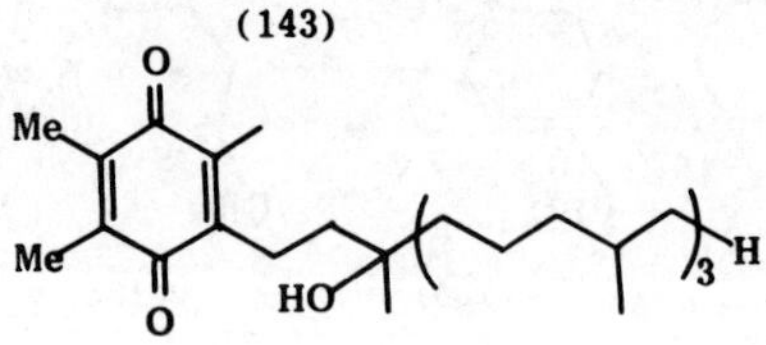

(144)

FIGURE 12.

cells of the moss hybrid of *Funaria hygrometrica* × *Physcomitrium pyriforme*.[111] Muromtsev et al.[112] noted the presence of a gibberellin-like hormone in the Musci. *Marchantia polymorpha, Pellia epiphylla, Atrichum undulatum,* and *Mnium hornum* produce α-tocopherol (vitamin E; **140**), vitamin K **(141)**, plastoquinone **(142)**, plastohydroquinone **(143)**, and α-tocoquinone **(144)** (Figure 12).[11] Some mosses are rich sources of vitamin B_2. Chickens and puppies fed with food, including the powder of *Barbella pendula, B. enervis, Floribundaria nipponica, Hypnum plumaeforme, Neckeropsis nitidula,* and *Ptychanthus stiatus* which contain vitamin B_2, gained more weight than the control, and no sickness or distaste resulted.[113] Most of the mosses produce highly unsaturated fatty acids. Ichikawa et al.[88,89] reported the presence of prostaglandin-like fatty acids **(145 to 150)** from the mosses *Dicranum scoparium, D. japonicum,* and *Leucobryum* spp. (Figure 13). These acids and the other unsaturated fatty acids are viscous liquids and are considered to be the significant components of herbivorous animals living in very cold places, protecting their bodies from the cold.[114]

XIII. CONCLUSION

Up to the present, more than 200 new compounds have been isolated from the bryophytes. Most of the compounds found in the Hepaticae are lipophilic terpenoids and aromatic

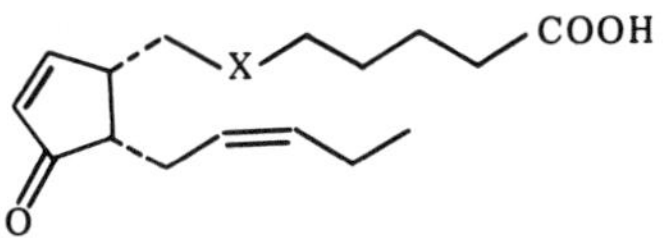

(145) X= -C≡C-
(146) X= $-(CH_2)_2-$

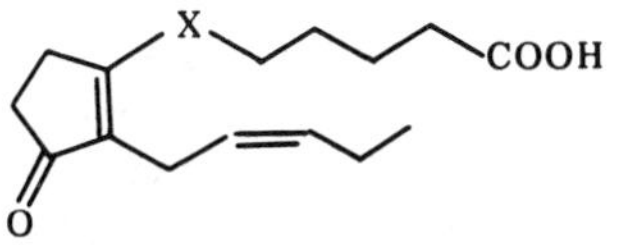

(147) X= $-CH_2-C\equiv C-$
(148) X= -CH=C=CH-
(149) X= $-CH_2-CH=CH-$
(150) X= $-(CH_2)_3-$

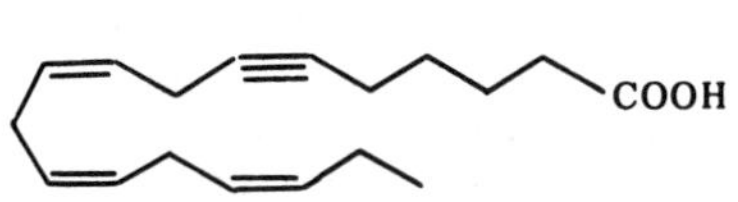

(151)

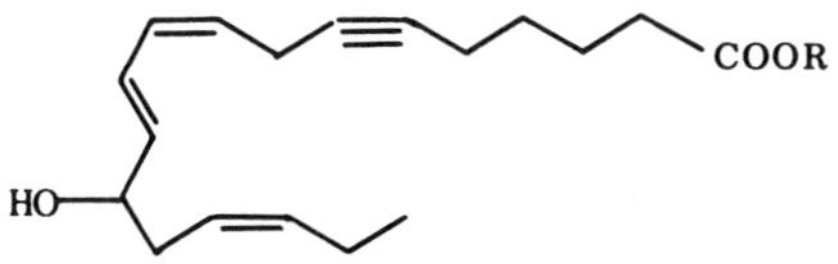

(152) R= $-CH_2-CH(OH)-CH_2OH$
(153) R=H

FIGURE 13.

compounds of which only two nitrogen-[9] and three sulfur-containing compounds have been found.[115,116] Mono- and sesquiterpenoids have not been found in the Musci and Anthocerotae. The major components of the former class are highly unsaturated fatty acids and their glycerides. As described above, the bryophytes contain some biologically active substances. However, the species of bryophytes studied chemically so far are only a small percentage of total bryophytes. Further, chemical and pharmacological investigation of the bryophytes listed in Table 1 as well as the other bryophytes may provide a number of different types of bioactive compounds for use as medicinal and agricultural drugs.

REFERENCES

1. **Ding, H.,** *Zhong guo Yao yun Bao zi Zhi wu,* Shanghai Kexue Jishu Chuban She, Shanghai, 1980, 1.
2. **Garnier, G., Bezanger-Beauquesne, L., and Debraux, G.,** *Resources Medicinales de la Flore Française,* Vol. 1, Vigot Frères Editeurs, Paris, 1961, 1.
3. **Ando, H. and Matsuo, A.,** Applied bryology, in *Advances in Bryology,* Vol. 2, Schultze-Motel, W., Ed., J. Cramer, Vaduz, West Germany, 1984, 133.
4. **Ri, Z.,** *Honzokomoku,* 21, 814, 1590.

5. **Müller, K.,** Beitrag zur Kenntnis der ätherischen Öle bei Lebermoosen, *Hoppe-Seylers Z. Physiol. Chem.*, 45, 299, 1905.
6. **Fujita, Y., Ueda, T., and Ono, T.,** Study on essential oils of Hepaticae. I. Essential oils of *Bazzania pompeana, Nippon Kagaku Zasshi,* 77, 400, 1956.
7. **Huneck, S. and Klein, E.,** Inhaltsstoffe der Mosse. III. Über die vergleichende gas- und dünnschicht-chromatographische Untersuchung der ätherischen Öle einiger Lebermoose und Isolierung von (−)-Longifolen und (−)-Longiborneol aus *Scapania undulata* (L.) Dum., *Phytochemistry,* 6, 383, 1967.
8. **Markham, K. R. and Porter, L. J.,** Chemical constituents of bryophytes, in *Progress in Phytochemistry,* Vol. 5, Reinhold, L. and Harborne, J. B., Eds., Pergamon Press, Oxford, 1978, 181.
9. **Asakawa, Y.,** Chemical constituents of the Hepaticae, in *Progress in the Chemistry of Organic Natural Products,* Vol. 42, Herz, W., Grisebach, H., and Kirby, G. W., Eds., Springer-Verlag, Vienna, 1982, 1.
10. **Huneck, S.,** Chemistry and biochemistry of bryophytes, in *New Manual of Bryology,* Vol. 1, Schuster, R. M., Ed., Hattori Botanical Laboratory, Nichinan, Japan, 1983, 1.
11. **Asakawa, Y.,** Physiologically active substances found in bryophytes, *Misc. Bryol. Lichenol.*, 7, 144, 1977.
12. **Asakawa, Y.,** Biologically active substances obtained from bryophytes, *J. Hattori Bot. Lab.*, 50, 123, 1981.
13. **Asakawa, Y.,** . Phytochemistry of Hepaticae: isolation of biologically active aromatic compounds and terpenoids, *Rev. Latinoam. Quim.*, 14, 109, 1984.
14. **Asakawa, Y.,** Biologically active substances of Hepaticae, *Kagaku To Seibutsu,* 22, 495, 1984.
15. **Asakawa, Y.,** Some biologically active substances isolated from Hepaticae: terpenoids and lipophilic aromatic compounds, *J. Hattori Bot. Lab.*, 56, 215, 1984.
16. **Asakawa, Y.,** Biologically active substances of Hepaticae, *Seiyakukojo,* 5, 284, 1985.
17. **Asakawa, Y.,** Biologically active compounds found in Hepaticae, *Farumashia,* 23, 455, 1987.
18. **Asakawa, Y.,** Biologically active substances found in Hepaticae, in *Studies in Natural Products Chemistry,* Rahman, Ed., Elsevier, Amsterdam, 1988, 277.
19. **Mizutani, M.,** Smell of some bryophytes, *Misc. Bryol. Lichenol.*, 6, 64, 1975.
20. **Suire, C., Asakawa, Y., Toyota, M., and Takemoto, T.,** Chirality of terpenoids isolated from the liverwort *Conocephalum conicum, Phytochemistry,* 21, 349, 1982.
21. **Asakawa, Y., Matsuda, R., and Takemoto, T.,** Mono- and sesquiterpenoids from *Wiesnerella denudata, Phytochemistry,* 18, 1007, 1980.
22. **Asakawa, Y.,** Comparative study of chemical constituents found in the thalli and female receptacles of *Wiesnerella denudata* and *Conocephalum conicum, J. Hattori Bot. Lab.*, 48, 277, 1980.
23. **Asakawa, Y., Toyota, M., and Cheminat, A.,** Terpenoids from the French liverwort *Targionia hypophylla, Phytochemistry,* 25, 2555, 1986.
24. **Herout, V.,** An approach to the isolation of the fragrant principle of liverwort *Lophocolea heterophylla, Flavour Fragrance J.*, 1, 43, 1985.
25. **Toyota, M., Kuwata, K., Asakawa, Y., and Frahm, J.-P.,** Fragrant component of the West German liverwort *Lophocolea heterophylla,* in Proc. 31st Symp. Chemistry of Terpenes, Essential Oils and Aromatics, Kyoto, Japan, September 10 to 12, 1987, 220.
26. **Asakawa, Y.,** unpublished data, 1987.
27. **Connolly, J. D., Harrison, L. J., and Rycroft, D. S.,** The structure of tamariscol, a new pacifigorgiane sesquiterpenoid alcohol from the liverwort *Frullania tamarisci, Tetrahedron Lett.*, 25, 1401, 1984.
28. **Matsuo, A., Nozaki, H., Nakayama, M., Kushi, Y., Hayashi, S., Komori, T., and Kamijo, N.,** (+)-Bicyclohumulenone, a novel sesquiterpene ketone of the humulane group from *Plagiochila acanthophylla* subsp. *japonica* (liverwort): X-ray crystal and molecular structure of the *p*-bromobenzoate derivative, *J. Chem. Soc. Chem. Commun.*, p. 174, 1979.
29. **Asakawa, Y.,** J.P. Patent 228409, 1985.
30. **Asakawa, Y.,** J.P. Patent 53273, 1986.
31. **Asakawa, Y.,** J.P. Patent 126013, 1986.
32. **Asakawa, Y.,** U.S. Patent 4659509, 1987.
33. **Asakawa, Y.,** European Patent 0185921 A2, 1986.
34. **Asakawa, Y., Wakamatsu, M., Takikawa, K., and Hattori, S.,** . Geographical difference of the chemical constituents of *Frullania tamarisci* subsp. *obscura,* in Proc. 31st Symp. Chemistry of Terpenes, Essential Oils and Aromatics, Kyoto, Japan, September 10 to 12, 1987, 223.
35. **Inoue, H.,** *Kokenosekai,* Idemitsu Ltd., Tokyo, 1978, 162.
36. **Mizutani, M.,** On the taste of some mosses, *Misc. Bryol. Lichenol.*, 2, 100, 1961.
37. **Asakawa, Y. and Aratani, T.,** Sesquiterpenes of *Porella vernicosa* (Hepaticae), *Bull. Soc. Chim. Fr.*, p.1469, 1975.
38. **Doskotch, R. W. and El-Feraly, F. S.,** The structure of tulipinolide and epitulipinolide. Cytotoxic sesquiterpenes from *Liriodendron tulipifera* L., *J. Org. Chem.*, 35, 1928, 1970.
39. **Huneck, S., Asakawa, Y., Taira, Z., Cameron, A. F., Connolly, J. D., and Rycroft, D. S.,** Gymnocolin, a new *cis*-clerodane diterpenoid from the liverwort *Gymnocolea inflata.* Crystal structure analysis, *Tetrahedron Lett.*, 24, 115, 1983.

40. **Hagiwara, H. and Uda, H.,** A total synthesis of (+)-perrottetianal A, *J. Chem. Soc. Chem. Commun.,* p. 1351, 1987.
41. **Toyota, M., Nagashima, F., and Asakawa, Y.,** Chemical constituents of the liverworts *Jungermannia.* II. New bitter diterpenoid glycosides from *Jungermannia infusca,* in Proc. 31st Symp. Chemistry of Terpenes, Essential Oils and Aromatics, Kyoto, Japan, September 10 to 12, 1987, 236.
42. **Huneck, S. and Overton, K. H.,** Neue Diterpenoids und andere Inhaltsstoffe aus Lebermoosen, *Phytochemistry,* 10, 3279, 1971.
43. **Connolly, J. D.,** New terpenoids from the Hepaticae, *Rev. Latinoam. Quim.,* 12, 121, 1982.
44. **Huneck, S., Baxter, G. A., Cameron, F., Connolly, J. D., Harrison, L. J., Phillips, W. R., Rycroft, D. S., and Sim, G. A.,** Dolabellane diterpenoids from the liverworts *Barbilophozia floerkei, B. lycopodioides,* and *B. attenuata:* spectroscopic and X-ray studies of structure, stereochemistry, and conformation. X-ray molecular structure of 3*S*,4*S*; 7*S*,8*S*-diepoxy-10*R*,18-dihydroxydolabellane,18-acetoxy-3*S*,4*S*,;7*S*,8*S*-diepoxydolabellane, and 10*R*,18-diacetoxy-3*S*,4*R*-epoxydolabella-7*E*-ene, *J. Chem. Soc. Perkin Trans. 1,* p. 807, 1986.
45. **Knoche, H., Ourisson, G., Perold, G. W., Foussereau, J., and Maleville, J.,** Allergenic component of a liverwort: a sesquiterpene lactone, *Science,* 166, 239, 1969.
46. **Perold, G. W., Muller, J.-C., and Ourisson, G.,** Structure d'une lactone allergisante: le frullanolide-I, *Tetrahedron,* 28, 5759, 1972.
47. **Asakawa, Y. Muller, J.-C., Ourisson, G., Foussereau, J., and Ducombs, G.,** Nouvelles lactones sesquiterpéniques de *Frullania* (Hépaticae). Isolement, structures, propriétés allergisantes, *Bull. Soc. Chim. Fr.,* p. 1465, 1976.
48. **Mitchell, J. C., Fritig, B., Singh, B., and Towers, G. H. N.,** Allergenic dermatitis from *Frullania* and Compositae, *J. Invest. Dermatol.,* 54, 233, 1970.
49. **Greene, A. E., Muller, J.-C., and Ourisson, G.,** A new approach to α-methylene γ-butyrolactones. Synthesis of (−)-frullanolide, *Tetrahedron Lett.,* p. 2489, 1972.
50. **Greene, A. E., Muller, J.-C., and Ourisson, G.,** Conversion of α-methyl to α-methylene γ-lactones. Synthesis of two allergenic sesquiterpene lactones, (−)-frullanolide and (+)-arbusculin B., *J. Org. Chem.,* 39, 186, 1974.
51. **Kido, F., Tsutsumi, R., Maruta, R., and Yoshikoshi, A.,** Lactone annulation of β-keto esters with β-vinylbutenolide and the total synthesis of racemic frullanolide, *J. Am. Chem. Soc.,* 101, 6240, 1979.
52. **Asakawa, Y., Matsuda, R., Toyota, M., Hattori, S., and Ourisson, G.,** Terpenoids and bibenzyls of 25 liverwort *Frullania* species, *Phytochemistry,* 20, 2187, 1981.
53. **Asakawa, Y., Takikawa, K., Toyota, M., and Takemoto, T.,** Novel bibenzyl derivatives and *ent*-cuparene-type sesquiterpenoids from *Radula* species, *Phytochemistry,* 21, 2481, 1982.
54. **Stampf, J.-C., Benezra, C., and Asakawa, Y.,** Stereospecificity of allergenic contact dermatitis (ACD) to enantiomers. III. Experimentally induced ACD to a natural sesquiterpene dialdehyde, polygodial in guinea pigs, *Arch. Dermatol. Res.,* 274, 277, 1982.
55. **Asakawa, Y., Masuya, T., Tori, M., and Campbell, E. O.,** Long chain alkyl phenols from the liverwort *Schistochila appendiculata, Phytochemistry,* 27, 735, 1987.
56. **Asakawa, Y., Toyota, M., Matsuda, R., Takikawa, K., and Takemoto, T.,** Distribution of novel cyclic bisbibenzyls in *Marchantia* and *Riccardia* species, *Phytochemistry,* 22, 1413, 1983.
57. **Tori, M., Toyota, M., Harrison, L. J., Takikawa, K., and Asakawa, Y.,** Total assignment of ^{1}H and ^{13}C NMR spectra of marchantins isolated from liverworts and its application to structure determination of two new macrocyclic bis(bibenzyls) from *Plagiochasma intermedium* and *Riccardia multifida, Tetrahedron Lett.,* 26, 4735, 1985.
58. **Tori, M., Masuya, T., Takikawa, K., Toyota, M., and Asakawa, Y.,** The structure and NMR of macrocyclic bis(bibenzyls) isolated from liverworts, in Proc. 28th Symp. Chemistry of Natural Products, Sendai, Japan, October 7 to 10, 1986, 9.
59. **Asakawa, Y., Tori, M., Takikawa, K., Krishnamurty, H. G., and Kar, S. K.,** Novel cyclic bis(bibenzyls) and related compounds from the Indian liverworts *Marchantia polymorpha* and *Marchantia palmata, Phytochemistry,* 27, 1811, 1987.
60. **Asakawa, Y., Okada, K., and Perold, G. W.,** Distribution of cyclic bis (bibenzyls) in the South African liverwort *Marchantia polymorpha, Phytochemistry,* 27, 161, 1988.
61. **Kodama, M., Shiobara, Y., Matsumura, K., and Sumitomo, H.,** Total synthesis of marchantin A, a cyclic bis(bibenzyl) isolated from liverworts, *Tetrahedron Lett.,* 26, 877, 1985.
62. **Asakawa, Y., Toyota, M., Taira, Z., Takemoto, T., and Kido, M.,** Riccardin A and riccardin B, two novel cyclic bis(bibenzyls) possessing cytotoxicity from the liverwort *Riccardia multifida* (L.) S. Gray, *J. Org. Chem.,* 48, 2164, 1983.
63. **Asakawa, Y. and Matsuda, R.,** Riccardin C, a novel cyclic bibenzyl derivative from *Reboulia hemisphaerica, Phytochemistry,* 21, 2143, 1982.
64. **Toyota, M., Nagashima, F., and Asakawa, Y.,** Fatty acids and cyclic bis(bibenzyls) from the New Zealand liverwort *Monoclea forsteri, Phytochemistry,* 27, 1988, 2603.

65. **Shiobara, Y., Sumitomo, H., Tsukamoto, M., Harada, C., and Kodama, M.,** Synthesis and structure confirmation of riccardin B, a macrocyclic bis(bibenzyl) from the liverwort, *Riccardia multifida, Chem. Lett.*, p. 1587, 1985.
66. **Iyoda, M., Sakaitani, M., Otsuka, H., and Oda, M.,** Synthesis of riccardin B by nickel-catalyzed intramolecular cyclization, *Tetrahedron Lett.*, 26, 4777, 1985.
67. **Nógrádi, M., Vermes, B., and Kajtárperedy, M.,** The constitution of riccardin B, *Tetrahedron Lett.*, 25, 2899, 1987.
68. **Toyota, M., Tori, M., Takikawa, K., Shiobara, Y., Kodama, M., and Asakawa, Y.,** Perrottetins E, F, and G from *Radula perrottetii* (Liverwort) — isolation, structure determination, and synthesis of perrottetin E, *Tetrahedron Lett.*, 26, 6097, 1985.
69. **Hashimoto, T., Tori, M., Asakawa, Y., and Fukazawa, Y.,** Plagiochins A, B, C, and D, a new type of macrocyclic bis(bibenzyls) having a biphenyl linkage between the ortho positions to the benzyl methylenes, from the liverwort *Plagiochila acanthophylla* subsp. *japonica, Tetrahedron Lett.*, 28, 6295, 1987.
70. **Asakawa, Y., Toyota, M., Tori, M., Fujiki, H., Suganuma, M., and Sugimura, T.,** Possible tumor promoters, and 5-lipoxygenase and calmodulin inhibitors isolated from some liverworts, in Proc. Int. Symp. Organic Chemistry of Medicinal Natural Products (IUPAC), Shanghai, China, November 10 to 14, 1985, B-021.
71. **Asakawa, Y. and Hashimoto, T.,** unpublished data.
72. **Belkin, M., Fitzgerald, D., and Felix, M. D.,** Tumor damaging capacity of plant materials. II. Plants used as diuretics, *J. Natl. Cancer Inst.*, 13, 741, 1952—1953.
73. **Yamada, K., Ichikawa, T., Yamashita, M., Tanimoto, M., Hikita, A., Kondo, K., and Sakai, K.,** Antitumor active substances of mosses: isolation and structure determination of a new maytansinoid, 15-methoxyansamitocin P-3, in Proc. 31st Symp. Chemistry of Terpenes, Essential Oils and Aromatics, Kyoto, Japan, September 10 to 12, 1987, 242.
74. **Asai, M., Mizuta, E., Izawa, M., Haibara, K., and Kishi, T.,** Isolation, chemical characterization and structure of ansamitocin, a new antitumor ansamycin antibiotic, *Tetrahedron*, 35, 1079, 1979.
75. **Kupchan, S. M., Kodama, Y., Thomas, G. J., and Hintz, H. P. J.,** Maytanprine and maytanbutine, new antileukemic ansa macrolides from *Maytenus buchananii, J. Chem. Soc. Chem. Commun.*, p. 1065, 1972.
76. **Powell, R. G., Weisleder, D., and Smith, C. R., Jr.,** Novel maytansinoid tumor inhibitors from *Trewia nudiflora:* trewiasine, dehydrotrewiasine, and demethyltrewiasine, *J. Org. Chem.*, 46, 4398, 1981.
77. **Goda, Y. and Sankawa, U.,** Thromboxane synthetase inhibitors from *Allium bakeri* Regel, in Proc. 105th Annu. Meet. Pharmaceutical Society of Japan, Kanazawa, Japan, April 3 to 5, 1985, 468.
78. **Hashimoto, T., Tori, M., and Asakawa, Y.,** A highly efficient preparation of lunularic acid and some biological activities of stilbene and dihydrostilbene derivatives, *Phytochemistry*, 27, 109, 1988.
79. **Asakawa, Y.,** unpublished data.
80. **Fujiki, H. and Sugimura, T.,** New potent tumor promoters: teleocidin, lyngbyatoxin A and aplysiatoxin, *Cancer Surv.*, 2, 539, 1983.
81. **Okamoto, H., Yoshida, D., and Saito, Y.,** Inhibition of 12-O-tetradecanoylphorbol-13-acetate-induced ornithine decarboxylase activity in mouse epidermis by sweetening agents and related compounds, *Cancer Lett.*, 21, 29, 1983.
82. **Madson, G. C. and Pates, A. L.,** Occurrence of antimicrobial substances in chlorophyllose plants growing in Florida, *Bot. Gaz. (Chicago)*, 113, 393, 1952.
83. **McCleary, J. A., Sypherd, P. S., and Walkington, D. L.,** Mosses as possible sources of antibiotics, *Science*, 131, 108, 1960.
84. **Pavletic, Z. and Stilinovic, B.,** Untersuchung über die antibiotische Wirkung von Moosextrakten auf einige Bakterien, *Acta Bot. Croat.*, 22, 133, 1963.
85. **McCleary, J. A. and Walkington, D. L.,** Mosses and antibiosis, *Rev. Bryol. Lichenol.*, 34, 309, 1966.
86. **Matsuo, A., Uohama, K., Hayashi, S., and Connolly, J. D.,** (+)-Acetoxyodontoschismenol, a new dolabellane diterpenoid from the liverwort *Odontoschisma denudatum, Chem. Lett.*, p. 599, 1984.
87. **Matsuo, A., Yoshida, K., Uohama, K., Hayashi, S., Connolly, J. D., and Sim, G. A.,** X-ray structures and conformation of 6β,12β-dihydroxy-3*E*,7*E*-dolabelladiene 6-*p*-bromobenzoate and acetoxyodontoschismenol 3*S*,4*S*-epoxide, *Chem. Lett.*, p. 935, 1985.
88. **Ichikawa, T., Namikawa, M., Yamada, K., Sakai, K., and Kondo, K.,** Novel cyclopentenoyl fatty acids from mosses, *Dicranum scoparium* and *Dicranum japonicum, Tetrahedron Lett.*, 24, 3337, 1983.
89. **Ichikawa, T., Yamada, K., Namikawa, M., Sakai, K., and Kondo, K.,** New cyclopentenoyl fatty acids from Japanese mosses, *J. Hattori Bot. Lab.*, 56, 209, 1984.
90. **Pryce, R. J.,** Metabolism of lunularic acid to a new plant stilbene by *Lunularia cruciata, Phytochemistry*, 11, 1355, 1972.
91. **Canonica, L. A., Corbella, A., Gariboldi, P., Jommi, G., Krepinsky, J., Ferrari, G., and Casagrande, C.,** Sesquiterpenoids of *Cinnamosma fragrans* Baillon. Structure of cinnamolide, cinnamosmolide and cinnamodial, *Tetrahedron*, 26, 3895, 1969.

92. **Nakanishi, K. and Kubo, I.,** Studies on warburganal, muzigadial, and related compounds, *Isr. J. Chem.,* 16, 28, 1977.
93. **Asakawa, Y., Toyota, M., Takemoto, T., Kubo, I., and Nakanishi, K.,** Insect antifeedant secoaromadendrane-type sesquiterpenes from *Plagiochila* species, *Phytochemistry,* 19, 2147, 1980.
94. **Wada, K. and Munakata, K.,** Insect feeding inhibitors in plants. III. Feeding inhibitory activity of terpenoids in plants, *Agric. Biol. Chem.,* 35, 115, 1971.
95. **Bernasconi, S., Gariboldi, P., Jommi, G., Montanari, S., and Sisti, M.,** Total synthesis of pinguisone, *J. Chem. Soc. Perkin Trans. 1,* p. 2394, 1981.
96. **Asakawa, Y., Harrison, L. J., and Toyota, M.,** Occurrence of a potent piscicidal diterpenedial in the liverwort *Riccardia lobata* var. *yakushimensis, Phytochemistry,* 24, 261, 1985.
97. **Kanasaki, T. and Ohta, K.,** Isolation and identification of costunolide as a piscicidal component of *Marchantia polymorpha, Agric. Biol. Chem.,* 40, 1239, 1976.
98. **Asakawa, Y., Toyota, M., and Aratani, T.,** (+)-Bornyl acetate from *Conocephalum conicum, Proc. Bryol. Soc. Jpn.,* 1, 155, 1976.
99. **Asakawa, Y., Huneck, S., Toyota, M., Takemoto, T., and Suire, C.,** Mono- and sesquiterpenes from *Porella arboris-vitae, J. Hattori Bot. Lab.,* 46, 163, 1979.
100. **Huneck, S. and Schreiber, K.,** Inhaltstoffe der Moose. XIII. Über die Inhaltsstoffe weiterer Lebermoose, *J. Hattori Bot. Lab.,* 39, 215, 1975.
101. **Gorham, J.,** Effect of lunularic acid analogues on liverwort growth and IAA oxidation, *Phytochemistry,* 17, 99, 1978.
102. **Benesova, V. and Herout, V.,** Components of liverworts. Their chemical structures and biological activity, *Bryophytorum Bibl.,* 13, 355, 1978.
103. **Arai, Y., Kamikawa, T., Kubota, T., Masuda, Y., and Yamamoto, R.,** Synthesis and properties of lunularic acid, *Phytochemistry,* 12, 2279, 1973.
104. **Baker, P. M., Fortes, C. C., Fortes, E. G., Gazzinelli, G., Gilbert, B., Lopes, J. N. C., Pellegrino, J., Tomassini, T. C. B., and Vichnewiski, W.,** Chemoprophylactic agents in schistosomaiasis: eremanthine, costunolide, α-cyclocostunolide and bisabolol, *J. Pharm. Pharmacol.,* 24, 853, 1972.
105. **Wilson, J. R. and Schwabe, W. W.,** Growth and dormancy in *Lunularia cruciata* (L.) Dum. III. The wavelengths of light effective in photoperiodic control, *J. Exp. Bot.,* 15, 368, 1964.
106. **Fries, K.,** Über einen genuinen keimungs- und streckungswachstumsactiven Hemmstoff bei *Marchantia polymorpha* L., *Beitr. Biol. Pflanz.,* 40, 177, 1964.
107. **Valio, I. F. M., Burdon, R. S., and Schwabe, W. W.,** New natural growth inhibitor in the liverwort *Lunularia cruciata* (L.) Dum., *Nature (London),* 223, 1176, 1969.
108. **Pryce, R. J.,** The occurrence of lunularic and abscisic acids in plants, *Phytochemistry,* 11, 1759, 1972.
109. **Gorham, J.,** Lunularic acid and related compounds in liverworts, algae, and *Hydrangea, Phytochemistry,* 16, 249, 1977.
110. **Schneider, M. J., Troxler, R. F., and Voth, P. D.,** Occurrence of indoleacetic acid in the bryophytes, *Bot. Gaz. (Chicago),* 128, 174, 1967.
111. **Beutelmann, P. and Bauer, L.,** Purification and identification of a cytokinin from moss callus cells, *Planta,* 133, 215, 1977.
112. **Muromtsev, G. S., Agnistikiva, V. N., Lupova, L. H., Dubovaya, L. P., and Lekareva, T. A.,** Gibberellin-like substances in ferns and mosses, *Invest. Akad. Nauk. SSSR Ser. Biol. Moskow,* 5, 727, 1964.
113. **Sugawa, S.,** Nutritive values of mosses as a food for domestic animals and fowls, *Hikobia,* 2, 119, 1960.
114. **Prins, H. H. Th.,** Why are mosses eaten in cold environments only?, *Oikos,* 38, 374, 1981.
115. **Asakawa, Y., Toyota, M., and Harrison, L. J.,** Isotachin A and isotachin B, two sulphur-containing acrylates from the liverwort *Isotachis japonica, Phytochemistry,* 24, 1505, 1985.
116. **Asakawa, Y., Takikawa, K., Tori, M., and Campbell, E. O.,** Isotachin C and balantiolide, two aromatic compounds from the New Zealand liverwort *Balantiopsis rosea, Phytochemistry,* 25, 2543, 1986.

INDEX

A

D

E

H

I

N

O

P

Q

R

S

T

U

V

W

X

Y

Z